T0289896

Understanding Geomorphology

Understanding Geomorphology

Edited by John Bailey

SYRAWOOD
PUBLISHING HOUSE

New York

Published by Syrawood Publishing House,
750 Third Avenue, 9th Floor,
New York, NY 10017, USA
www.syrawoodpublishinghouse.com

Understanding Geomorphology
Edited by John Bailey

International Standard Book Number: 978-1-64740-119-1 (Hardback)

Cataloging-in-Publication Data

Understanding geomorphology / edited by John Bailey.
    p. cm.
Includes bibliographical references and index.
ISBN 978-1-64740-119-1
1. Geomorphology. 2. Landforms. 3. Physical geography. I. Bailey, John.
GB401.5 .U53 2022
551.41--dc23

TABLE OF CONTENTS

PREFACE

Every book is initially just a concept; it takes months of research and hard work to give it the final shape in which the readers receive it. In its early stages, this book also went through rigorous reviewing. The notable contributions made by experts from across the globe were first molded into patterned chapters and then arranged in a sensibly sequential manner to bring out the best results.

The Earth's surface is affected by a combination of surface and geologic processes that shape landscapes and coastal geography. Such processes cause tectonic uplift and subsidence. The varied topographies of the Earth reflect the intersection of surface and subsurface action. The field concerned with the study of the origin and evolution of such bathymetric and topographic features of the Earth is known as geomorphology. Surface processes, such as wind, waves, groundwater movement, glacial action, volcanism, tectonism, etc. are responsible for most of Earth's topographic features. The history of Earth's landscapes, their shape and change in time are understood through a combination of physical experimentation, field observations and numerical modelling. This book brings forth some of the most innovative concepts and elucidates the unexplored aspects of geomorphology. Different approaches, evaluations, methodologies and advanced studies on geomorphology have been included herein. With state-of-the-art inputs by acclaimed experts of this field, this book targets students and professionals.

It has been my immense pleasure to be a part of this project and to contribute my years of learning in such a meaningful form. I would like to take this opportunity to thank all the people who have been associated with the completion of this book at any step.

Editor

Pluri-decadal (1955–2014) evolution of glacier–rock glacier transitional landforms in the central Andes of Chile (30–33° S)

Sébastien Monnier[1] and Christophe Kinnard[2]

[1]Instituto de Geografía, Pontificia Universidad Católica de Valparaíso, Valparaíso, Chile
[2]Département des Sciences de l'Environnement, Université du Québec à Trois-Rivières,
Trois-Rivières, Québec, Canada

Correspondence to: Sébastien Monnier (sebastien.monnier.ucv@gmail.com)

Abstract. Three glacier–rock glacier transitional landforms in the central Andes of Chile are investigated over the last decades in order to highlight and question the significance of their landscape and flow dynamics. Historical (1955–2000) aerial photos and contemporary (> 2000) Geoeye satellite images were used together with common processing operations, including imagery orthorectification, digital elevation model generation, and image feature tracking. At each site, the rock glacier morphology area, thermokarst area, elevation changes, and horizontal surface displacements were mapped. The evolution of the landforms over the study period is remarkable, with rapid landscape changes, particularly an expansion of rock glacier morphology areas. Elevation changes were heterogeneous, especially in debris-covered glacier areas with large heaving or lowering up to more than $\pm 1\,\mathrm{m\,yr^{-1}}$. The use of image feature tracking highlighted spatially coherent flow vector patterns over rock glacier areas and, at two of the three sites, their expansion over the studied period; debris-covered glacier areas are characterized by a lack of movement detection and/or chaotic displacement patterns reflecting thermokarst degradation; mean landform displacement speeds ranged between 0.50 and $1.10\,\mathrm{m\,yr^{-1}}$ and exhibited a decreasing trend over the studied period. One important highlight of this study is that, especially in persisting cold conditions, rock glaciers can develop upward at the expense of debris-covered glaciers. Two of the studied landforms initially (prior to the study period) developed from an alternation between glacial advances and rock glacier development phases. The other landform is a small debris-covered glacier having evolved into a rock glacier over the last half-century. Based on these results it is proposed that morphological and dynamical interactions between glaciers and permafrost and their resulting hybrid landscapes may enhance the resilience of the mountain cryosphere against climate change.

1 Introduction

Glacier–rock glacier interactions related to Holocene glacier fluctuations (e.g. Haeberli, 2005) and the current evolution of small debris-covered glaciers having survived to the post-Little Ice Age (LIA) warming (e.g. Bosson and Lambiel, 2016) are important issues in high-mountain studies. They may provide key insights into the mechanisms of rock glacier development (Dusik et al., 2015) and of cryosphere stability and resilience against climate changes; the latter topic is of societal importance in arid–semiarid mountain areas, where the potential permanence of underground solid water resources subsequent to deglaciation may constitute a non-negligible water resource (e.g. Rangecroft et al., 2013).

The most striking geomorphological expression of glacier–rock glacier interactions is large glacier–rock glacier transitional landforms, which are assemblages of debris-covered glaciers in their upper part and rock glaciers in their lower part (e.g. Kääb et al., 1997; Krainer and Mostler, 2000; Ribolini et al., 2007; Monnier et al., 2014; Janke et al., 2015).

Here, it is important to recall and highlight the differences between both types of landforms (Nakawo et al., 2000; Kääb and Weber, 2004; Haeberli et al., 2006; Degenhardt, 2009; Benn and Evans, 2010; Berthling, 2011; Cogley et al., 2011). Rock glaciers are perennially frozen homo- or heterogeneous ice–rock mixtures covered with a continuous and several-metre-thick ice-free debris layer that thaws every summer (known as the permafrost "active layer"); rock glacier movement is governed by gravity-driven permafrost creep. Debris-covered glaciers are glaciers covered with a thin (no more than several decimetres thick) and generally discontinuous debris layer; debris-covered glaciers movement is governed by gravity-driven ice creep and sometimes basal slip in response to a mass balance gradient; debris-covered glaciers do not require permafrost conditions. Rock glaciers and debris-covered glaciers exhibit distinct morphologies that are of critical importance in the surface energy balance and subsurface heat transfer. On their surface, rock glaciers exhibit "the whole spectrum of forms created by cohesive flows" (Barsch, 1992, p. 176) of "lava-stream-like … viscous material" (Haeberli, 1985, p. 92). These features vary for each case and study area; according to our field surveys in the Andes, they can be grouped into three main types: small-scale (< 1 m high) ripples or undulations resulting from deformations in the active debris layer moving together with the underlying perennially frozen core, medium-scale (1–5 m high) ridge-and-furrow assemblages resulting from the compression of the whole ice-debris mixture, and large-scale (5–20 m thick and > 100 m long) superimposed flow lobes upon which the first two feature types may naturally appear. Hereafter, we will simply refer to these features as "cohesive flow-evocative features". Contrarily, debris-covered glaciers are characterized by a chaotic distribution of features evocating surface instability such as hummocks, collapses, crevasses, meandering furrows, and thermokarst depressions and pounds. As a consequence, on rock glaciers the large- and fine-scale surface topography is rather smooth and convex, whereas on debris-covered glaciers it is rather rough and concave. Another morphological difference is the presence of ice visible from the surface: whereas ice is generally invisible from the surface of rock glaciers, it is frequently exposed on debris-covered glaciers due to the discontinuity of the debris cover or the occurrence of the aforementioned morphological features. Finally, and correlatively, over pluri-annual to pluri-decadal periods the morphology of well-developed rock glaciers is quite stable (besides cases of climate warming-related destabilizations, the geometry of surface features evolves but their overall pattern remains the same), while debris-covered glacier morphology is characterized by instability (surface features rapidly appear and disappear).

According to the literature at least three types of glacier–rock glacier interactions can be distinguished:

i. The readvance(s) and superimposition/embedding of glaciers or debris-covered glaciers onto/into rock glaciers, with related geomorphological and thermal consequences (Lugon et al., 2004; Haeberli, 2005; Kääb and Kneisel, 2006; Ribolini et al., 2007, 2010; Bodin et al., 2010; Monnier et al., 2011, 2014; Dusik et al., 2015). This is the sensu stricto significance of "glacier–rock glacier relationships" (Haeberli, 2005) as defined by what has been called the "permafrost school" in reference to the long-term "rock glacier controversy" (see Berthling, 2011).

ii. The continuous derivation of a rock glacier from a debris-covered glacier by evolution of the surface morphology (see above) together with the conservation and creep of a massive and continuous core of glacier ice (e.g. Potter, 1972; Johnson, 1980; Whalley and Martin, 1992; Potter et al., 1998; Humlum, 2000). This process was not initially called a "glacier–rock glacier relationship"; this view is indeed held by what has been called the "continuum school", in opposition to the permafrost school (Berthling, 2011). Nevertheless, such a phenomenon does belong, literarily, to the domain of glacier–rock glacier interactions.

iii. The transformation of a debris-covered glacier into a rock glacier not only by the evolution of the surface morphology but also by the evolution of the inner structure, i.e. the transformation of the debris-covered continuous ice body into a perennially frozen ice–rock mixture by addition from the surface of debris and periglacial ice and fragmenting of the initial glacier ice core. This has been described as an alternative to the dichotomous debate between the permafrost school and continuum school (Monnier and Kinnard, 2015); such phenomenon has been described as achievable over a human-life or historical timescale (Schroder et al., 2000; Monnier and Kinnard, 2015; Seppi et al., 2015).

In the present study, we aim to provide insights into the aforementioned issue using the variety of glacier–rock glacier transitional landforms encountered in the semiarid Andes of Chile and Argentina. These landforms have shown a particularly rapid evolution over the last decades which allow studying glacier–rock glacier interactions on an historical timescale. Three landforms with distinct morphologies have been chosen in the central Andes of Chile in an attempt to diagnose their geomorphological significance, especially in terms of glacier–rock glacier interactions and cryosphere persistence in the current climatic context. To this purpose, this study makes use of aerial and satellite imagery and remote sensing techniques in order to document the morphological and dynamical evolution of the studied landforms over a pluri-decadal time span.

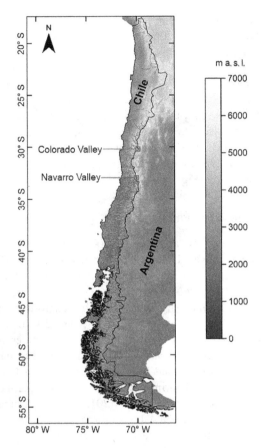

Figure 1. Location of the study sites. Drainage network, which reflects the variations in climatic–hydrologic conditions along the Chilean territory, is shown in blue.

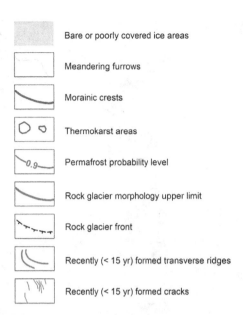

Figure 2. Geomorphological legend shared for all subsequent figures.

2 Study sites

We studied three glacier–rock glacier transitional landforms in the central Andes of Chile, respectively named Navarro, Presenteseracae, and Las Tetas (Fig. 1). Navarro and Presenteseracae are located in the Navarro Valley, in the upper Aconcagua River catchment (33° S). Las Tetas is located in the Colorado Valley, in the upper Elqui River catchment (30° S).

2.1 Upper Navarro Valley

The upper Navarro Valley belongs to the Juncal River catchment and Juncal Natural Park, which are part of the upper Aconcagua River catchment, in the Valparaíso region of Chile (32°53′ S, 70°02′ W; Fig. 1). In the Juncal River catchment (∼ 1400–6110 m a.s.l.), glaciers cover 14 % of the area (Bown et al., 2008; Ragettli et al., 2012) while active rock glaciers cover almost 8 % (Monnier and Kinnard, 2015). The climate is a mediterranean mountain climate. Brenning (2005) and Azócar and Brenning (2010) located the 0 °C isotherm of mean annual air temperature (MAAT) close to 3700 m a.s.l. and reported precipitations above

3000 m a.s.l. as ranging between 700 and 800 mm yr^{-1}. An automatic weather station located at 2800 m a.s.l. at the foot of the Juncal Glacier, 10 km SW from Navarro Valley, recorded a MAAT of 6.3 °C during the hydrological year 2013–2014. The upper Navarro Valley crosses, from west to east, the Albánico Formation (Upper Cretaceous; andesites, volcanic breccias), the San José Formation (Lower Cretaceous; limestones), and the Lagunilla Formation (Upper Jurassic; sandstones, lutites, gypsum). The glacial footprint is conspicuous through the Navarro Valley: the valley is U-shaped, with corries in the upper parts and latero-frontal moraines in the lower parts (Figs. 2 and 3).

2.1.1 Navarro

Navarro fills the major part of the upper Navarro Valley floor between ∼ 3950 and 3450 m a.s.l. (Fig. 3). The landform was described by Janke et al. (2015, p. 117) as a system composed of several classes of debris-covered glaciers and rock glaciers according to their presumed ice content. It is indeed a huge (> 2 km long and up to > 1 km wide) and complex assemblage with debris-covered glacier morphology in its upper parts and rock glacier morphology in its lower parts. The main presumed flow direction of the landform points towards 170° N. At least 10 conspicuous and sometimes > 15 m high morainic crests are visible at the surface of the landform, some of them being included in the rock glacier morphological unit. At one location (red circle in Fig. 3), the superposition of two series of morainic crests onto a rock glacier lobe suggests that the landform developed from a succession of glacier advances and rock glacier development phases.

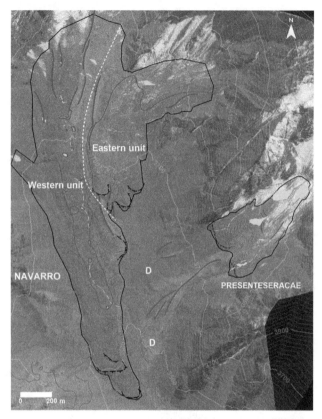

Figure 3. Map of the Navarro Valley. See Fig. 2 for legend. The background of the map is the 2014 Geoeye image draped over the Geoeye DEM (see the "Material and methods" section). Elevation contours are derived from the Geoeye DEM and the contour interval is 20 m. The boundary between the Navarro's western and eastern units is indicated with a dashed white line. The red circle indicates the location described in the text where morainic crests and rock glacier lobes are superimposed. Note also the decayed (D) rock glacier lobes in the area between Navarro and Presenteseracae.

Navarro is divided between an eastern and a western unit, with the two being separated by a central series of aligned morainic crests (Fig. 3). The eastern unit, which is located in the more shadowed northeastern part of Navarro Valley, is ∼ 1.2 km long, and about two-thirds of its area exhibits a rock glacier morphology. The terminal part exhibits three adjacent terminal lobes. The western unit is ∼ 2.4 km long and more complex. Sets of embedded morainic crests in the upper part delimit the retreat of a former glacier. The median part (∼ 1 km long) is peculiar, with the boundary between the debris-covered and rock glacier morphology extending far downslope and following the contour of an elongated central depression (10–15 m lower in altitude than the lateral margins) (Figs. 3 and 4). This central depression is characterized by numerous large (up to 50 m of diameter) thermokarst depressions with bare ice exposures, generally on their south-facing walls. The lower part of the western unit exhibits a rock glacier morphology and three superim-

posed fronts close to the terminus, the slope of the lowest front being gentler than that of the two upper fronts, which are almost at the same location.

Monnier and Kinnard (2015) provided an empirical model of permafrost probability based on logistical regression for the upper Aconcagua River catchment. According to this model, Navarro may be in a permafrost state. The permafrost probability is close to 1 in the upper parts; nevertheless, there is a marked decreasing gradient in permafrost probability from 0.9 to 0.7 between the central part and the terminus of the western unit (Fig. 3).

2.1.2 Presenteseracae

Presenteseracae is a small (∼ 600 m long and 300 m wide) debris-covered glacier located between ∼ 4080 and 3800 m a.s.l., in a narrow, SW-facing cirque, ∼ 300 m above and only 500 m east of Navarro (Fig. 3). The main presumed flow direction points towards 225° N. This landform has been thoroughly analysed by Monnier and Kinnard (2015). The debris-covered glacier exhibits rock glacier features in its lower part (Figs. 3, 4, and 7). The transverse and curved ridges (< 1.5 m high) and well-defined steep front (∼ 10 m high) have appeared during the last 15 years. The permafrost model of Monnier and Kinnard (2015) gave a permafrost probability of 1 for the whole Presenteseracae landform. The authors also correlated the development of the cohesive flow-evocative rock glacier morphology with the low estimated sub-debris ice ablation rates, and demonstrated that the sediment store on Presenteseracae and the potential formation times are in agreement with common rock wall retreat rates. They concluded that Presenteseracae is a debris-covered glacier currently evolving into a rock glacier. In the upper part of the landform, the debris cover is very thin (a few centimetres) and bare ice exposures are frequent. The debris cover thickens to more than 60 cm in the lower part, where the rock glacier morphology develops below a steeper sloping segment. Push moraine ridges (Benn and Evans, 2010) occur at the surface above 3780 m a.s.l. (Fig. 3). The lower part, which displays a rock glacier morphology, is clearly composed of two adjacent lobes, dividing away from a morainic crest overridden by the landform (Fig. 4). Depressed meandering furrows where buried ice is exposed are also present (Fig. 3). During hot summer days, the water flowing in the northernmost furrow sinks down a hole just before the front.

2.2 Las Tetas

Las Tetas is located in the Colorado Valley, which is the uppermost part of the Elqui River valley, in the Norte Chico region of Chile (30°10′ S, 69°55′ W; Fig. 1). Elevations in the Colorado Valley range between ∼ 3100 and 6255 m a.s.l. The landform is located on the south-facing side of Cerro Las Tetas (5296 m a.s.l.), less than 1 km south of Tapado Glacier

Figure 4. Photos of the lower **(a)** and upper part **(b)** of Navarro, seen from Presenteseracae; Presenteseracae seen from Navarro **(c)**; and the terminal part of Las Tetas **(d)** seen from its northeastern surrounding area. The white stars on photos **(a)** and **(b)** indicate the main location of the central depression and related thermokarst morphology on Navarro (see Fig. 3).

(e.g. Ginot et al., 2006; Pourrier et al., 2014). The climate of the area is a semiarid mountain climate. At the La Laguna artificial dam (~ 3100 m a.s.l., 10 km west of the study site), the mean annual precipitation was 167 mm during the 1970–2009 period, and the MAAT was 8 °C during the 1974–2011 period. The 0 °C isotherm is located near 4000 m a.s.l. (Brenning, 2005; Ginot et al., 2006). Materials composing the rock basement belong to the Pastos Blancos Formation (Upper Palaeozoic; andesitic to rhyolitic volcanic rocks). A set of embedded latero-frontal moraines is encountered ~ 700 m downslope from the front of Las Tetas, between ~ 4170 and 4060 m a.s.l.

Las Tetas is a ~ 1 km long landform located between 4675 and 4365 m a.s.l. (Fig. 5). The main presumed flow direction points towards 140° N. The boundary between debris-covered and rock glacier morphology is clear, in the form of a large and deep furrow, and divides the landform in two approximately equal units. The upper unit is characterized by a chaotic and hummocky morphology, and vast (up to more than 50 m of diameter) and deep (up to 20 m) thermokarst depressions exposing bare ice generally along their south-facing walls. The lower part of the landform exhibits tension cracks superimposed onto the ridge-and-furrow pattern. The front of Las Tetas is prominent; including the talus slope at the bottom, which may bury sediments or outcrops downward, it is almost 100 m high (Figs. 4 and 5). According to the logistic regression-based empirical permafrost model proposed by Azócar (2013) for the area, the 0.75 probability

level crosses the landform in its central part (Fig. 5). Permafrost favourability index (PFI) values proposed by Azócar et al. (2016, 2017) are > 0.7 in the upper part and between 0.6 and 0.7 in the lower part.

3 Material and methods

3.1 Satellite image and aerial photo processing

We acquired historical (prior to 2000) aerial photos and contemporary (after 2000) satellite images for the three study sites. Stereo pairs of aerial photos were inspected, selected, and scanned at the Geographic and Military Institute (IGM) of Chile. Scanning was configured in order to yield a ground resolution of 1 m. At Las Tetas, photos from 1978 and 2000 were selected; at Navarro and Presenteseracae, photos from 1955 and 2000 were selected. A stereo pair of Geoeye satellite images was also acquired for each site. The Geoeye imagery was acquired on 23 March 2012 and 14 February 2014 at Las Tetas and Navarro Valley, respectively, as panchromatic image stereo pairs (0.5 m of resolution) along with four bands in the near-infrared, red, green, and blue spectra (2 m of resolution).

Orthoimages, orthophotos, and altimetric information were generated from the data. The first step involved building a digital elevation model (DEM) from the stereo pair of Geoeye satellite images. The Geoeye images were triangulated using a rational polynomial camera (RPC) model sup-

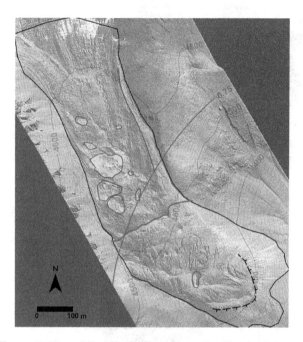

Figure 5. Map of the Las Tetas landform. See Fig. 2 for legend. The background of the map is the 2012 Geoeye image draped over the Geoeye DEM (see the "Material and methods" section).

Table 1. Errors generated during the aerial photo processing. The ground root mean square error (RMSE) relates to sets of ground control points (GCPs) extracted from the Geoeye orthoimage and used for the orthorectification of the aerial photos.

Site	Date	Horizontal ground RMSE (m)		Number of GCPs
		x	y	
Las Tetas	1978	1.13	1.16	10
	2000	0.33	0.54	8
Navarro Valley	1955	1.82	1.32	13
	2000	0.76	1.49	9

plied by the data provider. The exterior orientation was constrained using one or two (according to the site) ground control points (GCPs) acquired with a differential GPS system in the field in 2014 over bedrock outcrops visible on the images. Sets of three-dimensional (3-D) points were extracted automatically using standard procedures of digital photogrammetry (Kääb, 2005) and edited manually to remove errors. A 2×2 m DEM was generated using triangular irregular network interpolation of the 3-D points. The same processing scheme was followed for the aerial photo stereo pairs using control points visible both on the Geoeye image and the aerial photo stereo pairs. The vertical bias of the aerial photo DEMs was calculated by comparison with the Geoeye DEMs over flat and stable areas outside the landform studied and was removed from the subsequent calculations (see below). The automatic and manual extraction of 3-D points from aerial photo stereo pairs proved to be challenging in steep areas with unfavourable viewing geometry. The process failed for the 1955 stereo pair of Navarro Valley, with only a very sparse set of 3-D points extracted and including possible errors, ruling out the possibility to generate a reliable and complete DEM and to estimate the vertical bias.

The Geoeye images were pan-sharpened and orthorectified using the Geoeye DEM. The aerial photos were then orthorectified using the corresponding DEMs, except when no reliable DEM could be obtained (as for 1955 at Navarro); in that case, the Geoeye DEM was used. The orthorectification was constrained by the internal camera information, tie points, and GCPs extracted during the process. The accuracy

of the orthorectification was estimated using the GCPs. The root mean square error (RMSE) corresponding to the sets of GCPs at the different times is displayed in Table 1. The ground resolution of the orthophotos was then resampled at 0.5 m in order to equal that of the Geoeye products.

The altimetric information was used to calculate the elevation changes of the landforms between the different dates, after removal of the vertical bias. The total elevation change was further converted into annual rates of elevation change. As outlined by Lambiel and Delaloye (2004), elevation changes at the surface of rock glaciers may be explained by several and possibly concomitant factors: (i) downslope movement of the landform and advection of local topographic features, (ii) extensive or compressive flow, and (iii) melting or aggradation of internal ice. Therefore, it is difficult to unambiguously interpret elevation changes. Studying the Muragl rock glacier (Swiss Alps), Kääb and Vollmer (2000) highlighted how mass advection caused subtle elevation changes (between -0.20 and $+0.20$ m yr^{-1}), while surface lowering of up to -0.50 m yr^{-1} was considered as indicative of massive losses of ice. Accordingly, taking into account the range of values measured and the uncertainty (or detection threshold) on the measurements (see Table 2), we used an absolute value of 0.50 m yr^{-1} to generally discriminate between "moderate" and "large" vertical changes. The former were considered to relate primarily to the downslope expansion of the landform (including long profile adaptation and advection of topographic features) and, thus, to extensive flow; in the case of the latter, additional ice melting or material bulging by compression were considered necessary in the interpretation.

3.2 Image interpretation

The geomorphology of each landform was carefully interpreted from the orthoimages and orthophotos. First, we located and mapped the boundary between debris-covered and rock glacier morphology, according to the detailed criteria of differentiation presented in the Introduction. The thermokarst area was also monitored over time by mapping

Table 2. Uncertainty related to the measurement of annual elevation changes. Reported uncertainties correspond to 1 and 2-standard deviation (σ) probability of vertical errors for the generated DEMs. In Navarro Valley, no reliable DEM could be generated from the 1955 aerial photos, which explains the absence of data in the table for the 1955–2000 interval.

Site	Period	Vertical uncertainty (m yr^{-1})	
		1σ (66%)	2σ (95%)
Las Tetas	1978–2000	0.04	0.09
	2000–2012	0.22	0.43
Navarro Valley	1955–2000	–	–
	2000–2014	0.05	0.10

Table 3. Sizes of search template and search window used for the image feature tracking.

Site	Period	Search template size (pixels)	Search window size (pixels)
Las Tetas	1978–2000	100	250
	2000–2012	150	250
Navarro	1955–2000	300	550
	2000–2014	50	100
Presenteseracae	1955–2000	150	400
	2000–2014	80	180

the thermokarst depressions at the surface of the landforms as polygonal shapes, and their total area was calculated. Salient and recently appeared features such as cohesive flow-evocative ridges on Presenteseracae and cracks on Las Tetas were also mapped.

3.3 Image feature tracking

We used image feature tracking in order to measure horizontal displacements at the surface of the landforms. Computer-programmed image feature tracking is a sub-pixel precision photogrammetric technique that has been widely used for studying the kinematics of glaciers, rock glaciers, and other mass movements. We followed the principles and guidelines provided by Kääb and Vollmer (2000), Kääb (2005), Wangensteen et al. (2006), Debella-Gilo and Kääb (2011), and Heid and Kääb (2012). We used ImGRAFT, which is an open-source image feature tracking toolbox for MATLAB (Messerli and Grinsted, 2015) using two orthoimages (from spaceborne, airborne, or terrestrial sensors) of the same area and resolution but at different times. All the orthoimages were pre-processed in order to enhance their contrast. Two template matching methods were tested: normalized cross-correlation (NCC) and orientation correlation (OC). The NCC method was found to yield more consistent results at the different sites and was thus used for this study. NCC gives an estimate of the similarity of image intensity values between matching entities in the orthoimage at time 1 (I_1) and their corresponding entities in the orthoimage at time 2 (I_2). In I_1, a "search template" is defined around each pixel located manually or automatically inside a regular grid; the algorithm extracts this search template from I_1 and searches for it in I_2 within the area of a predefined search window (see, e.g., Fig. 2 in Debella-Gilo and Kääb, 2011, p. 132); the algorithm then computes the NCC coefficient between the search template in I_1 and that in I_2 and moves the search template until the entire search window is covered. The location that yields the highest correlation coefficient within the search window is considered as the likely best match for the original location in I_1. The sizes of the search template

and search window were first defined based on image quality and the time period considered; larger template and search windows were used for long periods, as only larger-scale morphological features were expected to be preserved over periods of several decades. The final choice of template and search window size was then set after several iterations of the algorithm (see Table 3). The NCC algorithm was performed over the whole area of the landforms using a 10 m spacing grid. Snow-covered areas were delineated on each image and excluded from the analysis, leaving an additional buffer of 10 m around the snow masks.

Results from feature tracking generally need to be filtered, especially when dealing with old orthophotos (Wangensteen et al., 2006). In this study, the following filtering procedure was followed. (1) We excluded displacements smaller than the orthorectification error (RMSE, Table 1). (2) We excluded displacements exhibiting a signal-to-noise ratio (SNR) < 2 (as recommended by Messerli and Grinsted, 2015); SNR is the ratio between the maximum NCC coefficient and the average of the NCC coefficient's absolute values in the search window, and can be used as an indicator of the "noise" in the results. (3) A directional filter was applied in order to eliminate vectors diverging excessively from one another, based on Heid and Kääb (2012). For that purpose, the mean displacements in the X ($\overline{du}$) and Y ($\overline{dv}$) directions were calculated in a 5 × 5 m running window centred on each displacement vector. The displacement vector was excluded if its du and dv component exceeded $\overline{du}$ and $\overline{dv}$, respectively, by more than 4 times the RMSE presented in Table 1. This last filtering step allowed excluding chaotic vectors with potential errors not removed by the first two filtering criteria, as well as highlighting areas with spatially coherent movement. Finally, the total displacements were converted to annual displacement rates and mapped.

Whereas all vectors obtained after filtering were mapped (see "Results and interpretations" and related figures), the final displacement statistics were calculated after removing upslope-pointing vectors (vectors deviating from more than ±45° from the landform longitudinal axis) (Table 4). These may include some remaining errors, but may also result from thermokarst degradation on debris-covered ice, as discussed later. As displacements statistics aimed at quantifying mean

Table 4. Summary statistics of horizontal displacements detected on the landform surfaces (see text for further details). The mean displacement ($\bar{d}$, $\mathrm{m\,yr^{-1}}$) and standard deviation (σ, $\mathrm{m\,yr^{-1}}$) are presented for the entire (snow-free) areas where movement was detected, and then only for the overlapping areas where movement was detected during both periods compared. The latter aimed at removing the spatial sampling bias when comparing movement statistics over time. n refers to the numbers of vectors retained and f to the corresponding fraction (%) of snow-free areas where movement was detected.

Site and period	Whole areas				Overlapping areas only		
	$\bar{d}$	σ	n	f	$\bar{d}$	σ	n
Navarro							
1955–2000	0.52	0.30	832	33	0.54	0.34	372
2000–2014	0.52	0.20	970	38	0.51	0.18	310
Presenteseracae							
1955–2000	1.04	0.46	219	15	1.10	0.41	73
2000–2014	0.96	0.47	162	12	0.82	0.40	63
Las Tetas							
1978–2000	0.88	0.35	79	15	0.69	0.21	18
2000–2012	0.86	0.45	163	31	0.65	0.37	30

downslope movement rates, these vectors were hence excluded from the calculation. Furthermore, for each landform, the mean annual displacement rate was re-calculated only over areas where movement was detected both during the historical (1955–2000 for Navarro Valley and 1978–2000 for Las Tetas) and recent (after 2000) periods in order to remove the spatial sampling bias (see Table 4). For this purpose, all the points present in a 20 m radius of each other's from one period to another were retained to estimate a mean displacement rate over a common area.

4 Results and interpretations

4.1 General performance of and insights provided by the methods

The methods used in this study first allowed to obtain series of images depicting at first sight conspicuous landscape evolutions: Figs. 6, 7, and 8 show the orthophotos and orthoimages obtained at each site together with the delineated boundary between debris-covered and rock glacier morphology areas and the front slope base at each time. These figures highlight how the landforms' landscape has changed over both historical (before 2000) and contemporary (after 2000) periods. Thermokarst areas could be easily mapped and calculated, except in 2000 at Las Tetas.

Reliable DEMs and related maps of elevation changes were obtained for the 2000–2014 period at Navarro (Fig. 10) and Presenteseracae (Fig. 12), and for both the 1978–2000 and 2000–2012 periods at Las Tetas (Figs. 13 and 14, re-

spectively). However, and as mentioned in the "Material and methods" section, no reliable and complete DEM could be obtained for the Navarro Valley in 1955, which explained the lack of elevation change measurements at Navarro and Presenteseracae.

The efficiency of the image feature tracking method varied according to the sites and periods but, on the whole, provided valuable information (Figs. 9–14 and Table 4). Filtering led to keeping between 12 and 38 % of the measured horizontal displacements according to the site and period (Table 4). The order of magnitude of the mean horizontal displacements is 0.50–1 $\mathrm{m\,yr^{-1}}$. Horizontal displacements were consistently detected in rock glacier areas, and much less in debris-covered glacier areas; Figs. 9–14 highlight spatially coherent flow vector patterns in the former, while the latter are characterized by a lack of movement detection and/or spatially chaotic patterns. This is consistent with the fact that the surface morphology of rock glaciers is more stable and preserved for longer times than the one of debris-covered glaciers, which is rather unstable and disrupt rapidly. Upslope-pointing vectors were kept in the figures in order to show that they frequently occur in sectors with thermokarst morphology where mass wasting processes are likely to occur. Finally, one will note that the most graphically striking results are obtained over the largest landform, i.e. Navarro (Figs. 9 and 10).

The interpretation of the main geomorphological evolution, elevation changes, and horizontal displacement patterns is summarized for each individual landform in Table 5a, b, and c, respectively, and the results are discussed jointly in the following section.

5 Discussion

The three cases studied have distinct significance in terms of glacier–rock glacier relationships and cryosphere persistence under ongoing climate change. Our results lead us to consider the following issues: (i) initial development of the landforms, (ii) differences between debris-covered and rock glacier areas, and (iii) current and future evolution of the landforms.

5.1 Initial landform development

Navarro and Las Tetas are composite landforms with a debris-covered glacier in their upper part and a rock glacier in their lower part. Considering the clear spatial organizations of surface features and the strong morphological boundaries, in particular the way the debris-covered glacier embeds into the rock glacier in the Navarro's western unit (Fig. 3) and the abrupt transition at Las Tetas (Fig. 5), these landforms most probably result from the (re)advance(s) of glaciers onto or in the back of pre-existing rock glaciers. Many other examples of such development of glacier–rock glacier assemblages have been studied and reported in the literature (Lugon et al., 2004; Haeberli, 2005; Kääb and Kneisel, 2006;

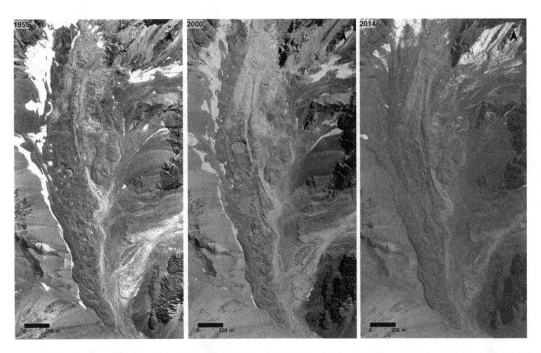

Figure 6. Sequence of orthophotos obtained for Navarro. The base of the landform front that could be reliably identified is indicated in colour (blue, magenta, and orange line in 1955, 2000, and 2014, respectively). At each date the boundary between debris-covered and rock glacier morphology is depicted with a red line (dotted in 1955, dashed in 2000, continuous in 2014).

Ribolini et al., 2007, 2010; Bodin et al., 2010; Monnier et al., 2011, 2014; Dusik et al., 2015). In the central part of the Navarro's western unit, the elevated lateral margins exhibit cohesive flow-evocative ridges, which probably resulted from the lateral compression exerted by the glacier during its advance ("composite ridges" of the glaciological terminology; Benn and Evans, 2010, p. 492). Also, the boundary between the debris-covered and rock glacier morphologies in 1955 (Fig. 6) gives a minimum indication of the lowest advance of the debris-covered glaciers onto the rock glaciers. However, the origin and age of the rock glaciers located in the lower part of the landforms are almost impossible to assess. Nonetheless, considering the context, they may have developed following several glacier advances and moraine deposition phases, suggesting the idea of a cycle in the landform development (see section "Study sites" and the red circle in Fig. 3). Such a development has led to the rock glacier being cut off from the main rock debris sources (i.e. the rock walls up-valley), resulting in the rock glacier being dependent on the ability of the debris-covered glacier to provide material (debris and ice) required for the sustainment of the rock glacier.

Presenteseracae is a completely distinct case. As studied by Monnier and Kinnard (2015) and the present work, in 1955 Presenteseracae was a debris-covered glacier and is now a debris-covered glacier transforming into a rock glacier. The initial development phase or, in this case, the "glacier–rock glacier transformation" has been occurring over the last decades. In less than 20 years, the surface debris cover

spread over almost all of the northern part; a front appeared at the terminus, and cohesive-flow evocative ridges appeared in the lower part, perpendicular to flow vectors (Figs. 7, 11, and 12). The latter ridges may be related to emergent, debris-rich shear planes (Monnier and Kinnard, 2015) bent by the landform movement. Displacement speeds were high ($> 1\,\mathrm{m\,yr^{-1}}$ on average) between 1955 and 2000, in agreement with the fast landscape evolution, before slowing down after 2000, which may reflect an acceleration of the transition towards a rock glacier. In the current state of our knowledge, what may have occurred in the internal structure in response to these drastic surface changes is uncertain: the continuous glacier core may, however, evolve into patches of buried ice progressively mixed with ice-mixed debris accumulated onto the surface.

5.2 Differences between debris-covered and rock glacier areas

Our study basically relied on the landscape differentiation between debris-covered and rock glacier areas. The criteria enounced and discussed in the Introduction section have been used to distinguish and partition the surface morphology of the landforms studied. Our subsequent results show that, at Navarro and Las Tetas, debris-covered and rock glacier areas are characterized by contrasting patterns of horizontal displacements and elevation changes. Flow patterns in rock glacier areas are conspicuous and spatially coherent and express the cohesive extensive flow of the landform in the di-

Table 5. (a) Summary of corresponding geomorphological evolution, elevation changes, horizontal displacements, and associated interpretations at Navarro for historical (1955–2000) and contemporary (2000–2014) periods. **(b)** Summary of corresponding geomorphological evolution, elevation changes, horizontal displacements, and associated interpretations at Presenteseracae for historical (1955–2000) and contemporary (2000–2014) periods. **(c)** Summary of corresponding geomorphological evolution, elevation changes, horizontal displacements, and associated interpretations at Las Tetas for historical (1978–2000) and contemporary (2000–2012) periods.

Geomorphological evolution	Elevation changes	Horizontal displacements	Interpretation
(a)			
Rock glacier morphology areas have expanded spatially between 1955 and 2014, both upward (eastern unit) and inward from the margin (western unit).	Elevation changes have been more pronounced and heterogeneous in debris-covered glacier morphology areas than in rock glacier morphology areas. In rock glacier areas, their moderate rates express the extensive flow of the landform (Fig. 10).	The area where movement was detected slightly increased (33 to 38 %), during 1955–2000 and 2000–2014. The mean displacement speed decreased over snow-free, overlapping areas (Table 4).	The progression of the rock glacier morphology correlates with a decrease in thermokarst areas, an expansion of coherent flow patterns, and a general deceleration of the landform movement. This reflects the expansion of slow, coherent downslope creep with minimal surface disturbance (typical of rock glacier) as the main geomorphic process. In the central depression of the western unit, general downslope movement occurred along with ice-loss-related downwasting.
The upward progression of the rock glacier morphology areas has been particularly strong in the eastern unit, especially between 1955 and 2000.	Elevation changes have been moderate in the eastern unit (Fig. 10).	Figures 9 and 10 highlight conspicuously spatially coherent patterns of flow vectors in the eastern unit, especially between 1955 and 2000.	
In the western unit, the progression has been more limited and occurred inward from the margins, toward the central depression (Figs. 6, 9, and 10).	Very large surface lowering (until more than $1\,\mathrm{m\,yr^{-1}}$) and moderate surface heaving alternate in the central depression (Fig. 10).	Many displacement spatially coherent vector patterns head towards the central depression (Figs. 9 and 10).	
Between 1955 and 2000, thermokarst area expanded from 11 950 to 16 520 m², before shrinking by a factor of 2 in less than 15 years (8560 m² in 2014).	The most pronounced surface lowering occurs at thermokarst locations (Fig. 10).	At thermokarst locations, displacement vectors are grouped in poorly organized, chaotic patterns, frequently pointing upward (Figs. 9 and 10).	
(b)			
The geomorphological evolution at the surface has been very fast, with the apparition of a rock glacier morphology in the lower half of the landform since 2000, in agreement with the description and analysis given by Monnier and Kinnard (2015) (Figs. 7, 11, and 12).	Elevation changes have been spatially very heterogeneous for such a small-size landform between 2000 and 2014. Nevertheless, the major part of the surface exhibits moderate elevation changes, which is seen as the expression of the extensive flow of the landform (Fig. 12).	The area where movement was detected slightly decreased (15 to 12 %) during 1955–2000 and 2000–2014, while the mean displacement speed decreased over snow-free, overlapping areas (Table 4). Horizontal flow vectors patterns are spatially more coherent in the lower than in the upper half of the landform (Figs. 11 and 12).	The geomorphological development, the distribution of flow vector patterns, and the deceleration of the landform movement point towards a transition towards rock glacier (Monnier and Kinnard, 2015); however, the decrease in the area where movement was detected does not corroborate such an interpretation. The absence of thermokarst at the surface of the landform for both periods studied may be explained by the small landform size, the cold conditions casted by the cirque topography (permafrost probability defined by the model in Fig. 3 is 1), the cirque floor slope, and/or even by the glacier–rock glacier transition phenomenon.
New morphological surface features, in the form of cohesive, flow-evocative downward (SE) convexly bent ridges, appeared in the lower-northern part of the landform (Figs. 3 and 7).	Elevation changes between 2000 and 2014 were generally moderate in the lower-northern part of the landform. Large surface heaving nevertheless occurred at the front of the landform (Fig. 12).	Horizontal displacement vectors in the lower part head towards SE (Figs. 11 and 12).	
No thermokarst.			

Table 5. Continued.

Geomorphological evolution	Elevation changes	Horizontal displacements	Interpretation
(c)			
The boundary between debris-covered and rock glacier morphology area has followed the overall displacement of the landform (Figs. 8, 13, and 14).	On the whole, elevation changes tended to decrease and be more spatially homogeneous from 1978 and 2000 to 2000 and 2012, with moderate rates expressing the extensive flow of the landform, especially in the lower rock glacier part (Figs. 13 and 14).	The area where movement was detected has strongly increased (15 to 31 %) between 1955 and 2000 as well as 2000 and 2014. The mean displacement speed decreased slightly (Table 4). However, between 2000 and 2012, the lower rock glacier part has displaced faster than the upper debris-covered glacier part (Fig. 14).	The decrease in thermokarst areas, the strong increase in movement detection areas, the apparition of coherent flow vectors patterns, and the deceleration of the whole landform support the idea that the rock glacier continues to develop. The higher displacement speed and the tension cracks in the lower rock glacier area nevertheless point towards an acceleration or even destabilization of the landform toward its terminus.
Tension cracks appeared in the lower part of the landform during the last decades (Figs. 5 and 8).		Whereas the mean displacement speed decreased slightly (Table 4), the lower part may be currently accelerating (Fig. 14).	
Thermokarst is striking by its aspect (depressions occur in the centre of coalescent mounds, reminiscent of impact craters) and its rapid evolution (Fig. 8). Between 1978 and 2012, thermokarst areas decreased 2-fold, from 23 248 m^2 in 1978 to 11 099 m^2 in 2012.	Large surface lowering occurred at the locations of thermokarst mounds and pounds, especially between 2000 and 2012 (Figs. 13 and 14).	Between 1978 and 2000, chaotic displacement patterns, with vectors frequently pointing upward, correlate with the thermokarst locations (Fig. 13). Between 2000 and 2012, very few vectors associated with thermokarst-related mass wasting are detected (Fig. 14).	

rection of the main longitudinal axis. Flow patterns in debris-covered glacier areas are either not detectable or, when detected, generally more chaotic. This low movement detection rate and chaotic organization of displacement patterns in debris-covered glacier areas can be explained by the inherently less cohesive mass flow and the unstable surface morphology resulting from the ablation of ice under a shallow debris layer. Elevation changes in debris-covered glacier areas have larger amplitudes and are spatially heterogeneous; in rock glacier areas, elevation changes are rather moderate and thus expressive of cohesive extensive flow. These different flow dynamics appear perfectly coherent with the definition of, and distinction made between, debris-covered and rock glaciers in the Introduction section.

5.3 Current evolution and its significance

5.3.1 Landscape evolution

All the landforms studied are characterized by a rapid landscape evolution over the last few decades. Changes occurred over the entire surface (Presenteseracae), in the contact/transition area between debris-covered and rock glaciers and in the debris-covered glacier area (Navarro), or even in both areas, though more subtly in the rock glacier area (Las

Tetas). This continuum in surface evolution perhaps best illustrates the process of glacial–periglacial transition. To our knowledge, an important result of our study not previously reported is the observed upward progression of the rock glacier areas which proceeds at the expense of the debris-covered glaciers on such composite landforms. At Presenteseracae, over a time span of a few decades, the rock glacier morphology has grown from being inexistent to covering approximately half the landform surface. At Navarro, rock glacier areas have subtly (in the western unit) or considerably (in the eastern unit) expanded, until, in the latter case, covering most parts of the essentially debris-covered glacier morphology present initially. As a first-order consideration, topoclimatic conditions seem to play a key role in this differentiated evolution: Presenteseracae and the eastern unit of Navarro are located in more shadowed and thus colder sites (see Figs. 3 and 5).

5.3.2 Dynamical evolution

The dynamical evolution correlates with the landscape evolution, to varying degrees according to the site. As stated in the Introduction, when areas with debris-covered glacier morphology evolve into areas with rock glacier morphology, changes occur in the surface energy balance and subsurface

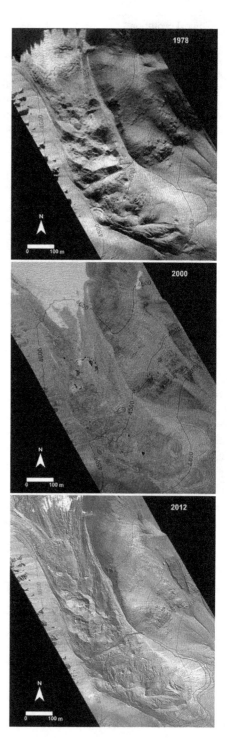

Figure 7. Sequence of orthophotos obtained for Presenteseracae. The base of the landform front that could be reliably identified is indicated in colour (blue, magenta, and orange line in 1955, 2000, and 2014, respectively). Note how the rock glacier morphology developed since 2000. In the southern part of the landform, it is nevertheless less well defined and more unstable; it is conspicuously cut by a central furrow and exhibits a few areas of bare ice over which debris slumps may occur. In the northern part of the landform, the rock glacier morphology is more developed; there is neither remaining bare ice area nor evidence of debris cover instability and sliding.

Figure 8. Sequence of orthophotos obtained for Las Tetas. The base of the landform front that could be reliably identified is indicated in colour (blue, magenta, and orange line in 1978, 2000, and 2012, respectively). At each date the boundary between debris-covered and rock glacier morphology is depicted with a red line (dotted in 1978, dashed in 2000, and continuous in 2012).

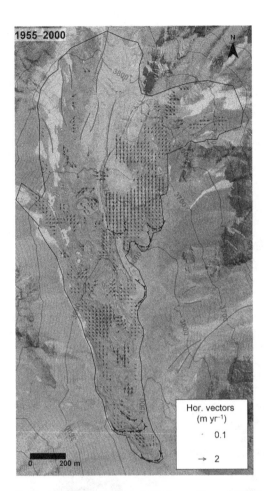

Figure 9. Horizontal displacements at the surface of Navarro between 1955 and 2000. The boundary between debris-covered and rock glacier morphology is depicted with a dotted red line in 1955 and with a dashed red line in 2014. Note that moraine crests and thermokarst depressions in 2000 are indicated. The background of the map is the 2000 orthophoto.

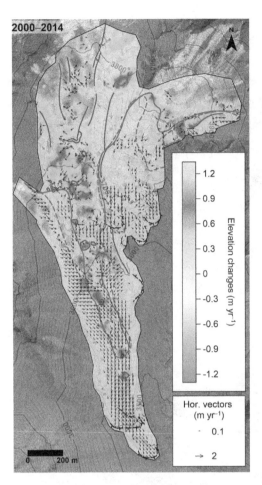

Figure 10. Horizontal displacements and elevation changes at the surface of Navarro between 2000 and 2014. The boundary between debris-covered and rock glacier morphology is depicted with a dashed red line in 2000 and with a continuous red line in 2014. Note that moraine crests and thermokarst depressions in 2014 are indicated. The background of the map is the 2014 Geoeye image.

heat transfers, which is likely to result in changes in flow dynamics depending upon the topography and the topoclimatic context. The displacement speed of the three studied landforms has decreased over the study period, at least over the overlapping areas where movement was detected in the different periods (Tables 4, 5a, b, and c). Whereas in other areas of the world many studies have reported rock glaciers to be accelerating under the current climate warming trend (e.g. Roer et al., 2005, 2008; Delaloye et al., 2010; Kellerer-Pirklbauer and Kaufmann, 2012), the decreased velocity highlighted in this analysis suggests an increasing stabilization of the landforms as they evolved from debris-covered glacier bodies to rock glaciers. As the transition from debris-covered to rock glacier seems to be proceeding mainly from the terminus upward, the increasingly debris-rich, lower rock glaciers may exert an increasing buttressing force on the remaining debris-covered glacier upslope, causing a general deceleration of the landform.

At Las Tetas, however, increasing displacement speeds downslope and the apparition of tension cracks in the lower rock glacier area during the recent (2000–2012) period point towards a possible acceleration or even destabilization of the landform terminus (Figs. 3, 8, and 14). Such evolution may be related to the observed decrease in modelled permafrost probability along the landform area (Fig. 5) and the climate evolution in this region: Rabatel et al. (2011) reported a warming trend of 0.19 °C decade^{-1} for the 1958–2007 period in the Pascua-Lama area, 80 km north of Las Tetas, and Monnier et al. (2014) also reported a trend of 0.17 °C decade^{-1} for the 1974–2011 period in the Río Colorado area. Such evolution is reminiscent of reports of acceleration and destabilization phenomena over rock glaciers in response to air and permafrost temperature increases (e.g. Roer et al., 2005, 2008; Delaloye et al., 2010; Kellerer-Pirklbauer and Kaufmann, 2012).

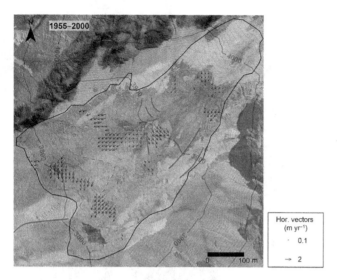

Figure 11. Horizontal displacements at the surface of Presenteser-acae between 1955 and 2000. The position of the base of the front at the two dates is indicated with dashed lines, as in Fig. 7; push moraine ridges in the upper part are also indicated. The background of the map is the 2000 orthophoto.

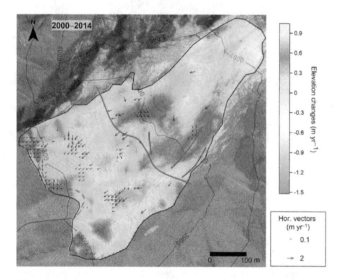

Figure 12. Horizontal displacements and elevation changes at the surface of Presenteseracae between 2000 and 2014. The position of the base of the front at the two dates is indicated with dashed lines, as in Fig. 7; the boundary between rock glacier and debris-covered glacier features and push moraine ridges in the upper part are indicated. The background of the map is the 2014 Geoeye image.

5.3.3 Final diagnostics and future evolution of the landforms

According to the results and interpretations presented for the Navarro's eastern part and Presenteseracae, rock glaciers can develop at the expense of debris-covered glaciers, by an upward progression of their morphology and correlative widespread development of cohesive mass flow. These

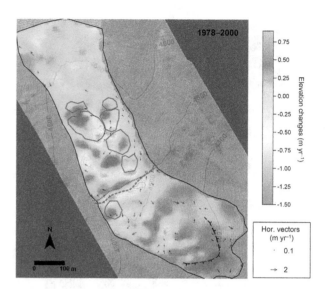

Figure 13. Horizontal displacements and elevation changes at the surface of Las Tetas between 1978 and 2000. The boundary between debris-covered and rock glacier morphology is depicted with a dotted red line in 1978 and with a dashed red line in 2000. Thermokarst depressions in 1978 are indicated. Thermokarst areas could not be accurately and reliably delineated on the 2000 orthophoto and are hence not mapped. The background of the map is the 2000 orthophoto.

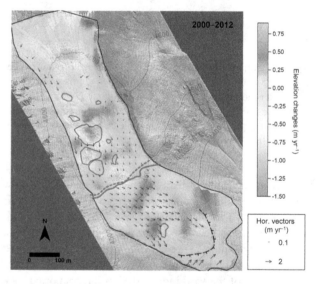

Figure 14. Horizontal displacements and altitudinal changes at the surface of Las Tetas between 2000 and 2012. The boundary between debris-covered and rock glacier morphology is depicted with a dashed red line in 2000 and with a continuous red line in 2012. Note that thermokarst depressions in 2012 are indicated. The background of the map is the 2012 Geoeye image.

are true cases of debris-covered glaciers evolving in rock glaciers (see Introduction: type iii). At Presenteseracae, however, the flow does not appear as strikingly cohesive as for the Navarro's western unit, possibly due to the smaller size of the landform as well as a steeper slope that may constitute a limiting dynamical parameter (Monnier and Kinnard, 2015). As these two landforms are located in favourable topoclimatic conditions, they should thus pursue their evolution towards rock glaciers. Despite the important insights presented by our study, it must be stressed that the evolution of the internal structure in response to morphological and dynamical changes at the surface remains unknown; it would require decades of borehole and geophysical survey monitoring to properly assess this. However, the transition may proceed by fragmentation of the glacier ice core and its mixing with debris and other types of ice (interstitial, intrusive) entrained from the surface. This is an alternative to the common and controverted model of the glacier ice-cored rock glacier where the evolution of the landform is controlled by the expansion and creep of a massive and continuous core of glacier ice (e.g. Potter, 1972; Whalley and Martin, 1992; Potter et al., 1998).

The Navarro's western unit and Las Tetas are more commonly known cases of assemblages that have formed and evolved in reaction to the superimposition/embedding of glaciers onto or in the back of rock glaciers and their subsequent dynamical interactions (see Introduction: type i). In both cases, the progression of the rock glacier at the expense of the debris-covered glacier is rather limited (Navarro's western unit) or null (Las Tetas). It is difficult to assert here whether the debris-covered glaciers are "pushing away" the rock glaciers or whether the latter are "pulling" the former; both processes probably occur (see also Sect. 5.3.2.). The dynamical links between both units certainly constitute a complex issue deserving more attention. Furthermore, as these whole landforms continue to advance, the rock glaciers could plausibly become entirely isolated from their main debris source in the upper cirques, while the increasingly warming conditions could cause the debris-covered glacier to become stagnant or disappear. Also, as the rock glaciers penetrate into areas with less favourable topoclimatic conditions, their future sustainment can be questioned.

6 Conclusion

We have used remote sensing techniques, including imagery orthorectification, DEM comparisons, and image feature tracking, in order to depict and measure the geomorphological evolution, elevation changes, and horizontal displacements of three glacier–rock glacier transitional landforms in the central Andes of Chile over a human-life timescale. Our study highlights how, as climate changes and mountain landscapes and their related dynamics shift, the glacial and periglacial realms can strongly interact. The pluri-decadal landscape evolution at the three studied sites is noticeable: rock glacier morphology areas expanded, as well

as the movement detection area in image feature tracking; thermokarst reduced; elevation changes tended to become more homogenous; and the mean horizontal displacement decreased and spatially coherent flow patterns enhanced. These overall results point toward the geomorphological and dynamical expansion of rock glaciers. However, the modalities and significance vary between sites. Navarro and Las Tetas are composite landforms resulting from the alternation between glacier (re)advance and rock glacier development phases; they currently exhibit an upward progression of the rock glacier morphology with associated cohesive mass flow and surface stabilization, or ice-loss-related downwasting and surface destabilization features. Presenteseracae is a special case of small debris-covered glacier that has evolved into a rock glacier during the last decades, with the rock glacier morphology having mostly developed ~ 15 years ago. Topoclimatic conditions appear to have been determinants in the landforms' evolution and, by extrapolation, could thus be expected to exert an important control on the development and conservation of underground ice in high-mountain catchments. From the latter point of view, our study stresses how spatial and dynamical interactions between glaciers and permafrost create composite landforms that may be more perennial than transitory: depending on the frequency of glacial–periglacial cycles, they participate in sustaining a hybrid cryospheric landscape that is potentially more resilient against climate change. This conclusion is of societal importance considering the location of the studied landforms in semiarid areas and the warming and drying climate predicted for the coming decades (Bradley et al., 2006; Fuenzalida et al., 2006).

We have furthermore provided new insights into the glacier–rock glacier transformation problem. Most of the common and previous glacier–rock glacier evolution models depicted a "continuum" process based on the preservation of an extensive core of buried glacier ice. However, our findings rather suggest that the transformation of a debris-covered glacier into a rock glacier may proceed from the upward progression of the rock glacier morphology at the expense of the debris-covered glacier, in association with an expanding cohesive mass flow regime and a probable fragmentation of the debris-covered glacier into an ice–rock mixture with distinct flow lobes. The highlighted importance of topoclimatic conditions and corresponding morphologic evolutions also supports the inclusion of the permafrost criterion within the rock glacier definition.

Competing interests. The authors declare that they have no conflict of interest.

Acknowledgements. This study is part of the Project Fondecyt Regular no. 1130566 entitled "Glacier-rock glacier transitions in shifting mountain landscapes: peculiar highlights from the central Andes of Chile". Fondecyt is the National Fund for Research and Technology in Chile. The authors want to thank Arzhan Surazakov, who performed the image processing, and Valentin Brunat, who was involved in the software handling and related data management in the framework of a master's thesis supported by the above-mentioned project. The authors also thank Andreas Kääb and Christophe Lambiel for their important help in improving this manuscript, as well as the associate editor for final edition and language corrections.

References

Azócar, G. F.: Modelling of permafrost distribution in the semiarid Chilean Andes, Master Thesis, University of Waterloo, Canada, 2013.

Azócar, G. F. and Brenning, A.: Hydrological and geomorphological significance of rock glaciers in the dry Andes, Chile (27°–33° S), Permafrost Periglac., 21, 42–53, 2010.

Azócar, G. F., Brenning, A., and Bodin, X.: Permafrost Favourability Index Map for the Chilean semi-arid Andes, Online data visualization, available at: www.andespermafrost.com (last access: 21 April 2017), 2016.

Azócar, G. F., Brenning, A., and Bodin, X.: Permafrost distribution modelling in the semi-arid Chilean Andes, The Cryosphere, 11, 877–890, https://doi.org/10.5194/tc-11-877-2017, 2017.

Barsch, D.: Permafrost creep and rock glaciers, Permafrost Periglac., 3, 175–188, 1992.

Benn, D. I. and Evans, D. J. A.: Glaciers and Glaciation, Routledge, London, 2010.

Berthling, I.: Beyond confusion: rock glaciers as cryo-conditioned landforms, Geomorphology, 131, 98–106, 2011.

Bodin, X., Brenning, A., Rojas, F.: Status and evolution of the cryosphere in the Andes of Santiago (Chile, 33.5° S), Geomorphology, 118, 453–464, 2010.

Bosson, J.-B. and Lambiel, C.: Internal structure and current evolution of very small debris-covered glacier systems located in alpine permafrost environments, Front. Earth Sci., 4, 1–17, 2016.

Bown, F., Rivera, A., and Acuña, C.: Recent glacier variations at the Aconcagua basin, central Chilean Andes, Ann. Glaciol., 48, 43–48, 2008.

Bradley, R. S., Vuille, M., Diaz, H. F., and Vergara, W.: Threats to water supply in the tropical Andes, Science, 23, 1755–1756, 2006.

Brenning, A.: Climatic and geomorphological controls of rock glaciers in the Andes of central Chile: combining statistical modelling and field mapping, PhD Thesis, Humboldt University, Berlin, 2005.

Cogley, J. G., Hock, R., Rasmussen, L. A., Arendt, A. A., Bauder, A., Braithwate, P., Jansson, P., Kaser, G., Möller, M., Nicholson, L., and Zemp, M.: Glossary of Glacier Mass Balance and Related Terms, UNESCO-IHP, Paris, 2011.

Debella-Gilo, M. and Kääb, A.: Sub-pixel precision image matching for measuring surface displacements on mass movements using normalized cross-correlation, Remote Sens. Environ., 115, 130–142, 2011.

Degenhardt, J. J.: Development of tongue-shaped and multilobate rock glaciers in alpine environments – Interpretations from ground penetrating radar surveys, Geomorphology, 109, 94–107, 2009.

Delaloye, R., Lambiel, C., and Gärtner-Roer, I.: Overview of rock glacier kinematics research in the Swiss Alps, Geogr. Helv., 65, 135–145, https://doi.org/10.5194/gh-65-135-2010, 2010.

Dusik, J.-M., Leopold, M., Heckmann, T., Haas, F., Hilger, L., Morche, D., Neugirg, F., and Becht, M.: Influence of glacier advance on the development of the multipart Riffeltal rock glacier, Central Austrian Alps, Earth Surf. Proc. Land., 40, 965–980, 2015.

Fuenzalida, H., Aceituno, P., Falvey, M., Garreaud, R., Rojas, M., and Sánchez, R.: Estudio de la variabilidad climática en Chile para el siglo XXI, Departamento de Geociencias, Universidad de Chile, Santiago, 2006.

Ginot, P., Kull, C., Schotterer, U., Schwikowski, M., and Gäggeler, H. W.: Glacier mass balance reconstruction by sublimation induced enrichment of chemical species on Cerro Tapado (Chilean Andes), Clim. Past, 2, 21–30, https://doi.org/10.5194/cp-2-21-2006, 2006.

Haeberli, W.: Creep of mountain permafrost: internal structure and flow of alpine rock glaciers, Mitteilungen der Versuchsanstalt für Wasserbau, Hydrologie und Glaziologie, Nr. 77, Zürich, 1985.

Haeberli, W.: Investigating glacier-permafrost relationships in high-mountain areas: historical background, selected examples and research needs, in: Cryospheric systems: glaciers and permafrost, edited by: Harris, C. and Harris, J. B., The Geological Society, London, 29–38, 2005.

Haeberli, W., Hallet, B., Arenson, L., Elconin, R., Humlum, O., Kääb, A., Kauffmann, V., Ladanyi, B., Matsuoka, M., Springman, S., and Vonder Mühll, D.: Permafrost creep and rock glacier dynamics, Permafrost Periglac., 17, 189–214, 2006.

Heid, T. and Kääb, A.: Evaluation of existing image matching methods for deriving glacier surface displacements globally from optical satellite imagery, Remote Sens. Environ., 118, 339–355, 2012.

Humlum, O.: The geomorphic significance of rock glaciers: estimates of rock glacier debris volumes and headwall recession rates in West Greenland, Geomorphology, 35, 41–67, 2000.

Janke, J. R., Bellisario, A. C., and Ferrando, F. A.: Classification of debris-covered glaciers and rock glaciers in the Andes of central Chile, Geomorphology, 241, 98–121, 2015.

Johnson, P. G.: Glacier–rock glacier transition in the Southwest Yukon territory, Canada, Arctic Alpine Res., 12, 195–204, 1980.

Kääb, A.: Remote sensing of mountain glaciers and permafrost, Zürich University, Switzerland, 2005.

Kääb, A., Haeberli, W., and Gudmundsson, G. H.: Analysing the creep of mountain permafrost using high precision aerial photogrammetry: 25 years of monitoring Gruben rock glacier, Swiss Alps, Permafrost Periglac., 8, 409–426, 1997.

Kääb, A. and Kneisel, C.: Permafrost creep within a recently deglaciated forefield: Muragl, Swiss Alps, Permafrost Periglac., 17, 79–85, 2006.

Kääb, A. and Vollmer, M.: Surface geometry, thickness changes and flow fields on creeping mountain permafrost: Automatic extraction by digital image analysis, Permafrost Periglac., 11, 315–326, 2000.

Kääb, A. and Weber, M.: Development of transverse ridges on rock glaciers: field measurements and laboratory experiments, Permafrost Periglac., 15, 379–391, 2004.

Kellerer-Pirklbauer, A. and Kaufmann, V.: About the relationships between rock glacier velocity and climate parameters in Central Austria, Austrian Journal of Earth Sciences, 105, 94–112, 2012.

Krainer, K. and Mostler, W.: Reichenkar rock glacier: a glacier-derived debris-ice system in the western Stubai Alps, Austria, Permafrost Periglac., 11, 267–275, 2000.

Lambiel, C. and Delaloye, R.: Contribution of real-time kinematic GPS in the study of creeping mountain permafrost: examples from the Western Swiss Alps, Permafrost Periglac., 15, 229–241, 2004.

Lugon, R., Delaloye, R., Serrano, E., Reynard, E., Lambiel, C., and González-Trueba, J. J.: Permafrost and Little Ice Age glacier relationships, Posets Massif, Central Pyrenees, Spain, Permafrost Periglac., 15, 207–220, 2004.

Messerli, A. and Grinsted, A.: Image Georectification and feature tracking toolbox: ImGRAFT, Geoscientific Instrumentation, Methods and Data Systems, 4, 23–34, 2015.

Monnier, S. and Kinnard, C.: Reconsidering the glacier to rock glacier transformation problem: new insights from the central Andes of Chile, Geomorphology, 238, 47–55, 2015.

Monnier, S., Kinnard, C., Surazakov, A., and Bossy, W.: Geomorphology, internal structure, and successive development of a glacier foreland in the semiarid Andes (Cerro Tapado, upper Elqui Valley, 30°08′ S, 69°55′ W), Geomorphology, 207, 126–140, 2014.

Monnier, S., Camerlynck, C., Rejiba, F., Kinnard, C., Feuillet, T., and Dhemaied A.: Structure and genesis of the Thabor rock glacier (Northern French Alps) determined from morphological and ground-penetrating radar survey, Geomorphology, 134, 269–279, 2011.

Nakawo, M., Raymond, C. F., and Fountain, A. (Eds.): Debris-covered Glaciers, IAHS Press, Wallingford, 2000.

Potter, N.: Ice-cored rock glacier, Galena Creek, Northern Absaroka Mountains, Wyoming, Geol. Soc. Am. Bull., 83, 3025–3058, 1972.

Potter, N., Steig, E. J., Clark, D. H., Speece, M. A., Clark, G. M., and Updike, A. U.: Galena Creek rock glacier revisited – new observations on an old controversy, Geogr. Ann. A, 80, 251–265, 1998.

Pourrier, J., H., Kinnard, C., Gascoin, S., and Monnier, S.: Glacier meltwater flow paths and storage in a geomorphologically complex glacial foreland: the case of the Tapado glacier, dry Andes of Chile (30° S), J. Hydrol., 519, 1068–1083, 2014.

Rabatel, A., Castebrunet, H., Favier, V., Nicholson, L., and Kinnard, C.: Glacier changes in the Pascua-Lama region, Chilean Andes (29° S): recent mass balance and 50 yr surface area variations, The Cryosphere, 5, 1029–1041, https://doi.org/10.5194/tc-5-1029-2011, 2011.

Ragettli, S., Cortés, G., McPhee, J., and Pellicciotti, F.: An evaluation of approaches for modelling hydrological processes in high-elevation, glacierized Andean watersheds, Hydrol. Process., 28, 5774–5695, 2012.

Rangecroft, S., Harrison, S., Anderson, K., Magrath, J., Castel, A. P., and Pacheco, P.: Climate change and water resources in arid mountains: an example from the Bolivian Andes, Ambio, 42, 852–863, 2013.

Ribolini, A., Chelli, A., Guglielmin, M., and Pappalardo, M.: Relationships between glacier and rock glacier in the Maritime Alps, Schiantala valley, Italy, Quaternary Res., 68, 353–363, 2007.

Ribolini, A., Guglielmin, M., Fabre, D., and Schoeneich, P.: The internal structure of rock glaciers and recently deglaciated slopes as revealed by geoelectrical tomography: insights on permafrost and recent glacial evolution in the Central and Western Alps (Italy–France), Quaternary Sci. Rev., 29, 507–521, 2010.

Roer, I., Kääb, A., and Dikau, R.: Rockglacier acceleration in the Turtmann valley (Swiss Alps): probable controls, Norsk Geografisk Tiddsskrift (Norwegian Journal of Geography), 59, 157–163, 2005.

Roer, I., Haeberli, W., Avian, M., Kaufmann, V., Delaloye, R., Lambiel, C., and Kääb, A.: Observations and considerations on destabilizing active rockglaciers in the European Alps, in: Proceedings of the Ninth International Conference on Permafrost, edited by: Kane, D. L. and Hinkel, K. M., University of Alaska, Fairbanks, 1505–1510, 2008.

Schroder, J. F., Bishop, M. P., Copland, L., and Sloan, V. F.: Debris-covered glaciers and rock glaciers in the Nanga Parbat Himalaya, Pakistan. Geogr. Ann. A, 82, 17–31, 2000.

Seppi, R., Zanozer, T., Carton, A., Bondesan, A., Francese, R., Carturan, L., Zumiani, M., Giorgi, M., and Ninfo, A.: Current transition from glacial to periglacial processes in the Dolomites (South-Eastern Alps), Geomorphology, 228, 71–86, 2015.

Wangensteen, B., Guðmundsson, A., Kääb, A., Farbrot, H., and Etzelmüller, B.: Surface displacements and surface age estimates for creeping slope landforms in Northern and Eastern Iceland using digital photogrammetry, Geomorphology, 80, 59–79, 2006.

Whalley, W. B. and Martin, H. E.: Rock glaciers. II. Models and mechanisms, Prog. Phys. Geog., 16, 127–186, 1992.

2

Geomorphic regulation of floodplain soil organic carbon concentration in watersheds of the Rocky and Cascade Mountains, USA

Daniel N. Scott and Ellen E. Wohl

Department of Geosciences, Colorado State University, Fort Collins, CO 80521, USA

Correspondence: Daniel N. Scott (dan.scott@colostate.edu)

Abstract. Mountain rivers have the potential to retain OC-rich soil and store large quantities of organic carbon (OC) in floodplain soils. We characterize valley bottom morphology, floodplain soil, and vegetation in two disparate mountain river basins: the Middle Fork Snoqualmie in the Cascade Mountains and the Big Sandy in the Wind River Range of the Rocky Mountains. We use this dataset to examine variability in OC concentration between these basins as well as within them at multiple spatial scales. We find that although there are some differences between basins, much of the variability in OC concentration is due to local factors, such as soil moisture and valley bottom geometry. From this, we conclude that local factors likely play a dominant role in regulating OC concentration in valley bottoms and that interbasin differences in climate or vegetation characteristics may not translate directly into differences in OC storage. We also use an analysis of OC concentration and soil texture by depth to infer that OC is input to floodplain soils mainly by decaying vegetation, not overbank deposition of fine, OC-bearing sediment. Geomorphology and hydrology play strong roles in determining the spatial distribution of soil OC in mountain river corridors.

1 Introduction

Terrestrial carbon storage plays an important role in regulating the global carbon cycle and the distribution of carbon between oceans, the atmosphere, long-term (10^5–10^9 years) storage in rock, and short- to moderate-term storage in the biosphere (10^1–10^4 years, including vegetation and soil) (Aufdenkampe et al., 2011; Battin et al., 2009). Soils, in particular, are a large organic carbon (OC) reservoir with significant spatial variability (Jobbágy and Jackson, 2000; Schmidt et al., 2011), making them difficult to characterize in the context of global carbon cycling. It is essential to quantify the spatial variability of OC stored in the biosphere to constrain the effects of climate change on feedbacks between biospheric and atmospheric carbon storage (Ballantyne et al., 2012). To provide a more complete understanding of how the biospheric carbon pool may change in the future and guide management of soil OC, we seek to provide a better con-straint on where carbon is stored in the biosphere and the processes that regulate that storage.

We focus here on river corridors, defined as channels, fluvial deposits, riparian zones, and floodplains (Harvey and Gooseff, 2015), which process, concentrate, transport, and store carbon (Wohl et al., 2017b). In the context of the carbon cycle, floodplains can act as a major component of the biospheric carbon pool (Aufdenkampe et al., 2011; Battin et al., 2009). Floodplain soils can act as a substantial pool of OC despite their relatively small aerial extent, indicating that floodplains may be disproportionately important compared to uplands in terms of carbon storage (D'Elia et al., 2017; Hanberry et al., 2015; Sutfin et al., 2016; Sutfin and Wohl, 2017; Wohl et al., 2012, 2017a). Mountainous regions, due to their high primary productivity (Schimel and Braswell, 2005; Sun et al., 2004), may play a substantial role in the freshwater processing and storage of OC whereby they retain sediment and water along the river network (Wohl et al., 2017b). Even laterally constrained floodplains in mountainous drainages can store significant quantities of OC that can be mobilized during floods (Rathburn et al., 2017). However, we lack a comprehensive characterization of how floodplain soil OC varies throughout watersheds. It is important to understand the spatial distribution of OC to predict its fate during floods and inform management to increase floodplain OC storage (Bullinger-Weber et al., 2014).

Floodplain OC enters river corridor soils via litterfall from vegetation and the erosion of OC-bearing bedrock (Hilton et al., 2011; Leithold et al., 2016; Sutfin et al., 2016). OC inputs are either allochthonous from upstream deposition of soil, particulate, and dissolved OC or autochthonous from riparian vegetation (Omengo et al., 2016; Ricker et al., 2013; Sutfin et al., 2016). As such, OC input can be regulated by vegetation dynamics and resulting litter input, hydrologic and sediment transport regimes, and water chemistry. However, it is unclear which OC inputs dominate under various conditions, hampering the prediction of changes in OC delivery to soils under a changing climate that may have significant effects on vegetation dynamics and hydrology.

OC concentration in soil is controlled by processes acting at multiple spatial scales. At broad, intra-basin scales, OC concentration in soil is regulated by the ability of carbon to sorb to soil particles and the ability of microbes and other organisms to respire soil OC, which can be controlled by rhizosphere dynamics, moisture, and temperature. Sorption of OC to soil particles reduces OC lability and is controlled by grain size and resulting available surface area as well as the availability of calcium, iron, and aluminum (Kaiser and Guggenberger, 2000; Rasmussen et al., 2018). Microbial respiration represents the primary pathway by which soil OC returns to the atmosphere. In general, low temperatures and frequent saturation inhibit microbial activity and promote OC storage (Falloon et al., 2011; Jobbágy and Jackson, 2000; Sutfin et al., 2016). At smaller, interbasin scales, the hydroclimatic regime controls vegetation dynamics, moisture, and temperature such that soil OC concentration in disparate regions can be approximately characterized by these predictors (Aufdenkampe et al., 2011; Schimel and Braswell, 2005). However, at the scale of a single watershed, hydrology, ecology, and geomorphology play strong roles in determining soil texture, moisture, and microbial dynamics, in turn controlling OC storage in valley bottoms (Scott and Wohl, 2017; Sutfin and Wohl, 2017; Wohl and Pfeiffer, 2018). We currently lack a comprehensive field-based examination of how processes acting at interbasin and intra-basin scales interact to regulate floodplain soil OC concentrations.

To address this knowledge gap, we quantify spatial variations in the OC concentration of floodplain soil across the entirety of two disparate mountain river networks. This allows us to examine interbasin hydroclimatic variation and intra-basin geomorphic and vegetation variation to understand the multi-scale controls on OC concentration. We use this multi-scale approach to draw inferences regarding the spatial distribution of floodplain OC, controls on that spatial distribution, and the dominant source of OC to mountain river floodplains.

Objectives and hypotheses

Across a basin, it is uncertain whether OC concentration in floodplain soils follows predictable longitudinal variation or is controlled by local factors. Similarly, in a vertical floodplain soil profile, it is uncertain whether OC concentration follows a trend similar to uplands, with declining OC concentration with depth, or exhibits vertical heterogeneity as a result of OC-rich layers deposited by floods. It is also unclear whether OC in floodplain soils is dominantly autochthonous or allochthonous. Floodplain soil OC source may be evident from the vertical heterogeneity of OC concentration: dominantly autochthonous OC profiles should decline with depth, whereas dominantly allochthonous OC profiles should exhibit vertical heterogeneity, reflecting episodic deposition. Our primary objective here is to understand spatial variations in OC concentration both with depth in a soil profile and across a basin. By quantifying these variations, we hope to infer the processes that regulate OC deposition in floodplain soil.

By examining two mountain river basins that differ in terms of hydroclimatic regime and vegetation characteristics, we can quantify both interbasin variation in OC storage and variation within each basin. We hypothesize that at an interbasin scale, the hydroclimatic regime and resulting rate of litterfall inputs in the riparian zone (Benfield, 1997) will dominantly regulate OC concentration (H1). We define hydroclimatic regime as the combination of precipitation and temperature dynamics that result in the vegetation characteristics of a basin. At an intra-basin scale, we expect that valley bottom geometry and river lateral mobility will regulate floodplain sediment characteristics and vegetation dynamics. Thus, we hypothesize that soil OC concentration does not vary along predictable longitudinal trends within mountain river basins, instead being more dominantly controlled by local fluvial processes and valley bottom form (H2a). We hypothesize that geomorphic process and form determine soil texture and moisture, which in turn set the boundary conditions that regulate the sorption of OC to mineral grains (promoting stabilization) and the potential of OC to be respired by microbes (H2b). In terms of OC inputs to floodplain soils, we hypothesize that the source of OC is dominated by autochthonous vegetation and litter inputs in these basins (H3). As such, we expect OC to dominantly decline with depth, only rarely exhibiting vertical heterogeneity that would represent allochthonous deposition from flooding.

2 Methods

This work was done alongside work presented in Scott and Wohl (2018b) and hence shares field sites, study design, GIS, and sampling techniques.

2.1   Field sites

We quantified soil organic carbon concentrations to a depth of approximately 1 m in the Big Sandy basin in the Wind River Range of Wyoming and the Middle Fork Snoqualmie basin in the central Cascade Mountains of Washington (Fig. 1). These basins represent distinct bioclimatic and geomorphologic regions ranging from the wet, high-relief Cascades to the semiarid, moderate-relief Middle Rockies.

The MF Snoqualmie has a mean annual precipitation of 3.04 m (Oregon State University, 2004), 2079 m of total relief over a 407 km^2 drainage area, and a mean basin slope of 60 %. Topography in the MF Snoqualmie is largely glaciogenic, with wide, unconfined valleys at both high and low elevations. Streams range from steep, debris-flow-dominated headwater channels to lower-gradient, wide, laterally unconfined channels in its lower reaches. The lower reaches of the MF Snoqualmie have been clear-cut extensively since the early 1900s, although there is little logging activity today. Vegetation follows an elevation gradient. The talus, active glaciers, and alpine tundra at the highest elevations transition to subalpine forests dominated by mountain hemlock (*Tsuga mertensiana*) (above approximately 1500 m), but also including Pacific silver fir (*Abies amabilis*) and noble fir (*Abies procera*) in the lower subalpine and montane zones (above approximately 900 m). Below the montane zone, uplands and terraces are covered by Douglas fir (*Pseudotsuga menziesii*) and western hemlock (*Tsuga heterophylla*), whereas active riparian zones are dominated by red alder (*Alnus rubra*) and bigleaf maple (*Acer macrophyllum*).

The semiarid Big Sandy is considerably drier than the MF Snoqualmie, but also exhibits broad, glacially carved valleys, especially in headwater reaches. It has a mean annual precipitation of 0.72 m (Oregon State University, 2004), 1630 m of total relief over a 114 km^2 drainage area, and a mean basin slope of 25 %. The lower reaches of the Big Sandy are anthropogenically impacted by moderate grazing use and an access road that crosses through part of the basin. Herbaceous alpine tundra dominates higher elevations (above approximately 3100 m), while the subalpine zone (approximately 2900 to 3100 m) is characterized by forests of whitebark pine (*Pinus albicaulis*), Engelmann spruce (*Picea engelmannii*), and subalpine fir (*Abies lasiocarpa*). The montane zone (approximately 2600 to 2900 m) is comprised dominantly of lodgepole pine (*Pinus contorta*). Only a small portion of this basin (approximately 1 %) resides below 2500 m, where shrub steppe begins to dominate (Fall, 1994). Parklands and meadows are abundant in this basin, creating a patchy forest structure. Comparing this basin to the MF Snoqualmie provides bioclimatic contrast that allows us to examine how floodplain soil OC concentrations vary across a range of stream morphologies and floodplain morphologic types in regions with differing precipitation, forest characteristics, and basin morphology.

To simplify comparison of these two basins, we henceforth refer to them by their dominant climate. The Middle Fork Snoqualmie is the wet basin and the Big Sandy is the semiarid basin (Fig. 1).

2.2   Study design and sampling

We sampled the semiarid basin in summer 2016 and the MF Snoqualmie in summer 2017. During each sampling campaign, no large floods occurred and we observed no floodplain erosion or deposition. Across both basins, we cored a total of 128 floodplain sites to determine soil OC concentration. Cores were collected as a series of individual soil samples at both regular and irregular depth increments.

2.2.1   Semiarid basin (Big Sandy)

The sparse vegetation in the semiarid basin enabled us to use a combination of a 10 m DEM and satellite imagery to manually map the extent of the valley bottom along the entire stream network and delineate valley bottoms based on confinement. We defined unconfined valley bottoms as those in which channel width occupied no more than half the valley bottom, and confined valley bottoms as those in which channel width occupied greater than half the valley bottom. Within each confinement stratum, we stratified the stream network by five drainage area classes to produce a total of 10 strata, ensuring even sampling across the basin. Within each of the resulting 10 strata, we randomly selected 5 reaches, producing a total of 50 sample sites throughout the basin. Due to access issues, we sampled 48 out of the 50 randomly located sites. We supplemented these with 4 subjectively located sites that we felt enhanced our ability to capture variation throughout the drainage based on observations in the field, resulting in a total of 52 sampled sites.

2.2.2   Wet basin (Middle Fork Snoqualmie)

The wet basin is larger than the semiarid basin and has extensive, low-gradient floodplains in its downstream reaches. These extensive floodplains display high spatial variability in vegetation, surface water, grain size, and estimated surface age based on aerial imagery and ground reconnaissance. To ensure an unbiased characterization of these heterogeneous floodplains, we used aerial imagery, a 10 m DEM, and pictures from field reconnaissance to delineate the floodplain into patch categories: fill channels (abandoned channels that have had enough sediment deposited to prevent an oxbow lake from forming), point bars (actively accreting surfaces on the inside of bends), wetlands (areas with standing water in imagery that are not obviously oxbows), oxbow lakes (abandoned channels dammed at the upstream and downstream ends to form a lake), and general floodplain surfaces (surfaces that cannot be classified into any of the above cat-

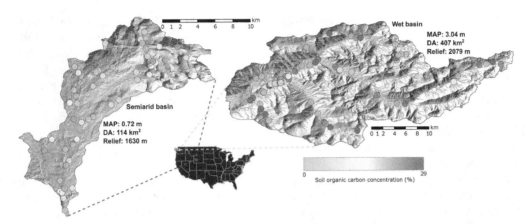

Figure 1. Hillshade showing the location, sampling sites, and stream network of the sampled basins. Big Sandy, Wyoming (42.69°, −109.25°), is on the left and MF Snoqualmie, Washington (47.53°, −121.52°), is on the right. Circles represent sampling locations at which floodplain soil OC was measured. Sample sites are colored by OC concentration. Mean annual precipitation (MAP), drainage area (DA), and relief are given for each basin.

egories). Within each of these five categories, we randomly selected six points at which to take soil cores.

We also stratified the entire wet basin stream network by channel slope into four strata. Within each channel slope strata (hereafter referred to as slope strata), we randomly selected 10 reaches to collect a single floodplain soil core, resulting in 40 randomly located sample sites.

To supplement randomly sampled sites and accommodate the infeasibility of accessing two of the randomly sampled sites along the stream network, we also subjectively selected sample sites in places that we felt enhanced the degree to which our sampling captured the variability present among streams in the basin. This resulted in a total of 46 sites stratified by slope, 38 of which were randomly sampled, in addition to 30 sites stratified by floodplain type.

2.3 Reach-scale field measurements

At each sampled reach (100 m or 10 channel widths, whichever was shorter), we measured channel geometry and other characteristics, although our measurements were not consistent across all basins because field protocol evolved during the course of the study. In both basins, we measured confinement, valley bottom width, and channel bed slope. We additionally measured bankfull width and depth in the wet basin. We did not measure channel characteristics for sites stratified by floodplain type in the wet basin, since they did not correspond to a single reach of channel as sites stratified by slope did in much more confined valleys.

In the wet basin, we also categorized channels by planform and dominant bedform (Montgomery and Buffington, 1997). We defined planforms as follows: straight, in which the channel was generally confined and significant lateral migration was not evident; meandering, in which lateral migration was evident but only a single channel existed; anastomosing, in which vegetated islands separated multiple channels; and

anabranching, in which a single dominant channel existed with relict channels separated by vegetated islands. We further classified channels as being either multithread (anastomosing or anabranching) or single thread (straight or meandering). Because logging records are inconsistent and likely inaccurate in the wet basin (based on the frequent observation of past logging activity that was not recorded in Forest Service records), we noted whether signs of logging, such as cut stumps, cable, decommissioned roads or railroads, or other logging-associated tools, were found near the reach.

We chose a representative location on the floodplain for each sampled site based on visual examination of vegetation type, soil surface texture, surface water presence, and elevation relative to the bankfull channel elevation (floodplain sites stratified by type in the wet basin were sampled as close to the randomly sampled point as possible). Once a location was chosen, we extracted a 32 mm diameter soil core using an open-sided corer (JMC Large Diameter Sampling Tube). Due to our adaptive methodology, we sampled soil OC slightly differently in the semiarid versus the wet basin. In the semiarid basin, we cored in irregular increments, generally 25–30 cm. After analyzing data from the semiarid basin, we realized that sampling in regular increments would make analysis more versatile. As a result, we switched to extracting soil samples at regular 20 cm increments in the wet basin. Cores were taken to refusal (i.e., coarse gravel or other obstructions preventing further soil collection) or a depth of approximately 1 m. Five cores in the semiarid basin, 12 cores in the wet basin sites stratified by slope, and 11 cores in the wet basin sites stratified by floodplain type did not reach refusal. When no sand or finer sediment was present in the valley bottom (only occurred in headwater channels of the wet basin), we recorded negligible OC concentration. Once soil samples were removed from the ground, they were placed in ziplock bags, frozen within 72 h (most samples were frozen within 8 h), and kept frozen until analysis.

2.4 Measuring soil OC and texture

To measure the concentration of organic carbon in soil samples, we used loss on ignition (LOI). We first defrosted samples for 24–48 h at room temperature. Once defrosted, we thoroughly mixed samples to ensure the most homogenous sample possible. We then subsampled 10–85 g of soil from each sample for analysis. Using crucibles in a muffle furnace, we dried samples in batches of 30 for 24 h at 105 °C to determine moisture content and remove all nonstructurally held water. Following the guidelines suggested by Hoogsteen et al. (2015), we then burned samples for 3 h at 550 °C to remove organic matter. By comparing the weight of the burned samples with that of the dried samples, we obtained an LOI weight.

After performing LOI, we used burned samples to perform texture by feel to determine the USDA soil texture class and estimated clay content (Thien, 1979). To convert LOI weight to OC concentration, we used the structural water loss correction of Hoogsteen et al. (2015) using clay content estimated from soil texture. This correction considers water held by clay that may not evaporate during drying but will evaporate during burning. It also estimates the proportion of the LOI weight that is OC. This correction, represented as a percent of the estimated sample OC content after the correction, ranged from 0.90 % to 495.76 % (95 % confidence interval of the median between 16.17 % and 24.25 %) for the wet basin, and ranged from 5.54 % to 570.53 % (95 % confidence interval of the median between 19.76 % and 32.11 %) for the semiarid basin.

One potential confounding factor in LOI is carbonates that may burn off during ignition, adding to the LOI weight while not being organic matter. In lithologies where carbonates are rare (e.g., granitoid rocks like those found in the upper part of the wet basin and entire semiarid basins), this is a relatively negligible issue. However, some of our soil samples came from parts of the wet basin draining rocks of the western mélange belt, including argillite, greywacke, and marble. We tested samples for the presence of carbonates to determine whether our LOI methods would be sufficient to accurately determine OC concentration. We randomly chose 10 soil samples of a total of 110 drained rocks that could include carbonates and submitted them to the Colorado State University soil testing laboratory for CHN furnace analysis (Sparks, 1996), which yielded data on the proportion of carbonates by mass in those samples. The median calcium carbonate concentrations of those 10 samples was 0.98 % (95 % confidence interval between 0.70 % and 1.03 %), and the median percentage of the total carbon in samples comprised of inorganic carbon was 5.05 % (95 % confidence interval between 1.97 % and 18.9 %). From this, we concluded that the amount of carbonate in the samples draining potentially carbonate-

bearing rocks was low enough that LOI was likely to still be accurate. Consequently, we analyzed all soil samples using LOI to obtain OC concentration. The median difference between the LOI and CHN OC estimates for these 10 samples was 0.45 % (95 % confidence interval between −0.83 % and 1.47 %), indicating no systematic bias in LOI estimates of OC concentration.

2.5 GIS and derivative measurements

After fieldwork in each basin, we collected the following data for each reach using a GIS platform: elevation, drainage area, land cover classification and canopy cover from the National Land Cover Database (Homer et al., 2015), and the mean slope of the basin draining to each reach (including hillslopes and channels). Utilizing drainage area at each reach and field-measured channel gradient, we calculated an estimated stream power as the product of drainage area, channel gradient, and basin-averaged precipitation. We utilized a 10 m DEM for all GIS topographic measurements. To estimate clay content for each sample, we used median values for assigned USDA texture classes. To obtain estimated clay content, moisture, and OC for each core, we calculated an average weighted by the percentage of core taken up by each soil sample. We categorized wet basin samples stratified by floodplain type into those with standing water (wetlands and oxbow lakes) and those with no standing water (all other types).

2.6 Statistical analyses

All statistical analyses were performed using the R statistical computing software (R Core Team, 2017). We conducted all analyses on three modeling groups based on the variables measured in each group. In the wet basin, we grouped observations by stratification type, separating observations stratified by channel slope from observations stratified by floodplain type. We separated these two groups from all observations in the semiarid basin, which were measured consistently. We modeled OC concentration and soil texture with a mixed-effects linear regression using individual soil samples (i.e., the individual samples that make up a core) as sample units ($n = 103$ for wet basin stratified by slope, 89 for wet basin stratified by floodplain type, and 101 for semiarid basin). We modeled the sampled site as a random effect, acknowledging that individual soil samples within a single core are likely nonindependent. We used profiled 95 % confidence intervals on effect estimates (β) for fixed effects to evaluate variable importance in mixed-effects models.

To gain further insight at the reach scale, we also modeled average OC concentration and soil moisture at each measured site using multiple linear regression. We modeled soil moisture at the reach scale because we felt that our single snapshot of moisture conditions was better represented as a site-level average. We first performed univariate analysis between each hypothesized predictor and the response, utilizing mainly comparative Wilcoxon rank-sum tests (Wilcoxon, 1945) or correlational Spearman correlation coefficient statistics. We utilized a Holm multiple-comparison correction (Holm, 1979) for pairwise comparisons. During this filtering, we also viewed box plots or scatterplots as appropriate to discern which variables appear to have anything other than a completely random relationship with the response. We then utilized the multiple linear regression of all subsets using the corrected Akaike information criterion as a model selection criterion (Wagenmakers and Farrell, 2004). We iteratively transformed response variables to ensure the homoscedasticity of error terms. To select a single best model, we utilized both Akaike-weight-based importance and parsimony to select a final reduced model. We considered sample sizes, p values, and effect magnitudes in determining variable importance.

We also analyzed each core to determine whether there were buried high-OC-concentration layers at depth. We compared each buried soil sample to the sample above it using the criterion that a peak in OC at depth should have an OC concentration 1.5 times that of the overlying sample and be above 0.5 % (Appling et al., 2014).

3 Results

Model results are presented in Table 1. Comparisons between basins and summaries of OC concentration, moisture, and estimated clay content are shown in Fig. 2.

3.1 OC concentration

Most cores display a decrease in OC concentration with depth (Fig. 3). Of cores with more than a single sample, 32 % (7/22) of cores stratified by slope in the wet basin, 32 % (8/25) of cores stratified by floodplain type in the wet basin, and 6 % (2/31) of cores in the semiarid basin exhibit OC concentration peaks at depth. Whether a soil sample was classified as an OC peak has no relation to estimated clay content in sites stratified by floodplain type ($p = 0.28$) or those stratified by slope ($p = 0.89$) in the wet basin. In the semiarid basin, soil samples classified as buried OC peaks have significantly higher estimated clay contents ($p = 0.05$) than those that were not classified as peaks.

In general, the floodplain-stratified sites in the wet basin store higher densities of OC than the semiarid basin (Fig. 2a, b). Figure 2a includes zero values (i.e., sites with no OC-bearing sediment; only present in the wet basin slope-stratified group), whereas Fig. 2b does not because sample units in Fig. 2b are individual soil samples. Comparing these two groups, it appears that soils in the wet basin exhibit much higher OC concentrations than those in the semiarid basin, but in general, there are many more reaches with no fine sediment available to store OC in the wet basin.

At the scale of individual soil samples, the depth below ground surface is by far the dominant control on OC concentration across all modeling groups. We used a cube root transform for all three mixed-effects models of OC concentration. For wet basin sites stratified by slope, deeper soil samples contain less OC ($\beta = -0.0084 \pm 0.0042$), whereas soil samples at higher elevations tend to contain more OC ($\beta = 0.0010 \pm 0.00099$). Depth is the only significant predictor of OC content for both wet basin soil samples stratified by floodplain type ($\beta = -0.0084 \pm 0.0042$) and soil samples in the semiarid basin ($\beta = -0.0037 \pm 0.0019$).

Modeling wet basin slope-stratified sites at the reach scale, we found that moisture ($\beta = 0.0078 \pm 0.0031$) and whether the reach was unconfined ($\beta = 0.77 \pm 0.49$) control soil OC (cube root transformation, model adjusted $R^2 = 0.54$, $p < 0.0001$). Modeling wet basin floodplain-stratified sites at the site scale, we found that canopy cover ($\beta = 0.012 \pm 0.011$) and moisture ($\beta = 0.0040 \pm 0.0011$) are controls on soil OC (cube root transformation, model adjusted $R^2 = 0.67$, $p < 0.0001$). Modeling Big Sandy sites at the reach scale, we found that soil depth ($\beta = -0.012 \pm 0.0071$) and moisture ($\beta = 0.014 \pm 0.0026$) are dominant controls on soil OC concentration (no transformation, model adjusted $R^2 = 0.69$, $p < 0.0001$).

In general, moister, deeper soils store more OC at the reach scale, whereas OC tends to vary dominantly with depth at the scale of individual soil samples. Although estimated clay content did not emerge as a significant predictor of OC concentration, it is used to calculate clay-held water to correct our LOI-based OC concentration measurements, making it important in determining OC for each sample.

3.2 Soil texture

In general, soil texture follows a predictable trend with river size between model groups (Fig. 2d). Floodplain type-stratified sites in the wet basin store the most clay, followed by slope-stratified sites and then sites in the semiarid basin.

Modeling soil texture at the individual soil sample scale across slope-stratified sites in the wet basin, we found whether the reach was confined ($\beta = 5.42 \pm 5.32$) and whether the bed material was dominantly sand ($\beta = 10.47 \pm 6.13$) to be dominant controls on estimated clay content. Modeling soil texture for sites stratified by floodplain type in the wet basin yielded no significant trends. In the semiarid basin, we found that either valley width ($\beta = 0.0050 \pm 0.0032$) or whether the stream was unconfined ($\beta = 0.41 \pm 0.33$) as well as depth below ground surface ($\beta = -0.0069 \pm 0.0033$ for model with valley width but not confinement) significantly control soil texture.

To summarize, sites from unconfined, lower-energy reaches in the wet basin and sites from reaches with wider valley bottoms and at lower depths in the semiarid basin exhibited more finely textured soils.

Table 1. Matrix of all models presented in the text. Each model is listed by model group, response variable, and scale. Scale refers to the sample unit of the model, and site refers to a core, with the response averaged over all the individual soil samples in the core. For each variable and model, an asterisk (*) indicates that the variable was included in model selection but not the full mixed-effects model. A minus (−) indicates that the variable was selected as important in predicting the response and denotes an indirect correlation, whereas a plus (+) indicates a direct correlation. In the case of confinement, a plus indicates that unconfined streams display a higher-magnitude response variable. In the case of bed material, a plus indicates that samples with sand exhibit a higher value of the response. NA indicates that either the variable was not measured for that basin or model group or that it is the model response.

Model group	Response (sample unit)	Scale	Confinement	Bedform	Channel slope	Bed material	Multithread	Valley width	Bankfull width	Bankfull depth	Stream power	Floodplain type	Standing water	Depth[1]	Clay content	Moisture	Logging nearby	Grasses present	Shrubs present	Trees present	Elevation	Basin slope	Canopy cover	NLCD	Drainage area
Wet basin stratified by slope	OC (%)	Soil sample	*		*		*				*	NA	NA	−	*	*	*				+		*		
	OC (%)	Site	+		*		*					NA	NA	*		+					*		*		
	Moisture (%)	Site	+		−						*	NA	NA	*	NA	NA					+		*		
	Texture (%)	Soil sample	+	*	*	+	*		*			NA	NA	*	NA						*	*			
Wet basin stratified by floodplain type	OC (%)	Soil sample	NA	NA	NA	NA	NA	NA	NA	NA	NA	*	*	−	*	*	*			*	*	*	*		NA
	OC (%)	Site	NA	NA	NA	NA	NA	NA	NA	NA	NA		*	*		+					*	*	+		NA
	Moisture[2] (%)	Site	NA	NA	NA	NA	NA	NA	NA	NA	NA	*	+	*	+	NA		*	*	*	*		*		NA
	Texture[2] (%)	Soil sample	NA	NA	NA	NA	NA	NA	NA	NA	NA		*	*	NA										NA
Semiarid basin	OC (%)	Soil sample	*	NA	*	NA	NA	*	NA	NA	*	NA	NA	−	*	*	NA	NA	NA	NA	*		*		
	OC (%)	Site	*	NA	*	NA	NA		NA	NA	*	NA	NA	+		+	NA	NA	NA	NA	*		*		
	Moisture (%)	Site	+	NA		NA	NA	+	NA	NA	*	NA	NA	+	NA	NA	NA	NA	NA	NA	+	*	*		
	Texture[3] (%)	Soil sample	+	NA	*	NA	NA		NA	NA	*	NA	NA	−	NA		NA	NA	NA	NA		*	*		

[1] Depth refers to either the soil sample depth below the ground or the total depth of the core, depending on the sample unit. [2] No significant results were observed for this model. [3] For this model, both valley width and confinement predict texture and can be interpreted interchangeably. However, including both in the same model would yield problems due to multicollinearity.

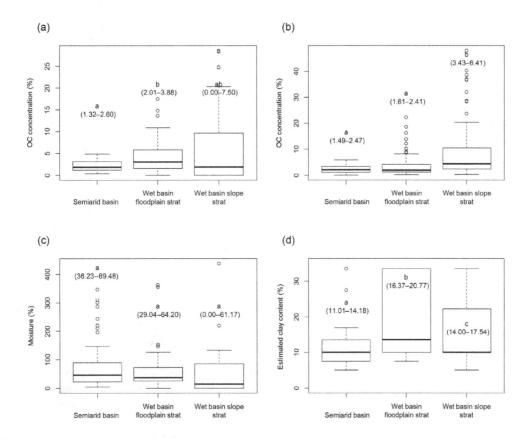

Figure 2. Box plots showing comparisons between model groups of OC concentration at the reach scale (**a**), OC concentration at the scale of individual soil samples (**b**), moisture at the reach scale (**c**), and estimated clay content at the scale of individual soil samples (**d**). Ends of dotted lines represent 1.5 times the interquartile range, which is represented by boxes. Bold line represents median. Circles represent outliers. Letters indicate probable differences between groups based on pairwise Wilcoxon (**a–c**) or t tests (**d**) with a Holm correction. Ranges in parentheses below letters show the 95 % confidence interval of the median value for the group (**a–c**) or the mean value for the group (**d**) in which median confidence intervals were overly constrained due to the categorical nature of our estimated clay content data.

3.3   Soil moisture

Soil moisture is less variable between basins than either texture or OC concentration (Fig. 2c). All model groups exhibit similar soil moisture conditions, although there was significant variability within each model group.

Soil moisture at wet basin sites stratified by slope is dominantly controlled by channel slope ($\beta = -13.15 \pm 9.11$), elevation ($\beta = 0.0046 \pm 0.0038$), and whether the stream is unconfined ($\beta = 3.89 \pm 2.67$; model adjusted $R^2 = 0.38$, $p < 0.0001$). At wet basin sites stratified by floodplain type, estimated clay content ($\beta = 0.060 \pm 0.042$) and whether the floodplain unit had standing water ($\beta = 1.15 \pm 0.71$) significantly control soil moisture (model adjusted $R^2 = 0.46$, $p < 0.0001$). In the semiarid basin, soil depth ($\beta = 0.012 \pm 0.012$), elevation ($\beta = 0.0023 \pm 0.0014$), and whether the reach was unconfined ($\beta = 0.91 \pm 0.75$) significantly control soil moisture (model adjusted $R^2 = 0.35$, $p < 0.0001$).

4   Discussion

4.1   Understanding spatial variability in OC concentration in floodplain soils (H1 and H2)

Comparing the wet basin to the semiarid basin shows that the wetter, higher primary productivity basin is capable of storing greater concentrations of OC in floodplain soils, but that both regions generally store similar OC concentrations in floodplain soils. This result partially agrees with the examination of subalpine lake deltas by Scott and Wohl (2017). In that study, subalpine lake deltas in the wet basin were compared to deltas in the drier Colorado Front Range. Subalpine lake deltas displayed similar OC concentrations, likely due to competing but complementary OC stabilization and loss mechanisms in each region. Those deltas represent a subset of the broader valley bottom soils studied here. This more expansive study points to both geomorphic controls, such as valley bottom geometry, and factors influenced by climate, such as canopy cover, as controls on OC storage in valley bottoms. These results also agree with the results of Lininger et

(a)

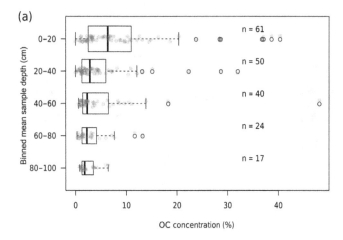

(b)

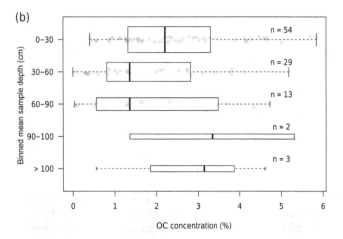

Figure 3. Box plots of sample OC concentration binned by mean sample depth for the wet (**a**) and semiarid (**b**) basins. Ends of dotted lines represent 1.5 times the interquartile range, which is represented by boxes. Bold line represents median. Black circles represent outliers. Transparent grey points show all data for each bin. Sample size for each bin is denoted by *n*.

al. (2018), which indicate that geomorphic context and vegetation dynamics control OC concentration on floodplain soils along large lowland rivers in Alaska, USA.

At the reach or site scale, wetter soil profiles consistently yield higher OC concentrations in all model groups. However, moisture does not differ significantly between model groups (Fig. 2c), indicating that this alone cannot explain differences between basins. Soils tend to be finer in the wet basin, but clay content was not an important predictor of OC concentration in studied soils. Clay content likely influences OC concentration based on previous research (Hoffmann et al., 2009), although the inclusion of coarse soil material (including particulate organic matter) in our samples may explain the lack of an observed correlation here. Variability in soil redox conditions (i.e., the presence of anaerobic microsites) may also introduce variability in respiration rate between basins and among individual soil samples (Keiluweit

et al., 2017). That is, variability in groundwater hydrology may result in a heterogeneous mix of aerobic (favoring microbial activity) and anaerobic (suppressing microbial activity) conditions within floodplain soils. We did not record redoximorphic features of our soil samples, so we are unable to determine whether this influence contributes to the error in our models. Although confinement plays a strong role in determining OC concentration in wet basin sites stratified by slope, it does not differ significantly between basins (52 % of semiarid basin reaches are unconfined compared to 63 % of wet basin reaches). The major differences between these basins are their hydroclimatic and disturbance regimes. The wet basin is at a lower elevation and has denser and higher biomass forests (Smithwick et al., 2002) compared to the sparser parkland forests of the semiarid basin, which likely also experiences more frequent fires based on fire histories of nearby regions (recurrence interval on the order of 10^1–10^2 years; Houston, 1973; Loope and Gruell, 1973). In addition, the volcanic soils present in the wet basin may suppress soil OC respiration (Matus et al., 2014), leading to a higher soil OC storage capacity there.

Between basins, it is likely that the hydroclimatic regime influencing primary production plays some role in the wet basin's higher maximum OC concentrations in floodplain soils compared to those of the semiarid basin. However, smaller-scale factors such as soil texture and moisture also likely play a role and are not related to drainage area (Table 1), indicating that neither OC concentration nor its controlling factors vary continuously along a river network, thus supporting H2a and H2b. This also indicates that local factors, set largely by geomorphic and hydrologic dynamics, play a significant role in modulating the effect of climate on OC concentrations. If the wet and semiarid basins displayed significantly different OC concentrations, our first hypothesis regarding the interbasin controls on OC concentration would be supported. However, we instead found that climate and primary productivity only partially determine OC concentrations, especially when viewed in the context of geomorphic and hydrologic variability. Thus, the results do not support H1.

Each basin (or model group) is slightly different in terms of the controls on soil OC concentration, moisture, and texture. In the wet basin sites stratified by slope, higher-elevation sites display higher OC concentrations. This is contrary to the general trend in primary productivity, which decreases with increasing elevation. However, it is important to note that the headwaters of the wet basin are dominated by lakes, deltas, and other depositional features in relatively broad, glacially carved valleys. Subalpine lake deltas have been shown to store high OC concentrations in this basin (Scott and Wohl, 2017), and many of the highest OC concentrations we measured were located in broad, wet meadows, subalpine lake deltas, or other unconfined high-elevation reaches. Such unconfined sites also likely have significantly cooler temperatures and tend to have higher soil moisture contents, as shown by our modeling (Table 1). As such, although high-elevation wet basin sites may receive less OC input, they likely have a low rate of OC respiration, resulting in higher OC concentrations on the whole, which agrees with the result of Bao et al. (2017). In the semiarid basin, our modeling suggests that the lower temperatures and higher moisture (Table 1) at higher elevations do not compensate for the lower primary productivity, as elevation does not correlate with OC concentration.

In both basins, unconfined reaches contain wetter and finer-textured soils, which may result in a higher soil OC capacity. Although confinement only relates directly to OC content in wet basin sites, it does play a strong role in determining moisture, which in turn plays a role in regulating OC concentration in both basins, likely via inhibiting microbial activity (Howard and Howard, 1993). The relevance of channel slope in determining soil moisture in the wet basin but not semiarid basin may reflect the prevalence of high-gradient, debris-flow-dominated channels in the wet basin that largely exhibit only gravel to boulder substrate, which we assume stores minimal fine sediment, moisture, or OC.

In the semiarid basin, higher soil depths correlate with moister and finer-textured soils, but less OC concentration. This indicates the trend in OC with depth likely dominates the signal of OC concentration, with deeper sites containing a higher proportion of OC-depleted deep samples.

4.2 Inferring sources of OC to floodplain soils (H3)

OC can be input to floodplain soils by two primary mechanisms. First, dissolved and particulate OC (including large wood) can be deposited on floodplain surfaces by overbank deposition, thus integrating fluvial sedimentary OC into the floodplain soil profile or, in the case of large wood, depositing discrete but potentially large concentrations of OC that can later be integrated into the soil profile. Second, litter and decomposing vegetation on the floodplain surface can act as autochthonous inputs of OC to floodplain soil.

Our modeling of OC concentration yielded results consistent with previous investigations of controls on soil OC storage capacity (Jobbágy and Jackson, 2000; Sutfin and Wohl, 2017). Sites in the heterogeneous floodplain of the wet basin display a direct correlation between canopy cover and OC concentration, indicating that increased litter inputs lead to increased floodplain soil OC concentration. Sediment inputs likely differ between floodplain depositional unit types (e.g., coarser sediment may deposit on point bars compared to filled secondary channels), although floodplain type does not predict OC concentration. This indicates that vegetation inputs may be more dominant than fluvial sediment inputs at these sites.

The finding that buried OC peaks in the wet basin do not have abnormally high clay contents supports the interpretation that wood and litter inputs to soil are the dominant source of OC in the floodplain soils we examined. Buried peaks can be layers created by overbank deposition and subsequent burial of fine, OC-bearing sediments (Blazejewski et al., 2009; Ricker et al., 2013), buried pieces of wood (Wohl, 2013), or buried organic horizons that are now capped by sediments that prevent OC respiration. If overbank deposition of fine sediment caused OC peaks, we would expect to see the soil samples classified as peaks exhibiting high clay contents, indicating finer sediment. Instead, our results suggest that in the wet basin, buried peaks are likely the result of either buried organic horizons or buried wood. We observed large pieces of decaying, buried wood in floodplain cut banks in the wet basin, supporting this inference. Overbank deposition of wood on the floodplain was only observed rarely in this basin, indicating that the OC measured in these soils is likely dominantly autochthonous.

In the semiarid basin, the two cores that exhibit peaks were collected from the same meadow, just downstream of a now-filled former lake that is a potential source of fine sediment. The channels draining this meadow exhibit an anabranching planform, indicating the potential to deposit and bury packets of potentially OC-rich, fine sediments. However, the majority of cores do not exhibit OC peaks, indicating OC input mainly from vegetation at the surface and continuing OC respiration at depth.

OC variation within each core is dominantly a function of depth. We observe a negative correlation between depth below ground surface and OC concentration, which has been observed in other studies, including mountain wetlands and floodplains (Jobbágy and Jackson, 2000; Scott and Wohl, 2017; Sutfin and Wohl, 2017; Zhao et al., 2017). In general, this indicates that at least in mountain river floodplains, OC is enriched at the surface and decomposes with depth, similar to upland soils. This fits with our finding that the majority of our cores do not exhibit significant OC peaks at depth and supports the dominance of litter and wood OC inputs to floodplain soils. These results support our hypothesis that decaying litter and wood, not overbank sediment deposition, dominate the input of OC to floodplain soils in our study basins (H3). We note that floodplain wood may also act as a trapping site for overbank fine organic matter, facilitating the deposition and input of OC from decaying vegetation. While these study basins likely accumulate soil OC mainly autochthonously, other basins that experience overbank flows, accompanying deposition of fine sediment, and burial of organic layers exhibit OC storage that is likely dominated by fluvial sediment deposition (e.g., Blazejewski et al., 2009; D'Elia et al., 2017; Ricker et al., 2013). Thus, it is likely that flow regime, lateral connectivity, and sediment transport dynamics regulate whether floodplain soil OC is dominantly input by the overbank deposition of fine material or litter and wood decay.

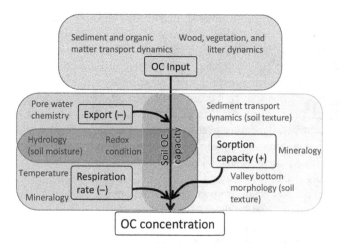

Figure 4. Conceptual model of physical processes that influence OC concentration in floodplain soils based on this study and other literature. Each box corresponds to a major factor that influences OC concentration. Colored text within each box denotes factors that influence OC inputs, sorption capacity, OC export, or respiration rate (note that some processes influence both OC export and respiration rate). As sorption capacity increases, so does the OC capacity of the soil. Conversely, as the rate of respiration and/or OC export increases, the soil OC capacity decreases. Floodplain soils can only develop high concentrations of OC if there are high rates of OC input. However, the capacity of the soils to store OC regulates that input, and is determined by the competing influences of sorption capacity and the combination of respiration rate and OC export. See the text for further details.

4.3 Conceptual model of soil OC concentration in floodplain soils

We present a conceptual model to summarize our results and place them in the context of work examining the controls on OC storage in soils (Fig. 4). OC is input to floodplains either through the decay of vegetation or the deposition of fine, OC-rich sediment. This input of OC only determines OC concentrations insofar as floodplain soils are capable of storing OC. That storage is effectively determined by a balance between processes that remove OC from floodplains, namely respiration or erosion followed by respiration (Berhe et al., 2007), and processes that regulate OC availability to microbes, namely the capability of the mineral fraction of the soil to sorb OC.

OC sorption capacity reflects a few specific processes. Although soil texture generally relates to the ability of OC to sorb to mineral grains and resulting OC availability, soil chemistry also plays a strong and potentially dominant role in regulating OC sorption capacity (Rasmussen et al., 2018). Soil texture is largely determined by valley morphology according to our modeling (Table 1), placing valley morphology and resulting sediment transport dynamics (Gran and Czuba, 2017; Wohl et al., 2017b) as indirect controls on sorption capacity.

Respiration rate is largely determined by microbial activity and the availability of OC to microbes. Erosion can rapidly expose soil OC to microbial respiration (Berhe et al., 2007), whereas soils that reside in largely anoxic conditions can exhibit low rates of microbial respiration (Boye et al., 2017). In addition, soil mineralogy, chemistry, and redox conditions (Keiluweit et al., 2017) can regulate microbial activity (e.g., andic soils may limit microbial respiration; Matus et al., 2014). Our results suggesting that moisture controls OC content support the idea that drier soils likely have higher rates of microbial respiration of OC. Moisture is a function of texture, valley bottom morphology, and elevation (a proxy for temperature) in our modeling (Table 1). Comparing floodplain types in the wet basin, we find that types with standing water exhibit significantly higher soil moisture contents than those without standing water. This indicates spatial variability in moisture content and likely microbial activity (Howard and Howard, 1993). In our modeling, this effect translates to spatial variability in OC concentration within floodplains and across entire basins.

OC export from soils refers specifically to leaching of dissolved OC, mainly from shallow soils. While we do not directly consider OC export in this study, it likely introduces variability in soil OC concentration and is regulated dominantly by hydrology (Ågren et al., 2014; Mcdowell and Likens, 1988), soil pore water chemistry (including pH and ionic strength; Brooks et al., 1999; Evans et al., 2006), and redox conditions (Knorr, 2013). OC export as DOC can act in conjunction with respiration to remove OC from soils and, like respiration rate, is countered by sorption capacity.

To summarize, OC inputs, regulated by the capacity of soils to store OC and suppress microbial respiration and OC export, determine OC concentrations in floodplain soils (Fig. 4). OC inputs to floodplain soils come from either autochthonous litter accumulation on the floodplain surface, allochthonous wood deposition, or allochthonous deposition of fine, OC-bearing sediments. Our results from the mountainous basins we studied suggest that deposition of fine material in overbank flows is rare, leading us to infer that autochthonous litter and allochthonous wood inputs to floodplains dominate OC input in mountain rivers. Where soils are more moist, microbial respiration is inhibited and more OC is stored. Although soil texture is likely not a limiting factor on OC concentration in these floodplains, finer-textured soils likely have a higher sorption capacity, retaining more of the OC input from decaying plant material. Our results indicate that geomorphic and hydrologic characteristics act as boundary conditions that regulate soil texture and moisture, in turn regulating sorption capacity, respiration rate, and resulting OC concentrations in floodplain soils.

5 Conclusions

We present floodplain soil OC concentration data from two disparate watersheds to compare how interbasin variability between the two watersheds compares with intra-basin variability in geomorphic and hydrologic characteristics in determining OC concentration. Our results indicate that OC concentration in mountain floodplain soils does not vary predictably along a longitudinal gradient, nor does it vary substantially between basins with differing climatic and vegetation characteristics. Instead, geomorphic and hydrologic characteristics, such as valley bottom morphology and soil moisture, dominantly determine floodplain OC concentration.

In our study basins, decaying litter and wood, and not overbank deposition of fine, OC-bearing sediment, is the main source of OC to floodplain soils. It is unclear whether that decaying vegetation is dominated by autochthonous litter inputs or transported downed wood. In comparing our basin to other studied floodplain soils, it seems that vegetation dynamics play a strong role in determining OC concentrations when fine sediment is not regularly deposited on floodplain surfaces. However, we suggest that floodplain soil characteristics, set by geomorphic and hydrologic conditions, regulate how OC inputs translate to the spatial distribution of OC along a river network.

This implies that OC storage in floodplains likely cannot be predicted using consistent downstream trends and that management prioritization designed to facilitate floodplain OC storage should be based on local geomorphic and hydrologic process variability within each basin. For instance, management to increase OC sequestration in floodplain soils will likely be more effective where floodplains are unconfined and soils already experience high moisture conditions for much of the year. Along these lines, our results show that modeling the floodplain biospheric OC pool to predict its response to warming and subsequent effects on climate based on regional factors such as climate and net primary productivity likely misses the substantial interbasin variability in OC concentration and storage resulting from variability in valley bottom geometry and both geomorphic and hydrologic processes (e.g., Doetterl et al., 2015).

Although our results provide some insights, the question of whether OC stored in floodplain soil comes dominantly from allochthonous versus autochthonous sources remains open. Our results imply that more productive, spatially heterogeneous floodplains likely input more OC to soils. Floodplain OC concentration, while mediated largely by moisture dynamics, likely depends mainly on OC inputs from productive riparian forests. This implies that management of OC storage in mountain river floodplains should focus on the restoration of riparian zones to maintain OC input to soil (e.g., Bullinger-Weber et al., 2014). More detailed studies in regions with varying sediment transport and hydrologic regimes are needed to determine what conditions favor autochthonous versus allochthonous OC inputs, but our results suggest that autochthonous sources dominate floodplain OC storage in basins with relatively low rates of vertical accretion and high channel–floodplain connectivity that promotes floodplain wetlands.

Author contributions. Field data collection was conducted primarily by DNS with assistance from EEW at some sites in the Middle Fork Snoqualmie basin. DNS performed data analysis and drafted the initial version of the paper, which was then edited and rewritten by DNS and EEW.

Competing interests. The authors declare that they have no conflict of interest.

Acknowledgements. This work was funded by NSF grant EAR-1562713. We thank Ellen Daugherty for extensive assistance in fieldwork and discussion that improved the conceptual model. We thank Katherine Lininger for stimulating discussion that improved the paper. Detailed and constructive comments from two anonymous reviewers and Robert Hilton improved the paper. We thank Sara Lowe for assistance in processing and analyzing soil samples.

References

Ågren, A. M., Buffam, I., Cooper, D. M., Tiwari, T., Evans, C. D., and Laudon, H.: Can the heterogeneity in stream dissolved organic carbon be explained by contributing landscape elements?, Biogeosciences, 11, 1199–1213, https://doi.org/10.5194/bg-11-1199-2014, 2014.

Appling, A. P., Bernhardt, E. S., and Stanford, J. A.: Floodplain biogeochemical mosaics: A multi-dimensional view of alluvial soils, J. Geophys. Res.-Biogeo., 119, 1538–1553, https://doi.org/10.1002/2013JG002543, 2014.

Aufdenkampe, A. K., Mayorga, E., Raymond, P. A., Melack, J. M., Doney, S. C., Alin, S. R., Aalto, R. E., and Yoo, K.: Riverine coupling of biogeochemical cycles between land, oceans, and atmosphere, Front. Ecol. Environ., 9, 53–60, https://doi.org/10.1890/100014, 2011.

Ballantyne, A. P., Alden, C. B., Miller, J. B., Tans, P. P., and White, J. W. C.: Increase in observed net carbon dioxide uptake by land and oceans during the past 50 years, Nature, 488, 70–73, https://doi.org/10.1038/nature11299, 2012.

Bao, H., Kao, S., Lee, T., Zehetner, F., Huang, J., Chang, Y., Lu, J., and Lee, J.: Distribution of organic carbon and lignin in soils in a subtropical small mountainous river basin, Geoderma, 306, 81–88, https://doi.org/10.1016/j.geoderma.2017.07.011, 2017.

Battin, T. J., Luyssaert, S., Kaplan, L. A., Aufdenkampe, A. K., Richter, A., and Tranvik, L. J.: The boundless carbon cycle, Nat. Geosci., 2, 598–600, https://doi.org/10.1038/ngeo618, 2009.

Benfield, E. F.: Comparison of Litterfall Input to Streams, J. N. Am. Benthol. Soc., 16, 104–108, https://doi.org/10.2307/1468242, 1997.

Berhe, A. A., Harte, J., Harden, J. W., and Torn, M. S.: The significance of the erosion-induced terrestrial carbon sink, Bioscience, 57, 337–346, https://doi.org/10.1641/B570408, 2007.

Blazejewski, G. A., Stolt, M. H., Gold, A. J., Gurwick, N., and Groffman, P. M.: Spatial Distribution of Carbon in the Subsurface of Riparian Zones, Soil Sci. Soc. Am. J., 73, 1733, https://doi.org/10.2136/sssaj2007.0386, 2009.

Boye, K., Noël, V., Tfaily, M. M., Bone, S. E., Williams, K. H., Bargar, J. R., and Fendorf, S.: Thermodynamically controlled preservation of organic carbon in floodplains, Nat. Geosci., 10, 415–419, https://doi.org/10.1038/NGEO2940, 2017.

Brooks, P. D., Mcknight, D. M., and Bencala, K. E.: The relationship between soil heterotrophic activity, soil dissolved organic carbon (DOC) leachate, and catchment-scale DOC export in headwater catchments, Water Resour. Res., 35, 1895–1902, 1999.

Bullinger-Weber, G., Le Bayon, R.-C. C., Thébault, A., Schlaepfer, R., and Guenat, C.: Carbon storage and soil organic matter stabilisation in near-natural, restored and embanked Swiss floodplains, Geoderma, 228–229, 122–131, https://doi.org/10.1016/j.geoderma.2013.12.029, 2014.

D'Elia, A. H., Liles, G. C., Viers, J. H., and Smart, D. R.: Deep carbon storage potential of buried floodplain soils, Sci. Rep.-UK, 7, 8181, https://doi.org/10.1038/s41598-017-06494-4, 2017.

Doetterl, S., Stevens, A., Six, J., Merckx, R., Van Oost, K., Pinto, M. C., Casanova-katny, A., Muñoz, C., Boudin, M., Venegas, E. Z., and Boeckx, P.: Soil carbon storage controlled by interactions between geochemistry and climate, Nat. Geosci., 8, 780–783, https://doi.org/10.1038/NGEO2516, 2015.

Evans, C. D., Chapman, P. J., Clark, J. M., Monteith, D. T., and Cresser, M. S.: Alternative explanations for rising dissolved organic carbon export from organic soils, Glob. Change Biol., 12, 2044–2053, https://doi.org/10.1111/j.1365-2486.2006.01241.x, 2006.

Fall, P. L.: Modern Pollen Spectra and Vegetation in the Wind River Range, Wyoming, USA, Arctic Alpine Res., 26, 383–392, 1994.

Falloon, P., Jones, C. D., Ades, M., and Paul, K.: Direct soil moisture controls of future global soil carbon changes: An important source of uncertainty, Global Biogeochem. Cy., 25, 1–14, https://doi.org/10.1029/2010GB003938, 2011.

Gran, K. B. and Czuba, J. A.: Sediment pulse evolution and the role of network structure, Geomorphology, 277, 17–30, https://doi.org/10.1016/j.geomorph.2015.12.015, 2017.

Hanberry, B. B., Kabrick, J. M., and He, H. S.: Potential tree and soil carbon storage in a major historical floodplain forest with disrupted ecological function, Perspect. Plant Ecol., 17, 17–23, https://doi.org/10.1016/j.ppees.2014.12.002, 2015.

Harvey, J. W. and Gooseff, M.: River corridor science: Hydrologic exchange and ecological consequences frombedforms to basins, Water Resour. Res., 51, 6893–6922, https://doi.org/10.1002/2015WR017617, 2015.

Hilton, R. G., Galy, A., Hovius, N., Horng, M. J., and Chen, H.: Efficient transport of fossil organic carbon to the ocean by steep mountain rivers: An orogenic carbon sequestration mechanism, Geology, 39, 71–74, https://doi.org/10.1130/G31352.1, 2011.

Hoffmann, T., Glatzel, S., and Dikau, R.: A carbon storage perspective on alluvial sediment storage in the Rhine catchment, Geomorphology, 108, 127–137, https://doi.org/10.1016/j.geomorph.2007.11.015, 2009.

Holm, S.: A Simple Sequentially Rejective Multiple Test Procedure, Scand. J. Stat., 6, 65–70, 1979.

Homer, C. G., Dewitz, J. A., Yang, L., Jin, S., Danielson, P., Xian, G., Coulston, J., Herold, N. D., Wickham, J. D., and Megown, K.: Completion of the 2011 National Land Cover Database for the conterminous United States-Representing a decade of land cover change information, Photogramm. Eng. Rem. S., 81, 345–354, 2015.

Hoogsteen, M. J. J., Lantinga, E. A., Bakker, E. J., Groot, J. C. J., and Tittonell, P. A.: Estimating soil organic carbon through loss on ignition: Effects of ignition conditions and structural water loss, Eur. J. Soil Sci., 66, 320–328, https://doi.org/10.1111/ejss.12224, 2015.

Houston, D. B.: Wildfires in Northern Yellowstone National Park, Ecology, 54, 1111–1117, https://doi.org/10.2307/1935577, 1973.

Howard, D. M. and Howard, P. J. A.: Relationships between CO_2 evolution, moisture content and temperature for a range of soil types, Soil Biol. Biochem., 25, 1537–1546, https://doi.org/10.1016/0038-0717(93)90008-Y, 1993.

Jobbágy, E. G. and Jackson, R. B.: The vertical distribution of soil organic carbon and its relation to climate and vegetation, Ecol. Appl., 10, 423–436, https://doi.org/10.1890/1051-0761(2000)010[0423:TVDOSO]2.0.CO;2, 2000.

Kaiser, K. and Guggenberger, G.: The role of DOM sorption to mineral surfaces in the preservation of organic matter in soils, Org. Geochem., 31, 711–725, https://doi.org/10.1016/S0146-6380(00)00046-2, 2000.

Keiluweit, M., Wanzek, T., Kleber, M., Nico, P., and Fendorf, S.: Anaerobic microsites have an unaccounted role in soil carbon stabilization, Nat. Commun., 8, 1–8, https://doi.org/10.1038/s41467-017-01406-6, 2017.

Knorr, K.-H.: DOC-dynamics in a small headwater catchment as driven by redox fluctuations and hydrological flow paths – are DOC exports mediated by iron reduction/oxidation cycles?, Biogeosciences, 10, 891–904, https://doi.org/10.5194/bg-10-891-2013, 2013.

Leithold, E. L., Blair, N. E., and Wegmann, K. W.: Source-to-sink sedimentary systems and global carbon burial: A river runs through it, Earth-Sci. Rev., 153, 30–42, https://doi.org/10.1016/j.earscirev.2015.10.011, 2016.

Lininger, K. B., Wohl, E., and Rose, J. R.: Geomorphic Controls on Floodplain Soil Organic Carbon in the Yukon Flats, Interior Alaska, From Reach to River Basin Scales, Water Resour. Res., 54, 1934–1951, https://doi.org/10.1002/2017WR022042, 2018.

Loope, L. L. and Gruell, G. E.: The ecological role of fire in the Jackson Hole area, northwestern Wyoming, Quatenary Res., 3, 425–443, https://doi.org/10.1016/0033-5894(73)90007-0, 1973.

Matus, F., Rumpel, C., Neculman, R., Panichini, M., and Mora, M. L.: Soil carbon storage and stabilisation in andic soils: A review, Catena, 120, 102–110, https://doi.org/10.1016/j.catena.2014.04.008, 2014.

Mcdowell, W. H. and Likens, G. E.: Origin, Composition, and Flux of Dissolved Organic Carbon in the Hubbard Brook Valley, Ecol. Monogr., 58, 177–195, 1988.

Montgomery, D. R. and Buffington, J. M.: Channel-reach morphology in mountain drainage basins, Bull. Geol. Soc. Am., 109, 596–611, 1997.

Omengo, F. O., Geeraert, N., Bouillon, S., and Govers, G.: Deposition and fate of organic carbon in floodplains along a tropical semiarid lowland river (Tana River, Kenya), J. Geophys. Res.-Biogeo., 121, 1131–1143, https://doi.org/10.1002/2015JG003288, 2016.

Oregon State University: PRISM Climate Group, 2004.

Rasmussen, C., Heckman, K., Wieder, W. R., Keiluweit, M., Lawrence, C. R., Berhe, A. A., Blankinship, J. C., Crow, S. E., Druhan, J. L., Hicks Pries, C. E., Marin-Spiotta, E., Plante, A. F., Schädel, C., Schimel, J. P., Sierra, C. A., Thompson, A., and Wagai, R.: Beyond clay: towards an improved set of variables for predicting soil organic matter content, Biogeochemistry, 137, 297–306, https://doi.org/10.1007/s10533-018-0424-3, 2018.

Rathbun, S., Bennett, G. L., Wohl, E., Briles, C., McElroy, B., and Sutfin, N.: The fate of sediment, wood, and organic carbon eroded during an extreme flood, Colorado Front Range, USA, Geology, 45, 1–14, https://doi.org/10.1130/G38935.1, 2017.

R Core Team: R: A Language and Environment for Statistical Computing, 2017.

Ricker, M. C., Stolt, M. H., Donohue, S. W., Blazejewski, G. A., and Zavada, M. S.: Soil Organic Carbon Pools in Riparian Landscapes of Southern New England, Soil Sci. Soc. Am. J., 77, 1070–1079, https://doi.org/10.2136/sssaj2012.0297, 2013.

Schimel, D. S. and Braswell, B. H.: The role of mid-latitude mountains in the carbon cycle: Global perspective and a Western U.S. case study, in: Global Change and Mountain Regions, edited by: Huber, U. M., Bugmann, H. K. M., and Reasoner, M. A., Springer, 449–456, 2005.

Schmidt, M. W. I. I., Torn, M. S., Abiven, S., Dittmar, T., Guggenberger, G., Janssens, I. A., Kleber, M., Kögel-Knabner, I., Lehmann, J., Manning, D. A. C. C., Nannipieri, P., Rasse, D. P., Weiner, S., and Trumbore, S. E.: Persistence of soil organic matter as an ecosystem property, Nature, 478, 49–56, https://doi.org/10.1038/nature10386, 2011.

Scott, D. N. and Wohl, E.: Evaluating Carbon Storage on Subalpine Lake Deltas, Earth Surf. Proc. Land., 42, 1472–1481, https://doi.org/10.1002/esp.4110, 2017.

Scott, D. and Wohl, E.: Dataset for Geomorphic regulation of floodplain soil organic carbon concentration in watersheds of the Rocky and Cascade Mountains, USA, https://doi.org/10.25675/10217/187762, 2018a.

Scott, D. N. and Wohl, E.: Natural and Anthropogenic Controls on Wood Loads in River Corridors of the Rocky, Cascade, and Olympic Mountains, USA, Water Resour. Res., https://doi.org/doi/10.1029/2018WR022754, 2018b.

Smithwick, E., Harmon, M. E., Remillard, S. M., Acker, S. A., and Franklin, J. F.: Potential upper bounds of carbon stores in forests of the Pacific Northwest, Ecol. Appl., 12, 1303–1317, https://doi.org/10.1890/1051-0761(2002)012[1303:PUBOCS]2.0.CO;2, 2002.

Sparks, D. L.: Methods of Soil Analysis. Part 3, Chemical Methods, edited by: Sparks, D. L., Page, A. L., Helmke, P. A., Loeppert, R. H., Soltanpour, P. N., Tabatabai, M. A., Johnston, C. T., Sumber, M. E., Bartels, J. M., and Bingham, J. M., Soil Science Society of America, Inc., Madison, Wisconsin, 1996.

Sun, O. J., Campbell, J., Law, B. E., and Wolf, V.: Dynamics of carbon stocks in soils and detritus across chronosequences of different forest types in the Pacific Northwest, USA, Glob. Change Biol., 10, 1470–1481, https://doi.org/10.1111/j.1365-2486.2004.00829.x, 2004.

Sutfin, N. A. and Wohl, E.: Substantial soil organic carbon retention along floodplains of mountain streams, J. Geophys. Res.-Earth, 122, 1325–1338, https://doi.org/10.1002/2016JF004004, 2017.

Sutfin, N. A., Wohl, E., and Dwire, K. A.: Banking carbon: A review of organic carbon storage and physical factors influencing retention in floodplains and riparian ecosystems, Earth Surf. Proc. Land., 60, 38–60, https://doi.org/10.1002/esp.3857, 2016.

Thien, S. J.: A flow diagram for teaching texture-by-feel analysis, J. Agron. Educ., 8, 54–55, 1979.

Wagenmakers, E.-J. and Farrell, S.: AIC model selection using Akaike weights, Psychon. B. Rev., 11, 192–196, https://doi.org/10.3758/BF03206482, 2004.

Wilcoxon, F.: Individual Comparisons by Ranking Methods, Biometrics Bull., 1, 80–83, https://doi.org/10.2307/3001946, 1945.

Wohl, E.: Floodplains and wood, Earth-Sci. Rev., 123, 194–212, https://doi.org/10.1016/j.earscirev.2013.04.009, 2013.

Wohl, E. and Pfeiffer, A.: Organic carbon storage in floodplain soils of the U.S. prairies, River Res. Appl., 34, 406–416, https://doi.org/10.1002/rra.3269, 2018.

Wohl, E., Dwire, K., Sutfin, N., Polvi, L. and Bazan, R.: Mechanisms of carbon storage in mountainous headwater rivers, Nat. Commun., 3, 1263, https://doi.org/10.1038/ncomms2274, 2012.

Wohl, E., Hall, R. O., Lininger, K. B., Sutfin, N. A., and Walters, D. M.: Carbon dynamics of river corridors and the effects of human alterations, Ecol. Monogr., 87, 379–409, https://doi.org/10.1002/ecm.1261, 2017a.

Wohl, E., Lininger, K. B., and Scott, D. N.: River beads as a conceptual framework for building carbon storage and resilience to extreme climate events into river management, Biogeochemistry, 1–19, https://doi.org/10.1007/s10533-017-0397-7, 2017b.

Zhao, B., Li, Z., Li, P., Xu, G., Gao, H., Cheng, Y., Chang, E., Yuan, S., Zhang, Y., and Feng, Z.: Spatial distribution of soil organic carbon and its influencing factors under the condition of ecological construction in a hilly-gully watershed of the Loess Plateau, China, Geoderma, 296, 10–17, https://doi.org/10.1016/j.geoderma.2017.02.010, 2017.

3

A lattice grain model of hillslope evolution

Gregory E. Tucker[1], Scott W. McCoy[2], and Daniel E. J. Hobley[3]

[1]Cooperative Institute for Research in Environmental Science (CIRES) and Department of Geological Sciences, University of Colorado, Boulder, CO 80305, USA
[2]Department of Geological Sciences and Engineering, University of Nevada, Reno, NV 89557, USA
[3]School of Earth and Ocean Sciences, Cardiff University, Cardiff, CF10 3AT, Wales

Correspondence: Greg Tucker (gtucker@colorado.edu)

Abstract. This paper describes and explores a new continuous-time stochastic cellular automaton model of hillslope evolution. The Grain Hill model provides a computational framework with which to study slope forms that arise from stochastic disturbance and rock weathering events. The model operates on a hexagonal lattice, with cell states representing fluid, rock, and grain aggregates that are either stationary or in a state of motion in one of the six cardinal lattice directions. Cells representing near-surface soil material undergo stochastic disturbance events, in which initially stationary material is put into motion. Net downslope transport emerges from the greater likelihood for disturbed material to move downhill than to move uphill. Cells representing rock undergo stochastic weathering events in which the rock is converted into regolith. The model can reproduce a range of common slope forms, from fully soil mantled to rocky or partially mantled, and from convex-upward to planar shapes. An optional additional state represents large blocks that cannot be displaced upward by disturbance events. With the addition of this state, the model captures the morphology of hogbacks, scarps, and similar features. In its simplest form, the model has only three process parameters, which represent disturbance frequency, characteristic disturbance depth, and base-level lowering rate, respectively. Incorporating physical weathering of rock adds one additional parameter, representing the characteristic rock weathering rate. These parameters are not arbitrary but rather have a direct link with corresponding parameters in continuum theory. Comparison between observed and modeled slope forms demonstrates that the model can reproduce both the shape and scale of real hillslope profiles. Model experiments highlight the importance of regolith cover fraction in governing both the downslope mass transport rate and the rate of physical weathering. Equilibrium rocky hillslope profiles are possible even when the rate of base-level lowering exceeds the nominal bare-rock weathering rate, because increases in both slope gradient and roughness can allow for rock weathering rates that are greater than the flat-surface maximum. Examples of transient relaxation of steep, rocky slopes predict the formation of a regolith-mantled pediment that migrates headward through time while maintaining a sharp slope break.

1 Introduction

Hillslopes take on a rich variety of forms. Their profile shapes may be convex-upward, concave-upward, planar, or some combination of these. Some slopes are completely mantled with soil, whereas others are bare rock, and still others draped in a discontinuous layer of mobile regolith. The processes understood to be responsible for shaping them are equally varied, ranging from disturbance-driven creep to dissolution to large-scale mass movement events.

Considerable research has been devoted to understanding the evolution of soil-mantled slopes that are primarily governed by disturbance-driven creep, such as downslope soil transport by biotic and abiotic soil-mixing processes. As a result, the geomorphology community has mathematical models that account well for observed slope forms and patterns of regolith thickness (e.g., Roering, 2008). Furthermore, stochastic-transport theory provides a mechanistic link between the statistics of particle motion, the resultant average rates of downslope transport, and the emergence of convex-upward, soil-mantled slope forms (Culling, 1963; Roering, 2004; Foufoula-Georgiou et al., 2010; Furbish et al., 2009; Furbish and Haff, 2010; Tucker and Bradley, 2010).

One gap that remains, however, lies in understanding steep, rocky slopes (Fig. 1). "Rocky" implies slopes that lack a continuous soil cover (e.g., Howard and Selby, 1994, and references therein); here, a transport law that assumes the existence of such a cover no longer applies. "Steep" implies angles approaching or exceeding the effective angle of repose for loose, granular material, so that ravel may be an important transport mode (e.g., Gabet, 2003; Roering and Gerber, 2005; Lamb et al., 2011; Gabet and Mendoza, 2012) and particles have the potential to fall as soon as they are released from bedrock. This type of relatively fast, long-distance transport does not fit comfortably in the framework of standard diffusion-based models of hillslope soil transport, which derive from an underlying assumption that the characteristic length scale of motion is short relative to the length of the slope.

Rocky slopes are rarely completely barren. More commonly, they have a patchy cover of loose material, which may either retard rock weathering by shielding the rock surface from moisture or temperature fluctuations, or enhance it by trapping water and allowing limited plant growth. A discontinuous cover does not fit easily within the popular exponential-decay regolith-production models (e.g., Heimsath et al., 2012; Lamb et al., 2013), which assume an essentially continuous soil mantle.

An additional issue, which pertains to both rocky and soil-mantled slopes, is the connection between sediment movement at the scale of individual "motion events," and the resulting longer-term average sediment flux, which forms the basis for continuum models of hillslope evolution. Recent theoretical and experiment work has begun to forge a mechanistic connection between these scales (Culling, 1963, 1965; Furbish et al., 2009; Furbish and Haff, 2010; Tucker and Bradley, 2010; Gabet and Mendoza, 2012; Lamb et al., 2013). However, the community's resources for computational analysis of particle-level dynamics remain limited, lagging behind developments in understanding sediment transport in coastal environments (Drake and Calantoni, 2001) and rivers (McEwan and Heald, 2001; MacVicar et al., 2006; Furbish and Schmeeckle, 2013; Schmeeckle, 2014).

To further our understanding of how grain-level weathering and transport processes translate into hillslope evolution, both for hillslopes in general and rocky slopes in particular, it would be useful to have a computational framework with which to conduct experiments. Ideally, such a framework should be sophisticated enough to capture the essence of weathering and granular mechanics, while remaining simple enough to involve only a small number of parameters and provide reasonable computational efficiency.

Our aim in this paper is to describe one such computational framework, test whether it is capable of reproducing commonly observed hillslope-profile forms, and examine how its parameters relate to the bulk-behavior parameters used in conventional continuum models of soil creep and regolith production. The model uses a pairwise, continuous-time stochastic (CTS) approach to combine a lattice grain model with rules for stochastic bedrock-to-regolith conversion ("weathering") and disturbance of surface regolith particles. One goal of this event-based approach is to study how bulk behavior, such as the diffusion-like net downslope transport of soil, can emerge from a large ensemble of stochastic events. In this paper, we present the "Grain Hill" model and examine its ability to reproduce three common types of slope profile: (1) convex-upward, soil-mantled slopes (Fig. 2a, b), (2) quasi-planar rocky slopes (Fig. 2c, d), and (3) cliff-rampart morphology in layered strata (Fig. 2e, f).

We begin with a description of the modeling technique. We then present results that illustrate the macroscopic behavior of the model under a variety of boundary conditions, and define the relationship between the cellular model's parameters and the parameters of conventional continuum mechanics models for hillslope evolution.

2 Model description

The model combines a cellular automaton representation of granular mechanics with rules for weathering of rock to regolith and for episodic disturbance of regolith. Cellular automata are widely used in the granular mechanics community, because they can represent the essential physics of granular materials at a reasonably low computational cost. Because the principles are often similar to those of lattice-gas automata in fluid dynamics (e.g., Chen and Doolen, 1998), cellular automata for granular mechanics are sometimes referred to as lattice grain models (LGrMs) (Gutt and Haff, 1990; Peng and Herrmann, 1994; Alonso and Herrmann, 1996; Károlyi et al., 1998; Károlyi and Kertész, 1999, 1998; Martinez and Masson, 1998; Désérable, 2002; Cottenceau and Désérable, 2010; Désérable et al., 2011).

2.1 CTS lattice grain model

Our approach starts with a two-dimensional (2-D) CTS lattice grain cellular automaton. A cellular automaton can be broadly defined as a computational model that consists of a lattice of cells, with each cell taking on one of N discrete states (represented by an integer value). These states evolve over time according to a set of rules that describe transitions from one state to another as a function of a particular cell's immediate neighborhood. A continuous-time stochastic model is one in which the timing of transitions is probabilistic rather than deterministic. Whereas transitions in traditional cellular automata occur in discrete time steps, in a

Figure 1. Examples of rocky hillslopes, sometimes referred to as Richter slopes. **(a)** Chalk Cliffs, Colorado, USA. **(b)** Canadian Rockies. **(c)** Grand Canyon, Arizona, USA. **(d)** Rocky Mountain National Park, Colorado, USA. **(e)** Guadeloupe Mountains, Texas, USA. **(f)** Waterton Lakes National Park, Canada (photos by Gregory E. Tucker).

CTS model, they are both stochastic and asynchronous. A CTS model can be viewed as a type of Boolean delay equation (Ghil et al., 2008), though the number of possible states is not necessarily limited to just two.

The method we present here, which we will refer to as the Grain Hill model, is implemented in the Landlab modeling framework (Hobley et al., 2017). The lattice grain component, on which Grain Hill builds, is described in detail by Tucker et al. (2016). Here, we present only a brief overview of the lattice grain model's rules and behavior. The framework is based on the pairwise ("doublet") method developed by Narteau and colleagues (Rozier and Narteau, 2014), which has been applied to problems as diverse as eolian dune dynamics (Narteau et al., 2009; Zhang et al., 2010, 2012) and the core-mantle interface (Narteau et al., 2001).

In the basic CTS lattice grain model, the domain consists of a lattice of hexagonal cells. Each cell is assigned one of eight states (Table 1, states 0–7). These states represent the

nature and motion status of the material: state 0 represents fluid (an "empty" cell into which a solid particle can move), states 1–6 represent a grain moving in one of the six lattice directions, and state 7 indicates a stationary grain (or aggregate of grains, as discussed below). For purposes of modeling hillslope evolution, we add an additional state (8) to represent rock, which is immobile until converted to granular material, representing regolith. An optional additional state (9) is used to model large blocks, as described below. Figure 3 shows several of these states in the form of a time sequence of transition events. Note that the timing of transition events is purely stochastic; there are no time steps in the usual sense.

Like other lattice grain models, the CTS lattice grain model is designed to represent, in a simple way, the motion and interaction of an ensemble of grains in a gravitational field. The physics of the material are represented by a set of transition rules, in which a given adjacent pair of states is assigned a certain probability per unit of time of un-

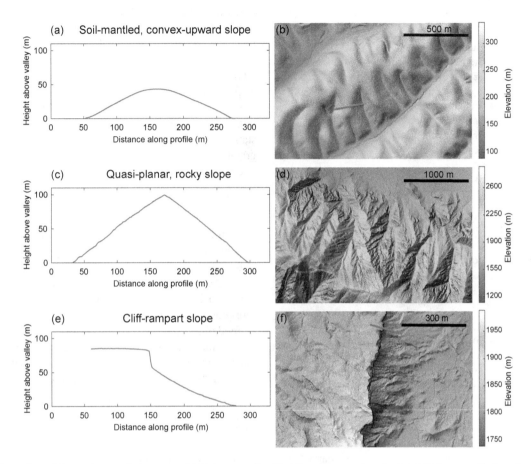

Figure 2. Examples of three characteristic types of hillslope profile. Red line in map view depicts hillslope profile location. **(a, b)** Soil-mantled, convex-upward slope (Gabilan Mesa, California, USA). **(c, d)** Quasi-planar, thinly mantled slope (Yucaipa Ridge, California, USA). **(e, f)** Cliff formed in resistant Tertiary laccolithic intrusive rocks overlying Jurassic sedimentary rocks (Cedar Mountain, Utah, USA).

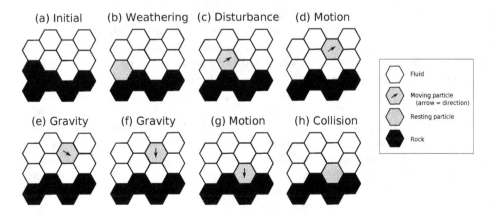

Figure 3. Hypothetical time sequence of transition events **(a–h)**, illustrating several of the states and transitions in the Grain Hill model. Note that although this example shows a single particle in motion, it is possible for multiple cells to exist in a state of motion at any given time.

Table 1. States in the Grain Hill model.

State	Description
0	Fluid
1	Grain moving upward
2	Grain moving up and right
3	Grain moving down and right
4	Grain moving down
5	Grain moving down and left
6	Grain moving up and left
7	Resting grain
8	Rock
(9)	Block (optional)

dergoing a transition to a different pair. For example, consider a vertically aligned pair of cells in which the top cell has state 4 (moving downward) and the bottom cell state 0 (empty/fluid) (Fig. 3f). Downward motion (falling) is represented by a transition in which the two states switch places (Fig. 3g).

The stochastic pairwise transitions in the CTS lattice grain model are treated as Poisson processes. The probability density function for the waiting time, t, to the next transition event at a particular pair is given by an exponential function with a rate parameter r, which has dimensions of inverse time:

$$p(t) = r e^{-rt}. \tag{1}$$

Each transition type is associated with a rate parameter that represents the speed of whichever process the transition is designed to represent. To implement these transitions, the CTS lattice grain model steps from one transition to the next, rather than iterating through time steps of fixed duration. Whenever the state of one or both cells in a particular pair changes, if the new pair is subject to a transition, the time at which the transition is scheduled to occur is added to a queue of pending events. The soonest among all pending events is chosen for processing, and the process repeats until either the desired run time has completed or there are no further events in the queue. Further details on the implementation and algorithms are provided in Tucker et al. (2016).

Grain motion through fluid is represented by a transition involving a moving grain and an adjacent fluid cell in the direction of the grain's motion: the two cells exchange states, representing the motion of the grain into the fluid-filled cell, and the replacement of the grain's former location with fluid (Fig. 3c, d and f, g). During this transition, the grain's motion direction remains unchanged (Fig. 4, top left). Note that the lattice itself never moves; rather, material motion is represented simply by an exchange of grain and fluid states between an adjacent pair of cells.

Gravity is represented by transitions in which a rising grain decelerates to become stationary, a stationary grain accelerates downward to become a falling particle, and a grain

moving upward at an angle accelerates downward to move downward at an angle (Fig. 5). An additional rule allows for acceleration of a particle resting on a slope: a stationary particle adjacent to a fluid cell below it and to one side may transition to a moving particle (Fig. 5, bottom row). Importantly for our purposes, this latter rule effectively imposes an angle of repose at 30°.

For gravitational transitions, the rate parameter, r_g, is determined by considering the time it would take for an initially stationary object to fall a distance of one cell width under gravitational acceleration without fluid drag. This works out to be

$$r_g = \sqrt{2\delta/g}, \tag{2}$$

where δ is cell width and g is gravitational acceleration. This rate parameter is used for all of the gravitational transitions illustrated in Fig. 5.

Because of the stochastic treatment of all transitions – including gravitational ones – it is possible for grains in the model to hover in mid-air for a brief period of time before plunging downward (e.g., Coyote, 1949). For purposes of modeling hillslope evolution, this is fine; what matters most is that there is a distinct timescale gap between "fast" (large rate constant) processes associated with grain motion and "slow" (small rate constant) processes associated with weathering and disturbance, which are described below. First, however, we must consider frictional interactions among moving particles.

We assume that biophysical disturbance events such as the growth of roots and burrowing by animals, and the settling motions that follow, tend to impart low kinetic energy, with "low" defined as ballistic displacement lengths that are short relative to hillslope length and comparable to or less than the characteristic disturbance-zone thickness. We consider such motions to be dominated by frictional dissipation rather than by transfer of kinetic energy by elastic impacts. This view is similar to the reasoning of Furbish et al. (2009) that the mean free path of mobile grains will typically be short relative to hillslope length, scaling with the grain radius and particle concentration. For this reason, unlike the original lattice grain model of Tucker et al. (2016), the present formulation includes only inelastic collisions (Fig. 4). These inelastic (frictional) collisions are represented by a set of rules in which one or both colliding particles become stationary, representing loss of momentum and kinetic energy as a result of the collision. The particular choices for frictional interaction are motivated simply by the geometry of the problem. They are non-unique in the sense that one could imagine reasonable alternatives to the rules illustrated in Fig. 4; however, the details of frictional interactions have little influence on the outcomes of the Grain Hill model. In the general lattice grain CTS model, the rate parameter for frictional transitions is set equal to the product of the gravitational parameter and a dimensionless friction factor, $f \in [0, 1]$ (there is also a corresponding elastic factor equal to $1 - f$). In the Grain Hill

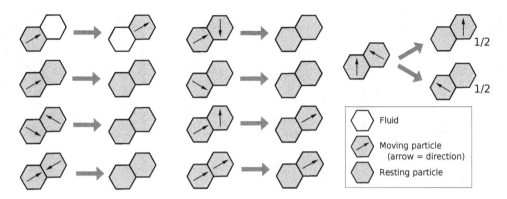

Figure 4. Rules for motion and frictional (inelastic) collisions, illustrated here for one of the six lattice directions.

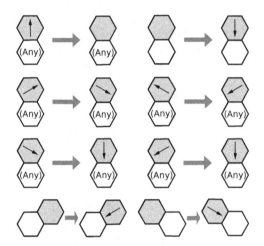

Figure 5. Illustration of gravitational rules. The bottom row shows the "falling on a slope" rule, which effectively imposes a 30° angle of repose. Modified from Tucker et al. (2016).

implementation explored in this paper, $f = 1$, such that particle collisions are purely frictional.

One limitation of the CTS lattice grain model is that falling grains do not accelerate through time; instead, they have a fixed transition probability that implies a statistically uniform downward fall velocity. This treatment is obviously unrealistic for particles falling in a vacuum, though it is consistent with a terminal settling velocity for grains immersed in fluid. Consistent with the above reasoning, the relatively short ballistic displacement lengths asserted for the modeled hillslopes also reduce the importance of this assumption, as a particle would typically have little time to accelerate before impacting another particle.

Tests of the CTS lattice grain model show that it reproduces several basic aspects of granular behavior (Tucker et al., 2016). For example, when gravity and friction are deactivated, the model conserves kinetic energy. When gravity and friction are active, the model reproduces some of the common behaviors observed with granular materials. For ex-

ample, Fig. 6 illustrates a simulation of the emptying of a silo to form an angle-of-repose grain pile. For our purposes, what matters most is simply that the model captures, in a reasonable way, the response of particles on a slope to episodic disturbance events.

2.2 Weathering and soil creep

Weathering of rock to form mobile regolith is modeled with a transition rule: when a rock cell lies adjacent to a fluid cell (which here is assumed to be air), there is a specified probability per unit of time, w $(1/T)$, of transition to a grain–air pair (Figs. 3a–b, and 7, top). In other words, w is the Poisson rate constant for the weathering transition process. This treatment means that the effective maximum expected weathering rate, in terms of the propagation of a weathering front, is cell diameter, δ, multiplied by w. An indirect consequence of this approach is that the weathering rate declines with increasing regolith thickness. As average regolith thickness increases, the fraction of the surface where rock is in contact with air diminishes, and consequently so does the average transition rate. A limitation of the approach is that when the rock is completely mantled, no further weathering can take place. We explore the consequences of this rule below, and compare it with the behavior of continuum regolith-production models.

Soil creep is modeled by a transition rule that mimics the process of episodic disturbance of the mobile regolith (which we use here as a generic term that includes various forms of unconsolidated granular material, such as soil, colluvium, and scree). For each resting grain that is adjacent to an air cell, there is a specified probability per unit of time, d $(1/T)$, that the regolith and air will exchange places, representing movement (Fig. 3b, c). The regolith cell is also converted from a stationary state to a state of motion (Fig. 7b). An advantage of this approach is that it mimics, in a general way, the effectively stochastic disturbance processes that are understood to drive soil creep.

Our definition of d is closely related to the activation rate, N_a, in the probabilistic theory for soil creep developed by

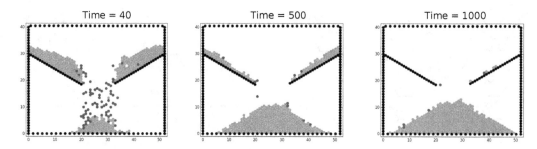

Figure 6. Lattice grain simulation of emptying of a silo. Light-shaded grains are stationary; darker-shaded ones are in motion. Black cells are walls (rock). Time units are indicated in seconds. From Tucker et al. (2016).

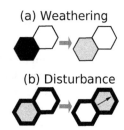

Figure 7. Transitions representing rock-to-regolith transformation by weathering **(a)** and regolith disturbance **(b)**, in which a stationary particle becomes mobile and switches position with a air cell. The illustration represents one of the six possible orientations.

Furbish et al. (2009). When combined with the lattice grain gravitational rules, the resulting cellular model captures both the scattering (disturbance) and settling (gravitational) behavior articulated by Furbish et al. (2009). In the Grain Hill cellular model, as in their theory, downslope regolith flux arises because, on average, scattering occurs perpendicular to the local surface while setting is vertical. The Grain Hill model includes an additional element not present in the Furbish et al. (2009) theory: an increase in (downward) scattering distance for particles on slopes steeper than 30°. This behavior, as illustrated below, promotes a non-linear relationship between gradient and flux, and leads to the possibility of threshold slopes.

Note that the weathering and disturbance rate constants (w and d, respectively) are understood to be considerably smaller than the gravitational rate constant, r_g. As noted above, a key concept here is that there are two distinct timescales: a short timescale associated with grain motion, and a much longer scale associated with weathering and disturbance frequency.

2.3 Cells as grain aggregates

Natural regolith disturbance events usually impact many grains at once. Raindrop impacts on bare sediment typically dislodge several grains at once (Furbish et al., 2007). Excavation of an animal burrow disturbs a volume of grains equal to the volume of the burrow, and the fall of a tree mobilizes a volume of regolith similar to the volume of the tree's root mound. Observations of such processes suggest that there may be a characteristic volume of disturbance that in some cases may be much larger than the volume of a single grain. For this reason, we envision regolith cells as being grain aggregates, with a length scale (width of a cell) δ and a volume scale δ^3.

2.4 Initial and boundary conditions

The 2-D model domain represents the cross section of a hypothetical hillslope, on which particles move within the cross-sectional plane. Any regolith cells that reach the model's side or top boundaries disappear. This treatment is meant to represent the presence of a stream channel at the base of each side of the model hillslope; particles reaching these channels are assumed to be eroded. Progressive lowering of the base level at the two model boundaries is treated by moving the interior cells upward away from the lower boundary, and adding a new row of rock or regolith cells along the bottom row. A new row of cells is added at time intervals of τ.

Cells around the lattice perimeter retain their initial states. If, for example, a transition occurs in which a grain "moves" into a fluid cell on the lattice perimeter, its former location will correctly transition to a fluid cell, but the perimeter cell itself will retain its status as a fluid cell. Effectively, this treatment means that grains or blocks reaching either of the two vertical boundaries are instantly eroded.

The initial condition for most runs presented here has the bottom two rows filled with regolith grains. The lower left and lower right cells are assigned to be rock, which represents the base-level (and incidentally helps keep a consistent color scheme among different model configurations, because the rock state is always present). The rest of the domain is initialized as air cells.

2.5 Scaling and non-dimensionalization

The basic model has four parameters: the disturbance rate, d (cells / T), weathering rate, w (cells / T), base-level lowering interval, τ (T), and width of domain, λ (cells). The base-level lowering timescale τ represents the time interval between episodes of relative uplift in which the interior domain is lifted by one cell relative to its side boundaries. The domain width might properly be considered a boundary condition rather than a parameter, but we include it here with an eye toward examining how slope width impacts hillslope properties such as mean height. Once we define the width of a cell, δ (L), we can define versions of these four parameters that explicitly incorporate this length scale:

$$D = d\delta, \tag{3}$$
$$W = w\delta, \tag{4}$$
$$U = \delta/\tau, \tag{5}$$
$$L = \lambda\delta. \tag{6}$$

Consider the case of dynamic equilibrium, in which the rate of base-level lowering is balanced by the hillslope's rate of erosion. The mean height of this steady state hillslope, H, is a function of the above four parameters plus the characteristic length scale δ, such that we end up with a total of six variables:

$$H = f(D, W, U, L, \delta). \tag{7}$$

Buckingham's Pi theorem dictates that these six variables, which collectively include dimensions of length and time, may be grouped into four dimensionless quantities:

$$\frac{H}{\delta} = f\left(\frac{D}{U}, \frac{W}{U}, \frac{L}{\delta}\right). \tag{8}$$

The ratio $d' = d\tau = D/U$ is a dimensionless disturbance rate. Similarly, $w' = w\tau = W/U$ is a dimensionless weathering rate. Noting the definitions above, Eq. (8) is equivalent to

$$h = f\left(d', w', \lambda\right), \tag{9}$$

where $h = H/\delta$ is dimensionless hillslope height. Hence, we have a dimensionless property of the hillslope, h, that depends uniquely on three other non-dimensional variables, representing disturbance rate, weathering rate, and length.

One can similarly define a dimensionless regolith thickness, $r = R/\delta$, where R is the dimensional equivalent; it too should depend on the three dimensionless parameters that represent disturbance rate d', weathering rate, w', and hillslope length, λ, respectively. For a hillslope composed entirely of regolith, r and h depend solely on d' and λ. Finally, we define a fractional regolith cover F_r. In the Grain Hill model, F_r is calculated as the number of air–regolith cell pairs divided by the total number of cell pairs that juxtapose air with either regolith or rock.

2.6 Blocks

The foregoing model is designed to represent regolith as grain aggregates composed of gravel-sized and finer grains: material fine enough that it is susceptible to being moved by processes such as animal burrowing, frost heave, tree throw, and so on. Some hillslopes, however, are adorned with grains that are simply too large to be displaced significantly by such processes. For example, Glade et al. (2017) presented a case study and model of slopes formed beneath a resistant rock unit that periodically sheds meter-scale or larger blocks. On at least some of these types of slope, the distance between surface blocks and their source unit is considerably greater than the distance that they could roll during an initial release event (Duszyński and Migoń, 2015). This observation implies that the blocks are transported downslope by a process of repeated undermining. Glade et al. (2017) hypothesized that erosion of soil beneath and immediately downhill can cause a block to topple and hence move a distance comparable to its own diameter in each such event.

We wish to capture this form of "too big to disturb" behavior in the Grain Hill model. The CTS approach, at least as it is defined here, does not lend itself to variations in grain size or geometry. Instead, we introduce an additional type of particle that represents the behavior of blocks rather than treating their difference in size explicitly. In a sense, the approach can be viewed as treating blocks as having greater density, rather than greater size, than other grains. A block particle differs from a normal regolith cell in that it cannot be scattered upward by disturbance. Motion of a block particle can only occur under two circumstances: when it lies directly above an air cell (in which case it falls vertically, trading places with the air cell) and when it lies above and to the side of an air cell (in which case it falls downslope at a 30° angle, with probability per time d). These rules mimic the undermining process discussed by Glade et al. (2017).

As in the Glade et al. (2017) model, block particles can also undergo weathering. Here, weathering is again treated in a probabilistic fashion: blocks form from weathering of bedrock, at probability per time w. Once created, a block can undergo a conversion to normal regolith with probability w when it sits adjacent to an air cell. This treatment of blocks captures, in a simple way, the weathering of blocks as they move downslope. For purposes of this paper, the block component is included simply to test whether a cellular automaton treatment produces results that are qualitatively consistent with observations, and also consistent with the hybrid continuum–discrete model of Glade et al. (2017) and Glade and Anderson (2017).

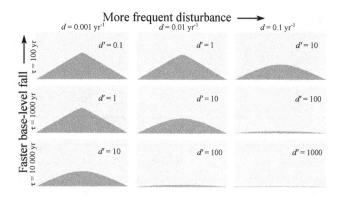

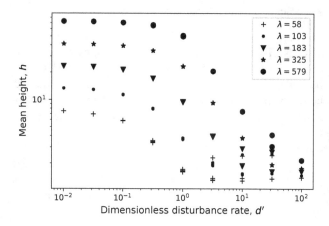

Figure 8. Equilibrium topographic cross sections using only regolith particles (no rock) and a variety of disturbance frequencies (d) and time interval between base-level fall events (τ). Fast basal incision and/or infrequent disturbance lead to planar threshold hillslopes; slow basal incision and/or frequent disturbance lead to parabolic hillslopes.

Figure 9. Dimensionless mean hillslope height, h, as a function of dimensionless disturbance rate d' for a range of hillslope lengths. Data points include 125 sensitivity analysis runs in which $d \in [10^{-3}, 10^{-2.5}, 10^{-2}, 10^{-1.5}, 10^{-1}]$, $\tau \in [10^2, 10^{2.5}, 10^3, 10^{3.5}, 10^4]$, and λ as shown in the legend.

3   Results

3.1   Fully soil-mantled hillslope

We start by considering the case of fully soil-mantled hillslopes, in which the supply of mobile regolith is effectively unlimited (Fig. 2a, b). Under this condition, the Grain Hill model represents a testable mechanistic hypothesis: that a transport limited, soil-mantled hillslope behaves essentially as a granular medium subject to periodic, quasi-random disturbance events. This concept was also the essence of the acoustic-disturbance experiments by Roering et al. (2001). To test the hypothesis, we run the Grain Hill model with a constant rate of material uplift relative to base-level until the system reaches quasi-steady state, to determine whether its steady form is smoothly convex upward (when the gradient is below the failure threshold) to planar (when the gradient lies at or near the failure threshold). Model runs were performed using a 251-row by 580-column lattice. Disturbance rates were varied from 0.001 to 0.1 yr^{-1} and intervals between relative-uplift events from 100 to 10 000 years.

Results show that the Grain Hill model produces parabolic to planar hillslope forms, depending on the ratio of disturbance to uplift rates, which is encapsulated in the dimensionless parameter d' (Fig. 8). At high d' (frequent disturbance and/or slow base-level fall), hillslope relief is low and the form is smoothly convex upward (Fig. 8, lower right panels). At somewhat lower d', the lower part of the slope approaches a threshold angle while the upper part remains smoothly convex (Fig. 8, middle diagonal panels). At low d', the form becomes predominantly planar and achieves a threshold relief that is insensitive to further increases in d' (Fig. 8, upper left panels).

Scaling of mean height as a function of d' is shown in Fig. 9. The figure shows results for 125 model runs spanning 2 orders of magnitude in each parameter (d, τ, and λ) in half-

decade intervals. The 125 runs represent a $5 \times 5 \times 5$ grid of experiments, in which each grid point represents a particular combination of the three parameters (d, τ, and λ).

For any given hillslope length, there are three regimes of behavior. Low d' (upper left of graph) leads to threshold hillslopes, in which relief depends only on hillslope length. Under moderate d', mean height scales inversely with d', as expected from linear diffusion theory. At high d', we have a finite-size regime in which dimensionless hillslope mean height is comparable to the disturbance scale, δ (cell size in the model); in other words, the hill is only one or a few cells high.

The behavior of the Grain Hill model in its simple, transport-limited configuration can be compared to diffusion theory, which relates volumetric sediment flux per unit contour length, $\mathbf{q}_s$, to topographic gradient:

$$\mathbf{q}_s = -D_s \frac{\partial \eta}{\partial x}, \tag{10}$$

where η is land-surface height, x is horizontal distance, and D_s is an effective transport coefficient. The Furbish et al. (2009) probabilistic theory for transport due to particle scattering and settling formulates D_s as

$$D_s = k r_p R_a N_a \overline{\left(1 - \frac{c}{c_m}\right)^2} \cos^2\theta, \tag{11}$$

where k is a dimensionless coefficient, r_p is particle radius, R_a is active regolith thickness, N_a is the activation rate, θ is slope angle, c is particle concentration, and c_m is a maximum concentration. The over-bar denotes an average over the active regolith thickness. For the Grain Hill model, R_a scales with the characteristic disturbance depth, δ. Further, because we treat grain aggregates, we may also assume $r_p \sim \delta$. Therefore, we have the prediction that

$$D_s = a\delta^2 N_a \cos^2\theta, \quad (12)$$

where a is a dimensionless proportionality constant.

The mean expected activation rate, N_a, is closely related to the Grain Hill model's disturbance frequency parameter, d. To relate the two quantitatively, one needs to make a trivial lattice-geometry correction. A straight-as-possible cut through the hex lattice exposes on average two faces per cell, both of which are susceptible to a disturbance event. Because d is the expected disturbance frequency per cell face, and because independent Poisson events are additive, the resultant disturbance frequency for each cell exposed along a quasi-horizontal surface is $N_a = 2d$.

A more important difference is that whereas N_a is defined as activation rate per unit horizontal area, d represents the rate per unit surface area regardless of orientation. For a given d, N_a will increase with surface roughness (because there is more exposed area of regolith–air contact), and with gradient (because the slope length increases).

An additional effect arises from the model's effective $30°$ angle of repose. On slopes steeper than this, the expected disturbance rate increases substantially because gravitational dislodgement becomes activated (Fig. 5, bottom row). Thus, the Grain Hill model incorporates an additional non-linear relationship between flux and gradient inasmuch as N_a depends on gradient.

We can derive an effective diffusivity, D_e, from the modeled topography by applying the expected relationship between mean elevation and diffusivity. Here, D_e is defined as that value which, if it were spatially uniform, would yield the same mean steady-state elevation as that produced by the particle model. Framing it this way allows us to interrogate how the effective transport coefficient varies as a function of mean slope gradient. At steady state, mass balance implies that

$$\mathbf{q}_s = Ex, \quad (13)$$

where E is the rate of erosion – equal to the rate of material uplift relative to base level – and x is horizontal distance from the ridge top. Substituting Eq. (10) and rearranging gives

$$\frac{d\eta}{dx} = -\frac{E}{D_s(x)}x \approx -\frac{E}{D_e}x. \quad (14)$$

Integrating and then averaging over x, we can solve for the average elevation, $\overline{\eta}$:

$$\overline{\eta} = \frac{E}{3D_e}L_h^2, \quad (15)$$

where $L_h = L/2$ is the length of the slope from ridge top to base (in other words, half the total length of the domain). We can then rearrange this to find D_e:

$$D_e = \frac{E}{3\overline{\eta}}L_h^2. \quad (16)$$

To examine how D_e scales, we can define a dimensionless form, normalizing by the disturbance frequency, d, and the square of active regolith thickness (equal to particle diameter), δ^2:

$$D_e' = \frac{D_e}{d\delta^2} = \frac{EL_h^2}{3\overline{\eta}d\delta^2}. \quad (17)$$

Noting that $E = \delta/\tau$, $L = 2L_h$, and $L/\delta = \lambda$, this is equivalent to

$$D_e' = \frac{\lambda^2}{12\overline{h}d\tau}, \quad (18)$$

where $\overline{h}$ is the mean hillslope height in particle diameters.

As expected, D_e' increases with hillslope gradient (Fig. 10). The effective diffusivity approaches an asymptote at $30°$ (mean gradient ≈ 0.6), representing an angle of repose. The pattern resembles the family of non-linear flux–gradient curves introduced by Andrews and Bucknam (1987) and explored further by Howard (1994) and Roering et al. (1999). At low gradients, D_e' approaches a value of about 60. (This method of estimating D_e' is similar to fitting the standard theoretical parabolic curve to the experimental profiles, except that here we use the integral of the profiles.)

The link between D_e and d provides a way to scale the Grain Hill model to field-derived estimates of D_s and R_a. Here, we equate the theoretical effective diffusivity, D_e, with the definition of the transport coefficient D_s of Furbish et al. (2009). Noting that, at low gradients, $\cos^2\theta$ in Eq. (12) approaches unity, and using the prior relation $N_a = 2d$, we may write D_s for low slope angle as

$$D_s(\theta \to 0) = 2a\delta^2 d. \quad (19)$$

In the Grain Hill model, the fact that low-angle $D_e' \approx 60$ implies that the dimensional equivalent $D_e(\theta \to 0) \approx 60\delta^2 d$. Equating D_s (the transport coefficient derived by Furbish et al., 2009) and D_e (the effective transport coefficient derived from the Grain Hill model),

$$D_s(\theta \to 0) \approx 60\delta^2 d. \quad (20)$$

This relation can be used to scale the parameters in the Grain Hill model with field data. For example, if one were to assume an active regolith thickness of 0.4 m and a low-gradient transport coefficient of $D_s = 0.01\,\mathrm{m}^2\,\mathrm{yr}^{-1}$, and set δ to the active regolith thickness, then

$$d = \frac{D_s}{60\delta^2} \approx 0.001\,yr^{-1}. \quad (21)$$

Here, d represents the frequency with which a given exposed patch of regolith of width and depth δ is disturbed upward. With the above values, the simulated hills in Fig. 8 would be 232 m long (valley to valley), with height ranging from 1.6 to 57.6 m and base-level lowering rate from 0.04 to 4 mm yr^{-1}.

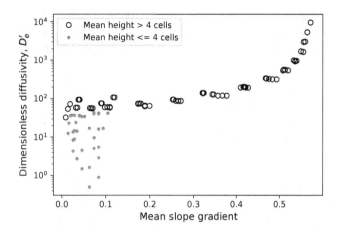

Figure 10. Relationship between dimensionless diffusivity and mean gradient, from the series of 125 model runs of which a subset is shown in Fig. 8.

3.2 Hillslope with regolith production from rock

Having established that the Grain Hill model reproduces classic soil-mantled hillslope forms and has parameters that can be related to the parameters in commonly used continuum hillslope transport theories, we turn now to the case in which regolith is generated from bedrock with a production rate that may (or may not) limit the rate of erosion. We explore the role of regolith production with a series of model runs in which w' varies from 0.4 to 40. The upper end of this range represents a condition in which the potential maximum rate of regolith production greatly exceeds the rate of base-level lowering. The lower end, 0.4, is less than the rate of base-level fall, and would seem to be insufficient to allow for equilibrium to occur, and yet nonetheless it does. Examples of equilibrium hillslope forms found in this parameter space are shown in Fig. 11.

Relationships among mean gradient, fractional regolith cover, dimensionless disturbance rate d', and dimensionless weathering rate w' are illustrated in Fig. 12. For $w' > 1$, the gradient–d' relation (Fig. 12a) has the same shape as in the purely regolith models: a threshold regime at lower d' transitioning to an inverse gradient–d' relation at higher d'. This indicates that when the maximum weathering rate (for a flat surface) is substantially greater than the rate of base-level fall, we recapture transport-limited conditions. With $w' < 1$, however, the hillslope achieves an equilibrium gradient that is greater than that for the transport-limited case, and at lower d', is greater than the threshold angle (Fig. 12a, b).

We can also examine the fractional regolith cover, which is defined here as the number of rock–air cell pairs divided by the total number of cell pairs at which air meets either regolith or rock (Fig. 12c, d). The fractional regolith cover shows relatively little sensitivity to d' (Fig. 12c). The cover hovers around unity for high w' and d' but systematically declines with w' when w' is below about 10. (Note that the

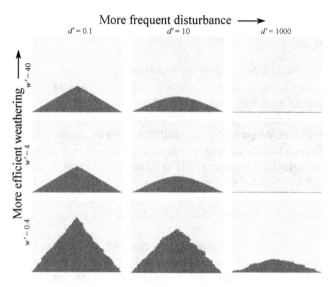

Figure 11. Final equilibrium profiles from Grain Hill runs with rock and weathering. Domain size is 222 rows by 257 columns, and uplift interval ranges from 100 to 10 000 years.

data points representing $d' = 1000$ and $w' > 1$ have hillslope heights of only a few particles and are therefore sensitive to finite-size effects.)

The models with $w' < 1$ present a seeming paradox: how is it possible to achieve an equilibrium form when the maximum weathering rate appears to be lower than the rate of uplift relative to base level? The solution to the paradox lies in surface area. The surface area of rock that is exposed to weathering is not fixed but rather depends on the overall slope length, the terrain roughness, and the fractional regolith cover. To appreciate the first effect, consider a planar slope at angle θ with no regolith cover. If $w\delta$ represents the maximum slope-normal bedrock weathering rate, then the vertical rate is simply $w\delta/\cos\theta$. All else equal, increasing gradient will increase vertical weathering rate, thereby providing a feedback between gradient and rock lowering rate. A second feedback relates to topographic roughness: all else equal, a rougher surface will experience a greater weathering rate because it provides more surface area. The third feedback, which is embedded in the depth-dependent regolith production hypothesis (Ahnert, 1967) lies in regolith cover: the greater the exposure of rock (or the thinner the cover), the faster the average rate of rock-to-regolith conversion. In the Grain Hill model, this third feedback is represented by fractional bedrock exposure (since weathering only occurs when rock cells are juxtaposed with air cells).

To test whether these are indeed the feedbacks responsible for equilibrium topography in the Grain Hill model, we can compare the rate of material influx (uplift relative to base-level) with the expected rate of rock-to-regolith conversion. In the Grain Hill model, the expected rate of regolith production, P, in cross-sectional area per time, is the product of weathering rate per cell face, w, the cross-sectional area of a

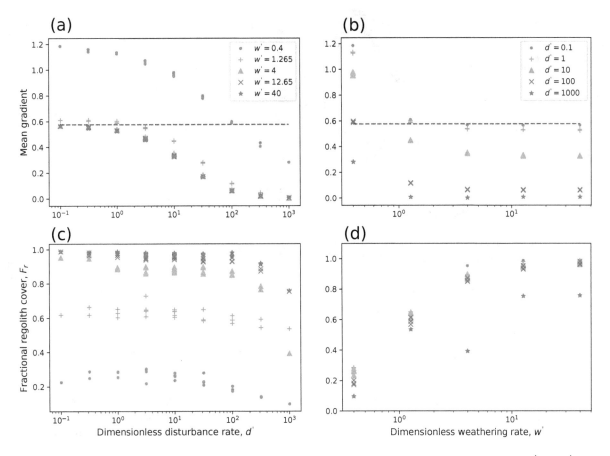

Figure 12. Mean equilibrium gradient and regolith thickness for models with rock and weathering, as a function of d' and w'. Data represent 125 runs with $d \in [10^{-3}, 10^{-2.5}, 10^{-2}, 10^{-1.5}, 10^{-1}]$ yr^{-1}, $w' \in [10^0, 10^{0.5}, 10^1, 10^{1.5}, 10^2]$, and $\tau \in [10^2, 10^{2.5}, 10^3, 10^{3.5}, 10^4]$ yr. Horizontal dashed lines show model's angle of repose for regolith.

cell, A, and the number of rock–air cell faces, n_{ra},

$$P = wAn_{ra}. \tag{22}$$

The rate of material addition due to uplift relative to base-level, U, again in cross-sectional area per time, is the area of a cell, A, multiplied by the horizontal width of the domain in cells, n_H, divided by the interval between uplift events, τ:

$$U = n_H A/\tau. \tag{23}$$

Equality between rock uplift and weathering can be expressed as

$$\frac{1}{\tau} = w\frac{n_{ra}}{n_H}. \tag{24}$$

The ratio on the right side represents the surface-area effect, in the form of surface area exposed to weathering per unit horizontal area. The balance is illustrated in Fig. 13, which compares the left-hand and right-hand terms for each of the 125 model runs with weathering. Each data point represents a single snapshot in time, and so scatter is to be expected. To help diagnose the scatter around the 1 : 1 line, the

data are divided into quintiles by fractional regolith cover, F_r (note that some of the points in the lower quintiles are obscured by being over-plotted along the 1 : 1 line). Many of the points that fall off the 1 : 1 line, especially at the high end (higher $1/\tau$), come from runs with $F_r > 80\%$; with very few exposed rock–air pairs, a small fluctuation in the n_{ra} can produce a relatively large change in predicted weathering rate. At the low end, many of the points above the 1 : 1 line come from runs with a maximum height of only a few cells, which are subject to finite-size effects.

The main message of Fig. 13 is that the Grain Hill model demonstrates an equilibrium adjustment between rock uplift and rock weathering. The weathering rate does not have a fixed upper "speed limit," but rather is set by the exposed surface area, which in turn is a function of gradient, roughness, and regolith cover. Solutions with a discontinuous regolith cover are indicative of this adjustment. Slopes can grow arbitrarily steep, with weathering and erosion increasingly attacking from the sides as the gradient rises.

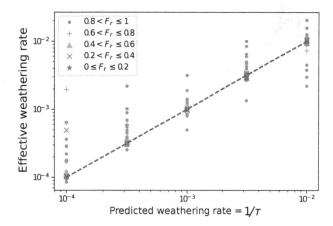

Figure 13. Comparison between rate of material input, $1/\tau$ (cells / year), with effective rate of weathering, $w n_{\mathrm{ra}}/n_{\mathrm{H}}$, from 125 model runs (see text).

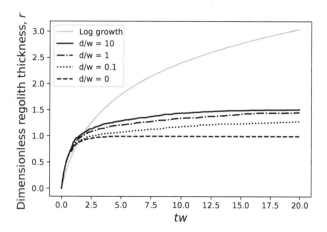

Figure 14. Regolith thickness versus time, as predicted by inverse-exponential theory (log growth; solid cyan curve) and the Grain Hill model with a range of ratios of disturbance rate (d) to weathering rate (w). Time (horizontal axis) is non-dimensionalized by multiplying by w.

3.3 Comparison between weathering rule and inverse-exponential model

The most popular function to describe regolith production from bedrock is the decaying exponential formula proposed by Ahnert (1967), which has proved consistent with estimates of production rate obtained using cosmogenic radionuclides (Heimsath et al., 1997; Small et al., 1999). The production rate is given by

$$P = P_0 \exp(-R/R_*), \qquad (25)$$

where P_0 is the maximum (bare bedrock) production rate, R is regolith thickness, and R_* is a depth-decay scale on the order of decimeters. On a flat surface, assuming no erosion or deposition, the expected rate of change of R over time is

$$\frac{dR}{dt} = \frac{\rho_{\mathrm{r}}}{\rho_{\mathrm{s}}}(1-\omega)P_0 \exp(-R/R_*), \qquad (26)$$

where ρ_{r} and ρ_{s} are the bulk densities of parent material and regolith, respectively, and ω is the fraction of parent material removed in solution upon weathering. Starting from a bare surface, and assuming isovolumetric weathering (in which case $\rho_{\mathrm{s}} = (1-\omega)\rho_{\mathrm{r}}$), the expected regolith thickness as a function of time can be found by integrating Eq. (26):

$$\frac{R}{R_*} = \ln\left[\frac{P_0}{R_*}t + 1\right]. \qquad (27)$$

We can compare this with the behavior of the cellular weathering rule by running the case of a flat, initially bare-rock surface from which weathered material may neither enter nor leave (Fig. 14, case $d w^{-1} = 0$). When the disturbance rate is zero, the cellular weathering model asymptotically approaches a steady regolith thickness of exactly one cell (thickness equal to δ). This is so because the model allows weathering to occur only when rock cells are exposed to air

cells, and there is no disturbance process that would juxtapose rock and air once the initial weathered layer has formed. When disturbance rate is non-zero, however, regolith continues to form even after the mean thickness r exceeds unity (representing one characteristic disturbance depth). Continuation of regolith production occurs because the disturbance process intermittently exposes rock, at which point it becomes subject to weathering. The greater the disturbance rate, the more frequent the exposure and hence the more rapid weathering (Fig. 14). For any ratio $d w^{-1}$, the model's weathering behavior clearly differs from the logarithmic growth in thickness predicted by exponential theory. This represents both a strength and a weakness in the Grain Hill model. On the one hand, the model under its present configuration cannot account for rock-to-regolith conversion resulting from processes that penetrate more than one characteristic disturbance depth δ into the subsurface. For example, the model neglects the possibility that some plant roots may penetrate deeply and contribute to disaggregation, or that an unusually deep freezing front in a cold winter might cause rock fracture and displacement of the resulting fragments (e.g., Anderson et al., 2012). On the other hand, the model honors the likelihood that soil disturbance and regolith production are closely linked processes, rather than independent: all else equal, a greater disturbance rate will tend to produce faster rates of both regolith production and downslope soil movement.

3.4 Rock collapse and vertical cliffs

Some rock slopes display a cliff-and-rampart morphology in which a vertical or near-vertical rock face stands above an inclined, often sediment-mantled buttress (Figs. 1 and 15). Although common in sedimentary rocks where a resistant unit forms the cliff and a weaker unit the buttress, the same

Figure 15. Two examples of cliff-and-rampart morphology. **(a)** Near Palisade, Colorado, USA, after recent rock-fall event (photo courtesy of D. Nathan Bradley and Dylan Ward). **(b)** Colorado Plateau, Utah, USA. Note that contact between lower rampart and subvertical slopes, both of which have formed in a gray shale unit, occurs without any apparent break in lithology.

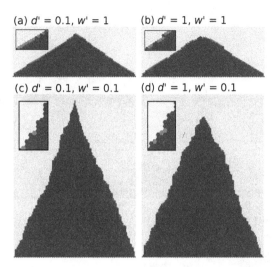

Figure 16. Quasi-steady model hillslope profiles created using a collapse rule, under four different combinations of d' and w'. Insets show magnified views of a portion of each hillslope.

morphology is sometimes found in apparently homogeneous lithology (Fig. 15b). The cliff portion of such slopes suggests a process of undermining and collapse, with the cliff-forming material being cohesive enough to maintain a vertical face but too weak to support overhangs.

To explore the origins of ramp-and-cliff morphology, we consider a version of the Grain Hill model that adds an extra rule to represent collapse: any rock particle that directly overlies air has the possibility to transition to a falling regolith particle, with the same rate as gravitational transition from resting to falling; in other words, as soon as a rock particle has been undermined, it behaves like cohesionless material.

Under dynamic equilibrium, this rule produces a morphology with slopes that are roughly planar, with alternating vertical and sloping sections and patchy regolith cover (Fig. 16). With $w' \leq 1$, gradient and regolith cover depend strongly on w' and show little or no sensitivity to d'. When $w' \ll 1$, the hillslope forms resemble pinnacles. These examples demonstrate two combined feedbacks between weathering and base-level fall: the surface area susceptible to weath-

ering, and the frequency and magnitude of material collapse through undermining.

The case of transient evolution under a stable base level leads to the formation of a regolith-mantled, angle-of-repose ramp (Fig. 17). The slope break remains relatively sharp as it retreats headward. The ramp forms as a transport slope. The angle of repose is an attractor state: if the angle were steeper, weathered material would be rapidly removed as a result of gravitational instability; if it were substantially lower, material would accumulate, because transport would be limited to the (much lower) rate of disturbance-driven creep motions. Hence, the Grain Hill model predicts that formation of a sediment-mantled ramp beneath a steeper, actively weathering rock slope is an expected outcome for a steep rock slope under stable base level.

3.5 Blocks

Weathering and erosion in landscapes underlain by relatively massive, fracture- or joint-bounded rock can sometimes produce large "blocks" of rock, defined here as clasts that are too large to be displaced upward by normal hillslope processes. The release of blocks from dipping sedimentary or volcanic strata can alter both the shape and relief of hillslopes (Glade et al., 2017). When blocks are delivered to streams, they can influence the channel's roughness, gradient, erosion rate, and longitudinal profile shape (Shobe et al., 2016).

As discussed in Sect. 2.6, the Grain Hill model can be modified to honor blocks by defining an additional cell type that represents blocks. The weathering process is modified such that a rock cell now weathers into a block, and the block in turn may weather to form regolith. When a block is undermined directly from below, it will fall just as a normal regolith particle would. When a block particle lies adjacent

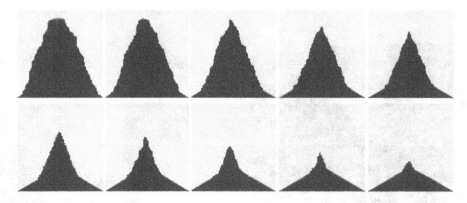

Figure 17. Time series showing transient erosion of a steep rock slope under a stable base-level, highlighting formation of ramp-and-cliff morphology. Simulation shows 20 000 years of slope evolution under $d = w = 10^{-3}$ yr^{-1}. Nominal width, assuming $\delta = 0.1$ m, is 12 m.

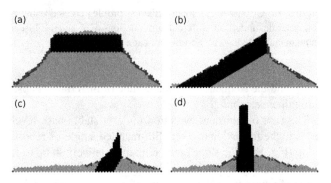

Figure 18. Examples of models that include blocks. Rock (black) weathers to blocks (dark red), which can only move downward or downward plus laterally. Blocks in turn weather to regolith (light brown) (GIF-format animations of similar runs are available as an online resource; see Tucker, 2018a).

to and above an air cell, a disturbance event may occur that causes the block to shift downward on the slope. By these means, blocks in the model may move downward, or downward and laterally, but never upward. An implicit assumption in this treatment is that blocks do not roll long distances (further than their own diameter) upon release.

We examine model runs in which a resistant rock layer is embedded in a weak sedimentary material that is soft enough to be treated as regolith (Fig. 18). The modeled hillslopes are qualitatively consistent both with field observations and with the mixed continuum–discrete model of Glade et al. (2017) and Glade and Anderson (2017) in that block-mantled slopes are generally concave upward, reflecting a downslope decrease in the flux of blocks as weathering progressively transmutes them into regolith.

4   Comparison to field sites

We perform a basic validation of the Grain Hill model by comparing its output to real field sites, testing whether the

model is capable of reproducing realistic hillslope forms at the correct spatial scale under known boundary conditions. Field sites were chosen such that model boundary conditions could be derived from independent field estimates of rate parameters such as D_s and the rate of base-level fall. To perform this test, we consider two examples: a convex-upward, soil-mantled hillslope in Gabilan Mesa, California, USA (Fig. 2a, b), and a steep, quasi-planar, discontinuously mantled hillslope in the Yucaipa Ridge, California, USA (Fig. 2c, d). For each of these two case studies, the hillslopes appear to be approximately at steady state, and independent estimates exist for the rate of base-level fall, U (Binnie et al., 2007; Perron et al., 2009, 2012). We estimated the effective transport coefficient, D_s, for the profiles shown in Fig. 2a, c by measuring the second derivative of the one-dimensional hillslope elevation profiles, $\frac{\partial^2 \eta}{\partial x^2}$, and solving for D_s using

$$D_s = -\frac{U}{\frac{\partial^2 \eta}{\partial x^2}}. \tag{28}$$

For the Gabilan Mesa profile, we estimated the profile-averaged effective transport coefficient as 0.0345 m^{2}yr^{-1}. The effective rate of base-level lowering has been estimated at $U \approx 1.47 \times 10^{-4}$ m yr^{-1} (Perron et al., 2012). To construct a Grain Hill model for the Gabilan profile, we begin by assuming a characteristic disturbance depth of $\delta = 1$ m. This value was chosen to be consistent with measured soil depths that typically range between 0.2 and 1.2 m (Johnstone et al., 2017). We treat the system as transport-limited, consisting of mobile material, so that weathering is not explicitly modeled. The disturbance parameter, d, is then calculated from the independently estimated value of D_s using Eq. (20). The interval between uplift events is $\tau = \delta/U \approx 6800$ years. The resulting modeled equilibrium profile provides a reasonably good match to the observed Gabilan profile, with a convex-upward shape and a hilltop height of about 45 m above the slope base (Fig. 19a).

For Yucaipa Ridge, we estimated the transport coefficient at $D_s \sim 0.028$ m^2 yr^{-1} on the basis of hilltop curva-

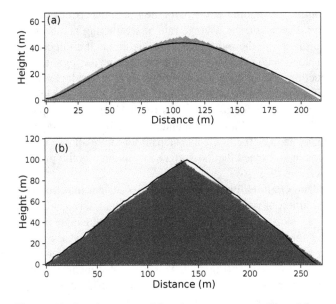

Figure 19. Steady state models using parameters estimated from observed hillslope profiles. **(a)** Parameters are based on Gabilan Mesa, with the profile shown in Fig. 2a, b for comparison. **(b)** Parameters based on Yucaipa Ridge, with the profile shown in Fig. 2c, d for comparison.

ture and an estimated effective rate of base-level lowering of $\approx 0.0027\,\mathrm{m\,yr^{-1}}$ (Binnie et al., 2007). Using Eq. (20), this equates to a disturbance-rate parameter $d = 0.00468\,\mathrm{yr^{-1}}$ and an uplift interval of 370 years in the Grain Hill model. Bedrock outcrops are common on the Yucaipa Ridge hillslopes, implying a thin, discontinuous regolith cover. We therefore treat the system as consisting of bedrock that must be weathered before it can become mobile. Because we do not have independent information on the effective maximum rock weathering rate, the Yucaipa case is a somewhat weaker test: we can only ask whether there exists a geologically reasonable value of w such that the model reproduces the observed relief and shape of the slope profile. Through trial and error, we find that with a weathering rate parameter $w = 0.002\,\mathrm{yr^{-1}}$ (which corresponds to a maximum regolith production rate of $2\,\mathrm{mm\,yr^{-1}}$), the model does a credible job of capturing the shape and size of the Yucaipa profile (compare Fig. 2c with 19b). Although this particular value was obtained through a simple calibration process, it is at least both geologically reasonable and, as one might expect, somewhat lower than the rate of base-level lowering.

To test the sensitivity of the Yucaipa example to the assumed characteristic disturbance depth, we ran a second experiment in which δ was reduced to 0.75 m, and the weathering, disturbance rate, and uplift parameters were rescaled accordingly. The relief and mean gradients of the two cases are nearly identical, with planar slopes and a relief in both cases of $\sim 100\,\mathrm{m}$.

These two examples demonstrate that the Grain Hill model parameters are not arbitrary but instead can be linked through straightforward reasoning to field estimates of transport efficiency and base-level lowering. When one does so, the model successfully reproduces both the shape and scale of observed slopes.

5 Discussion

With just three parameters – disturbance frequency (d), characteristic disturbance depth (δ), and base-level fall frequency (u) – the Grain Hill algorithm can reproduce the convex-upward to quasi-planar forms associated with soil-mantled hillslopes (Fig. 8). With the addition of a parameter that represents rock-to-regolith conversion rate, the algorithm accommodates partly mantled, rocky hillslopes (Figs. 11, 16, 17). By adding a rule for detachment of blocks from resistant rock, the model reproduces hillslope forms associated with hogbacks and ledge-forming escarpments (Fig. 18).

A common criticism of cellular automaton models is that they involve arbitrary rules and/or parameters that can neither be measured nor verified in the real world. That is not the case for the Grain Hill model, for which the parameters are tied to measurable physical quantities. For example, the disturbance frequency d is directly related to the frequency parameter N_a in statistical theory of soil transport developed by Furbish et al. (2009), and through that theory to the diffusion-like transport coefficient D_s that is commonly estimated in field studies. This connection between model parameters and field measurements is illustrated by the model's ability to reproduce the correct shape and scale of observed hillslope forms when estimates of D_s and U are available (Figs. 2, 19). In the transport-limited case, there are no tunable parameters: given independent estimates of D_s and U, the correct morphology is recovered (Figs. 2a, 19a). In the case where rock weathering appears to play a role, and an independent estimate of P_0 is not available, the model requires an estimation of maximum weathering rate w. Nonetheless, a plausible value of w ($0.002\,\mathrm{m\,yr^{-1}}$), somewhat smaller than the rate of base-level fall ($0.0027\,\mathrm{m\,yr^{-1}}$), reproduces the observed shape and relief in the Yucaipa Ridge case study.

The transport dynamics predicted by the Grain Hill model are consistent with continuum soil-transport theory, which treats soil as a fluid with a downslope flow rate that depends on slope gradient. Like the popular Andrews–Bucknam nonlinear transport law (e.g., Andrews and Bucknam, 1987; Howard, 1994; Roering et al., 1999), the transport-limited form of the Grain Hill model predicts diffusion-like behavior in which the effective diffusivity increases with slope gradient, with an asymptote at a threshold angle (Fig. 10). In one sense, the Grain Hill model is actually closer to the process level than fluid-like continuum models, because net downslope mass flux arises from a sequence of stochastic disturbance events rather than being dictated by a macroscopic transport law.

One limitation of the Grain Hill model is that its threshold-like behavior arises from the lattice geometry: regolith cells perched at a 30° angle above and to one side of an air cell are treated as unconditionally unstable. Whereas the timing of motion is treated as a stochastic process, the occurrence of motion is inevitable (unless some other event occurs first). This treatment neglects the possibility of frictional locking among noncohesive grains at angles somewhat above 30°, as well as the possibility of cohesion. This limitation could be overcome by introducing a probabilistic treatment of grain stability: a grain aggregate will be stable with a given probability p, and unstable with probability $1-p$. Such a treatment would introduce an additional parameter, but this parameter could in principle be estimated from physical experiments. The addition of a "sticking rule" like this might also make it possible for models with alternative lattice geometries to manifest the same dynamics, thereby decoupling the basic model framework from the geometry of the lattice on which it is implemented.

The inclusion of rock-to-regolith conversion enables the Grain Hill model to predict a continuum of slope forms from fully soil mantled to intermittently covered to bare. However, there are several limitations in the treatment of regolith production that could be improved on. The weathering rule assumes that regolith production can only occur when rock is exposed to air, which obviously neglects the role of shallow subsurface processes such as root or frost wedging. The effective weathering depth scale is the same as the disturbance scale, and equal to the cell size. This assumption is probably reasonable if the processes responsible for weathering and disturbance were one and the same, but not if they are distinct processes with different length scales. The Grain Hill model also does not account explicitly for chemical weathering, which in some cases can extend well below the surface. Finally, the model's effective regolith-production behavior does not follow the log-growth curve predicted by inverse-exponential theory for a stable surface (Fig. 14). With these caveats in mind, one advantage of the stochastic model of regolith production is that it effectively treats the disturbance and regolith-production processes as being closely linked: all else equal, the production rate is higher when disturbance is more frequent.

The popular inverse-exponential model for regolith production implies the existence of a speed limit to landscape evolution: in the absence of rock landsliding, erosion rate cannot exceed the maximum rate of rock-to-regolith conversion. Moreover, the model implies the existence of a bare landscape once the rate of erosion exceeds the maximum rate of regolith production. Heimsath et al. (2012) found evidence, however, that in fact there are additional stabilizing mechanisms, and that these manifest in landscapes with thin, patchy soils. The Grain Hill model is consistent with these observations in that it predicts the natural emergence of a dis-

continuous regolith cover, with the fractional cover exerting an influence on the average rate of weathering and erosion. Furthermore, the model behavior highlights the importance of slope length and roughness in modulating the regolith production rate: all else equal, steeper or rougher slopes allow higher production rates, leading to an additional feedback between relief and erosion rate for rocky hillslopes. The possibility of rock collapse upon undermining by weathering provides another feedback mechanism that may allow rates of erosion to exceed the flat-surface maximum regolith production rate (Fig. 16).

The Grain Hill model also provides insight into transient evolution of rocky slopes. Experiments on the relaxation of rocky slopes that are steeper than the threshold angle predict the formation of a regolith-mantled pediment at the angle of repose, which extends upslope as the steep upper slope gradually recedes (Fig. 17). This scarp–pediment morphology emerges without any variation in material strength, requiring only a period of base-level stability.

As a computational framework for exploring hillslope forms, the Grain Hill model has the advantage that it provides a mechanistic link between events (disturbance and weathering) and long-term morphologic evolution, without the need to specify a flux law. The model has the further advantage of being fully two dimensional, allowing disturbance and weathering events to initiate from the side as well as vertically. A further key element is that the model can mix timescales: a short timescale associated with grain motion, an intermediate timescale associated with disturbance events, and a much longer timescale for slope evolution. Mixing these disparate timescales in a single computer model is made possible by the fact that most of the time grains are stationary: the algorithm operates on small (stochastic) time steps during those moments when grains are moving, and on much longer steps when no grains are in motion (for further information on the discrete-event algorithm behind the model, see Tucker et al., 2016).

The Grain Hill framework has several important limitations. It is not practical to simulate motion of individual grains unless the spatial scale is quite limited (e.g., Fig. 6) or the grains are unusually large (Fig. 18). If one wished to model individual grains (of order, say, 10^{-3} m) at the scale of a hillslope (of order 10^2 m), a much more efficient solution algorithm would be needed. Furthermore, the nature of a cellular automaton is such that physical interactions are limited to adjacent cells only; long-distance effects such as stress transmission cannot easily be represented. In one sense, the restriction to short-range influence could be seen as an advantage, in that it forces one to think about how it is that mass or energy is actually transmitted in a granular medium. But the restriction means that well-known principles such as solid-state stress cannot easily be represented. On the other hand, the model does capture non-local transport, in which particles set in motion can travel a distance comparable to the slope length (Foufoula-Georgiou et al., 2010; Tucker and

Bradley, 2010; Furbish and Roering, 2013). Non-local transport emerges in the Grain Hill model when the slope angle is near or above 30°, such that there is a high probability that a disturbed particle will land in an unstable location and continue moving without the need for a second disturbance event.

A further limitation concerns the fixed cell size. Because the model is restricted to a fixed cell size, the Grain Hill framework does not lend itself to treatment of multiple grain sizes (apart from the simple "aggregates and blocks" approach illustrated in Fig. 18). Despite these limitations, the Grain Hill model provides a useful framework for exploring hillslope process and form in the context of stochastic events.

6 Conclusions

A continuous-time stochastic cellular automaton model known as the Grain Hill model allows for computational simulation of two-dimensional slope forms that arise from stochastic disturbance and (possibly) weathering events. The model operates on a hexagonal lattice, with cell states representing fluid, rock, and grain aggregates that are either stationary or in a state of motion in one of the six cardinal lattice directions. An optional additional state represents unusually large grains ("blocks") that cannot be displaced upward by disturbance events.

The Grain Hill model is able to reproduce a range of common slope forms, from fully soil mantled to rocky and partially mantled. The bestiary of forms that the model can produce includes convex-upward soil mantled slopes, planar slopes (bare, soil mantled, or in between), and cliffs with basal ramparts. When the model is configured to include a resistant rock layer that decomposes into blocks, the model reproduces observed hogback-like slope forms and qualitatively matches the behavior predicted by a recent continuum–discrete model (Glade et al., 2017; Glade and Anderson, 2017).

In its simplest guise, the model has only three process parameters, which represent disturbance frequency, characteristic disturbance depth, and base-level lowering rate, respectively. Incorporating physical weathering of rock adds one additional parameter, representing the characteristic rock weathering rate. These parameters are not arbitrary but rather have a direct link with corresponding parameters in continuum theory. Comparison between observed and modeled slope forms demonstrates that the model can reproduce both the shape and scale of real hillslope profiles.

Experiments with the Grain Hill model highlight the importance of regolith cover fraction in governing both the downslope mass transport rate and the rate of physical weathering. Equilibrium rocky hillslope profiles are possible even when the rate of base-level lowering exceeds the nominal bare-rock weathering rate, because increases in both slope gradient and roughness can allow for rock weathering rates that are greater than the flat-surface maximum. Finally, experiments in transient relaxation of steep, rocky slopes predict the formation of a regolith-mantled pediment that migrates headward through time while maintaining a sharp slope break.

Author contributions. The idea to develop a 2-D cellular rock-slope model arose from conversations among all three authors. The model code was written in Landlab by GT. Both GT and DEJH contributed to the underlying grid data structures and Python code. SWM extracted the hillslope profiles and estimated the parameters for the two field sites. GT performed the computational experiments and wrote the paper, with input and editing from SWM and DEJH.

Competing interests. The authors declare that they have no conflict of interest.

Acknowledgements. This research was supported by the US National Science Foundation (EAR-1349390 and ACI-1450409 to Gregory E. Tucker and Daniel E. J. Hobley, and EAR-1349229 to Scott W. McCoy). Daniel E. J. Hobley's participation was also supported in part by the National Center for Earth Surface Dynamics (EAR-1246761). Support for high-performance computing and software development was provided by the Community Surface Dynamics Modeling System (CSDMS) (EAR-1226297). High-resolution topographic data were downloaded from Open Topography (http://www.opentopography.org/, last access: 17 October 2017).

References

Ahnert, F.: The role of the equilibrium concept in the interpretation of landforms of fluvial erosion and deposition, Proc. symp. l'évolution des versants (Liége), 50, 23–51, 1967.

Alonso, J. and Herrmann, H.: Shape of the tail of a two-dimensional sandpile, Phys. Rev. Lett., 76, 4911, https://doi.org/10.1103/PhysRevLett.76.4911, 1996.

Anderson, R. S., Anderson, S. P., and Tucker, G. E.: Rock damage and regolith transport by frost: An example of climate modulation of the geomorphology of the critical zone, Earth Surf. Proc. Land., 38, 299–316, 2012.

Andrews, D. and Bucknam, R. C.: Fitting degradation of shoreline scarps by a nonlinear diffusion model, J. Geophys. Res., 92, 12857–12867, 1987.

Binnie, S. A., Phillips, W. M., Summerfield, M. A., and Fifield, L. K.: Tectonic uplift, threshold hillslopes, and denudation rates in a developing mountain range, Geology, 35, 743–746, https://doi.org/10.1130/G23641A.1, 2007.

Chen, S. and Doolen, G. D.: Lattice Boltzmann method for fluid flows, Annu. Rev. Fluid Mech., 30, 329–364, 1998.

Cottenceau, G. and Désérable, D.: Open Environment for 2d Lattice-Grain CA, in: Cellular Automata, ACRI, edited by: Bandini S., Manzoni S., Umeo H., and Vizzari G., Lecture Notes in Computer Science, 6350, Springer, Berlin, Heidelberg, 2010.

Coyote, W. E.: Fast and furry-ous: exploring the links between gravitational forces and situational awareness, PhD thesis, Acme Technical College, Tombstone, Arizona, USA, 1949.

Culling, W.: Soil creep and the development of hillside slopes, J. Geol., 71, 127–161, 1963.

Culling, W.: Theory of erosion on soil-covered slopes, J. Geol., 73, 230–254, 1965.

Désérable, D.: A versatile two-dimensional cellular automata network for granular flow, SIAM J. Appl. Math., 62, 1414–1436, 2002.

Désérable, D., Dupont, P., Hellou, M., and Kamali-Bernard, S.: Cellular automata in complex matter, Aip. Conf. Proc., 20, 67, 2011.

Drake, T. G. and Calantoni, J.: Discrete particle model for sheet flow sediment transport in the nearshore, J. Geophys. Res.-Oceans, 106, 19859–19868, 2001.

Duszyński, F. and Migoń, P.: Boulder aprons indicate long-term gradual and non-catastrophic evolution of cliffed escarpments, Stołowe Mts, Poland, Geomorphology, 250, 63–77, 2015.

Foufoula-Georgiou, E., Ganti, V., and Dietrich, W.: A nonlocal theory of sediment transport on hillslopes, J. Geophys. Res., 115, F00A16, https://doi.org/10.1029/2009JF001280, 2010.

Furbish, D. and Haff, P.: From divots to swales: Hillslope sediment transport across divers length scales, J. Geophys. Res., 115, F03001, https://doi.org/10.1029/2009JF001576, 2010.

Furbish, D. J. and Roering, J. J.: Sediment disentrainment and the concept of local versus nonlocal transport on hillslopes, J. Geophys. Res.-Earth, 118, 937–952, 2013.

Furbish, D. J. and Schmeeckle, M. W.: A probabilistic derivation of the exponential-like distribution of bed load particle velocities, Water Resour. Res., 49, 1537–1551, 2013.

Furbish, D., Hamner, K., Schmeeckle, M., Borosund, M., and Mudd, S.: Rain splash of dry sand revealed by high-speed imaging and sticky paper splash targets, J. Geophys. Res., 112, F01001, https://doi.org/10.1029/2006JF000498, 2007.

Furbish, D., Haff, P., Dietrich, W., and Heimsath, A.: Statistical description of slope-dependent soil transport and the diffusion-like coefficient, J. Geophys. Res., 114, F00A05, https://doi.org/10.1029/2009JF001267, 2009.

Gabet, E.: Sediment transport by dry ravel, J. Geophys. Res., 108, 2049, https://doi.org/10.1029/2001JB001686, 2003.

Gabet, E. J. and Mendoza, M. K.: Particle transport over rough hillslope surfaces by dry ravel: Experiments and simulations with implications for nonlocal sediment flux, J. Geophys. Res., 117, F01019, https://doi.org/10.1029/2011JF002229, 2012.

Ghil, M., Zaliapin, I., and Coluzzi, B.: Boolean delay equations: A simple way of looking at complex systems, Physica D., 237, 2967–2986, 2008.

Glade, R. and Anderson, R.: Quasi-steady evolution of hillslopes in layered landscapes: An analytic approach, J. Geophys. Res., 123, 26–45, https://doi.org/10.1002/2017JF004466, 2017.

Glade, R. C., Anderson, R. S., and Tucker, G. E.: Block-controlled hillslope form and persistence of topography in rocky landscapes, Geology, 45, 311–314, 2017.

Gutt, G. and Haff, P.: An automata model of granular materials, in: Proceedings of the fifth distributed memory computing conference, Charleston, SC, USA, 1990.

Heimsath, A., Dietrich, W., Nishiizumi, K., and Finkel, R.: The soil production function and landscape equilibrium, Nature, 388, 358–361, 1997.

Heimsath, A., DiBiase, R., and Whipple, K.: Soil production limits and the transition to bedrock-dominated landscapes, Nat. Geosci., 5, 210–214, 2012.

Hobley, D. E. J., Adams, J. M., Nudurupati, S. S., Hutton, E. W. H., Gasparini, N. M., Istanbulluoglu, E., and Tucker, G. E.: Creative computing with Landlab: an open-source toolkit for building, coupling, and exploring two-dimensional numerical models of Earth-surface dynamics, Earth Surf. Dynam., 5, 21–46, https://doi.org/10.5194/esurf-5-21-2017, 2017.

Howard, A. D.: A detachment-limited model of drainage basin evolution, Water Resour. Res., 30, 2261–2285, 1994.

Howard, A. D. and Selby, M. J.: Rock Slopes, in: Geomorphology of Desert Environments, 123–172, Springer, Dordrecht, 1994.

Hutton, E., Hobley, D. E. J., Tucker, G. E., Nudurupati, S. S., Adams, J. M., Gasparini, N. M., Knuth, J. S., Strauch, R., Shobe, C. M., Barnhart, K. R., Rengers, F. K., and Istanbulluoglu, E.: Landlab version 1.0., https://doi.org/10.5281/zenodo.154179, 2016.

Johnstone, S. A., Chadwick, K. D., Frias, M., Tagliaro, G., and Hilley, G. E.: Soil Development over Mud-Rich Rocks Produces Landscape-Scale Erosional Instabilities in the Northern Gabilan Mesa, Geol. Soc. Am. Bull., 129, 1266–79, 2017.

Károlyi, A. and Kertész, J.: Lattice-gas model of avalanches in a granular pile, Phys. Rev. A., 57, 852, https://doi.org/10.1103/PhysRevE.57.852, 1998.

Károlyi, A. and Kertész, J.: Granular medium lattice gas model: the algorithm, Comput. Phys. Commun., 121, 290–293, 1999.

Károlyi, A., Kertész, J., Havlin, S., Makse, H. A., and Stanley, H. E.: Filling a silo with a mixture of grains: friction-induced segregation, Europhys. Lett., 44, 386, 1998.

Lamb, M. P., Scheingross, J. S., Amidon, W. H., Swanson, E., and Limaye, A.: A model for fire-induced sediment yield by dry ravel in steep landscapes, J. Geophys. Res.-Earth, 116, F03006, https://doi.org/10.1029/2010JF001878, 2011.

Lamb, M. P., Levina, M., DiBiase, R. A., and Fuller, B. M.: Sediment storage by vegetation in steep bedrock landscapes: Theory, experiments, and implications for postfire sediment yield, J. Geophys. Res.-Earth, 118, 1147–1160, 2013.

MacVicar, B., Parrott, L., and Roy, A.: A two-dimensional discrete particle model of gravel bed river systems, J. Geophys. Res.-Earth, 111, F3, https://doi.org/10.1029/2005JF000316, 2006.

Martinez, J. and Masson, S.: Lattice grain models, in: Silos, edited by: Brown, C. and Nielsen, J., London, CRC Press, 1998.

McEwan, I. and Heald, J.: Discrete particle modeling of entrainment from flat uniformly sized sediment beds, J. Hydraul. Eng., 127, 588–597, 2001.

Narteau, C., Le Mouël, J., Poirier, J., Sepúlveda, E., and Shnirman, M.: On a small-scale roughness of the core–mantle boundary, Earth Planet. Sc. Lett., 191, 49–60, 2001.

Narteau, C., Zhang, D., Rozier, O., and Claudin, P.: Setting the length and time scales of a cellular automaton dune model from the analysis of superimposed bed forms, J. Geophys. Res.-Earth, 114, F03006, https://doi.org/10.1029/2008JF001127, 2009.

Peng, G. and Herrmann, H. J.: Density waves of granular flow in a pipe using lattice-gas automata, Phys. Rev. A., 49, R1796, https://doi.org/10.1103/PhysRevE.49.R1796, 1994.

Perron, J. T., Kirchner, J. W., and Dietrich, W. E.: Formation of evenly spaced ridges and valleys, Nature, 460, 502–505, 2009.

Perron, J. T., Richardson, P. W., Ferrier, K. L., and Lapôtre, M.: The Root of BBranching River Networks, Nature, 492, 100–103, 2012.

Roering, J.: Soil creep and convex-upward velocity profiles: Theoretical and experimental investigation of disturbance-driven sediment transport on hillslopes, Earth Surf. Proc. Land., 29, 1597–1612, 2004.

Roering, J.: How well can hillslope evolution models "explain" topography? Simulating soil transport and production with high-resolution topographic data, Geol. Soc. Am. Bull., 120, 1248–1262, 2008.

Roering, J., Kirchner, J., and Dietrich, W.: Evidence for nonlinear, diffusive sediment transport on hillslopes and implications for landscape morphology, Water Resour. Res., 35, 853–870, 1999.

Roering, J., Kirchner, J., Sklar, L., and Dietrich, W.: Hillslope evolution by nonlinear creep and landsliding: An experimental study, Geology, 29, 143–146, 2001.

Roering, J. J. and Gerber, M.: Fire and the evolution of steep, soil-mantled landscapes, Geology, 33, 349–352, 2005.

Rozier, O. and Narteau, C.: A real-space cellular automaton laboratory, Earth Surf. Proc. Land, 39, 98–109, 2014.

Schmeeckle, M. W.: Numerical simulation of turbulence and sediment transport of medium sand, J. Geophys. Res.-Earth, 119, 1240–1262, 2014.

Shobe, C. M., Tucker, G. E., and Anderson, R. S.: Hillslope-derived blocks retard river incision, Geophys. Res. Lett., 43, 5070–5078, 2016.

Small, E., Anderson, R., and Hancock, G.: Estimates of the rate of regolith production using 10Be and 26Al from an alpine hillslope, Geomorphology, 27, 131–150, 1999.

Tucker, G. and Bradley, D.: Trouble with diffusion: Reassessing hillslope erosion laws with a particle-based model, J. Geophys. Res., 115, F1, https://doi.org/10.1029/2009JF001264, 2010.

Tucker, G. E., Hobley, D. E. J., Hutton, E., Gasparini, N. M., Istanbulluoglu, E., Adams, J. M., and Nudurupati, S. S.: CellLab-CTS 2015: continuous-time stochastic cellular automaton modeling using Landlab, Geosci. Model Dev., 9, 823–839, https://doi.org/10.5194/gmd-9-823-2016, 2016.

Tucker, G. E.: GrainHill cellular hillslope model: GIF animations of hillslope evolution, https://doi.org/10.6084/m9.figshare.6720476, 2018a.

Tucker, G. E.: GrainHill version 1.0, https://doi.org/10.5281/zenodo.1306961, 2018b.

Zhang, D., Narteau, C., and Rozier, O.: Morphodynamics of barchan and transverse dunes using a cellular automaton model, J. Geophys. Res., 115, F3, https://doi.org/10.1029/2009JF001620, 2010.

Zhang, D., Narteau, C., Rozier, O., and du Pont, S. C.: Morphology and dynamics of star dunes from numerical modelling, Nat. Geosci., 5, 463–467, 2012.

4

Morphology of bar-built estuaries: empirical relation between planform shape and depth distribution

Jasper R. F. W. Leuven, Sanja Selaković, and Maarten G. Kleinhans

Faculty of Geosciences, Utrecht University, Princetonlaan 8A, 3584 CB, Utrecht, the Netherlands

Correspondence: Jasper R. F. W. Leuven (j.r.f.w.leuven@uu.nl)

Abstract. Fluvial–tidal transitions in estuaries are used as major shipping fairways and are characterised by complex bar and channel patterns with a large biodiversity. Habitat suitability assessment and the study of interactions between morphology and ecology therefore require bathymetric data. While imagery offers data of planform estuary dimensions, only for a few natural estuaries are bathymetries available. Here we study the empirical relation between along-channel planform geometry, obtained as the outline from imagery, and hypsometry, which characterises the distribution of along-channel and cross-channel bed levels. We fitted the original function of Strahler (1952) to bathymetric data along four natural estuaries. Comparison to planform estuary shape shows that hypsometry is concave at narrow sections with large channels, while complex bar morphology results in more convex hypsometry. We found an empirical relation between the hypsometric function shape and the degree to which the estuary width deviates from an ideal convergent estuary, which is calculated from river width and mouth width. This implies that the occurring bed-level distributions depend on inherited Holocene topography and lithology. Our new empirical function predicts hypsometry and along-channel variation in intertidal and subtidal width. A combination with the tidal amplitude allows for an estimate of inundation duration. The validation of the results on available bathymetry shows that predictions of intertidal and subtidal area are accurate within a factor of 2 for estuaries of different size and character. Locations with major human influence deviate from the general trends because dredging, dumping, land reclamation and other engineering measures cause local deviations from the expected bed-level distributions. The bathymetry predictor can be used to characterise and predict estuarine subtidal and intertidal morphology in data-poor environments.

1 Introduction

Estuaries develop as a result of dynamic interactions between hydrodynamic conditions, sediment supply, underlying geology and ecological environment (Townend, 2012; de Haas et al., 2017). One model for the resulting morphology is that of the "ideal estuary" that is hypothesised to have along-channel uniform tidal range, constant depth and current velocity, and a channel width that exponentially converges in the landward direction such that the loss of tidal energy by friction is balanced by the gain in tidal energy by convergence (Savenije, 2006, 2015; Townend, 2012; Dronkers, 2017). One would expect that in this case the along-channel variation in hypsometry is also negligible. However, natural estuaries deviate from ideal ones as a result of a varying degree of sediment supply, lack of time for adaptation and sea-level rise (Townend, 2012; de Haas et al., 2017), and locations wider than ideal are filled with tidal bars (Leuven et al., 2018a) (Fig. 1). Differences in bed-level profiles between ideal and non-ideal estuaries are further enhanced by damming, dredging, dumping, land reclamation and other human interference (e.g. O'Connor, 1987; Wang and Winterwerp, 2001; Lesourd et al., 2001; Jeuken and Wang, 2010; Wang et al., 2015). All these natural deviations from the ideal estuary mean that there is no straightforward relation

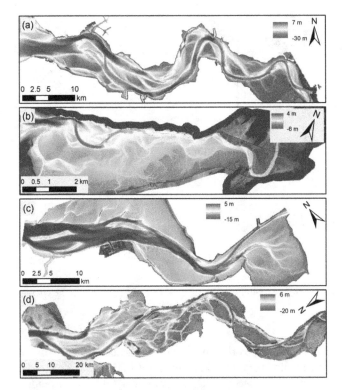

Figure 1. Bathymetry from **(a)** the Western Scheldt (NL), **(b)** the Dovey (Wales), **(c)** the Eems–Dollard (NL) and **(d)** the Columbia River Estuary (USA). Source: **(a, c)** Rijkswaterstaat (NL), **(b)** Natural Resources Wales, **(d)** Lower Columbia Estuary Partnership.

between the planform geometry of the estuary and the hypsometry or distribution of depths.

Hypsometry captures key elements of geomorphological features (Strahler, 1952; Boon and Byrne, 1981; Dieckmann et al., 1987; Kirby, 2000; Toffolon and Crosato, 2007; Townend, 2008, 2010; de Vet et al., 2017) (Fig. 2). The hypsometric method was developed by Strahler (1952) and Boon and Byrne (1981) to relate the planform area of a basin to elevation. Later the resulting functions were used to predict the influence of basin morphology on the asymmetry of the horizontal and vertical tides, to predict flood or ebb dominance and maturity of an estuary (Boon and Byrne, 1981; Wang et al., 2002; Moore et al., 2009; Friedrichs, 2010), and to characterise the trend of salt marsh development (Gardiner et al., 2011; Hu et al., 2015). Furthermore, hypsometry was used as a data reduction method to characterise entire reaches spanning bars and channels in estuaries (Toffolon and Crosato, 2007) and shapes of individual tidal bar tops (de Vet et al., 2017). Hypsometry has also been used to describe the dimensions of channels and tidal flats in an idealised model (Townend, 2010). In the latter, a parabolic shape was prescribed for the low water channel, a linear profile for the intertidal low zone and a convex profile for the intertidal high zone. While these profiles are valid for perfectly con-

verging channels, it is unknown to what extent they are applicable to estuaries with irregular planforms and whether the currently assumed profiles are valid to assess flood or ebb dominance. Hypsometric profiles and derived inundation duration are also relevant indicators for habitat composition and future transitions from mudflat to salt marsh (Townend, 2008). To predict and characterise the morphology and assess habitat area, we need along-channel and cross-channel bed-level predictions for systems without measured bathymetry (Wolanski and Elliott, 2015).

For only a few natural estuaries is bathymetry available, which leaves many alluvial estuaries with irregular planforms from all around the world underinvestigated. However, many estuaries are visible in detail on satellite imagery, which raises the question of whether there is a relation between planform geometry and depth distribution. Such a relation is known to exist in rivers in the form of hydraulic geometry depending on bar pattern and meander pool depths depending on planform channel curvature (e.g. Kleinhans and van den Berg, 2011; van de Lageweg et al., 2016). Therefore, it seems likely that such a relation between the horizontal and vertical dimensions exists for sandy estuaries as well, but this is not reported in the literature. Morphological models can simulate 3-D bed levels with considerable accuracy (van der Wegen and Roelvink, 2008, 2012; van Maren et al., 2015; Braat et al., 2017), but these models are computationally intensive and need calibration and specification of initial and boundary conditions. To study unmapped systems for which only aerial photography is available, it would be useful to be able to estimate bed-level distributions from planform geometry. Here we investigate this relation.

Previously, we showed in Leuven et al. (2018a) how locations and widths of tidal bars can be predicted from the excess width, which is the local width of the estuary minus the ideal estuary width. The summed width of bars in each cross section was found to approximate the excess width. This theory describes bars as discrete recognisable elements truncated at low water level on what is essentially a continuous field of bed elevation that changes in the along-channel direction (Leuven et al., 2016). However, to predict morphology in more detail, predictions of along-channel and cross-channel bed elevations are required. While hypsometry can summarise bed elevation distribution as a cumulative profile, it is unknown whether the shape of the profile is predictable. Our hypothesis is that the along-channel variation in hypsometry depends on the degree to which an estuary deviates from its ideal shape. Therefore, we expect that locations with large excess width and thus a large summed width of bars have a more convex hypsometry (Fig. 3c, d). In the case of an ideal estuary with (almost) no bars (Fig. 3a, b), we expect concave hypsometry.

The aim of this paper is to investigate the relation between estuary planform outline and along-channel variation in hypsometry. To do so, first hypsometric curves are used to summarise the occurring bed elevations in a cumulative profile.

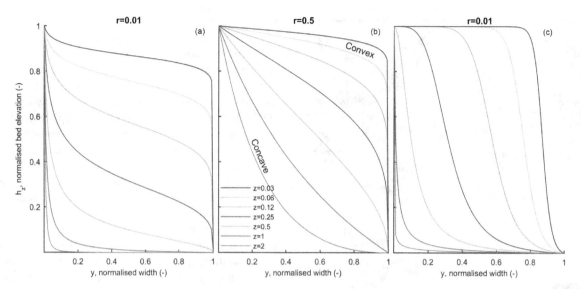

Figure 2. Hypsometric functions to describe morphological systems; modified from Strahler (1952). **(a, b)** Effect of z values with the r value kept constant. **(c)** Inverted version of the Strahler function (see Eq. 4).

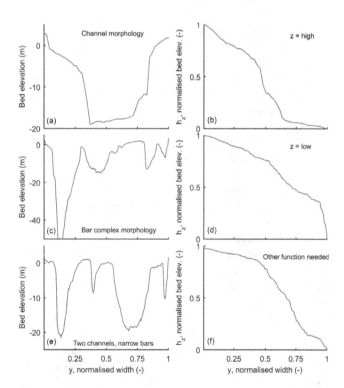

Figure 3. Example cross sections and hypsometry, suggesting that channel-dominated morphology **(a, b)** generally results in concave hypsometric functions (high z value in Fig. 2), while bar complexes **(c, d)** generally result in convex hypsometry (low z value in Fig. 2). In the case of narrow bars with a flat top and relatively steep transition from bar top to channel, the original hypsometry function by Strahler (1952) is less appropriate and the inverted function fits better.

Then, we use the original function of Strahler (1952) to fit the data obtained from the bathymetry of four estuaries (Fig. 1). In the results, we develop an empirical function to predict hypsometry. The quality and applications of the predictor are assessed in the discussion.

2 Methods

In this section, first, the definition of an ideal estuary is given with a description of how we derived a geometric property that characterises deviation from an ideal shape. Second, the general form of a hypsometric curve is described. Then, the available datasets that were used for curve fitting are given. Last, the methodology to fit a hypsometric function to bathymetry in systems is presented.

2.1 Deviation from "ideal"

A useful model to describe the morphology of estuaries is that of the "ideal estuary", in which the energy per unit width remains constant along-channel. The ideal state can be met when tidal range and tidal current are constant along-channel such that the loss of tidal energy by friction is balanced by the gain in tidal energy per unit width by channel convergence (Pillsbury, 1956; Dronkers, 2017). In the case that the depth is constant along-channel, the ideal estuary conditions are approximately met when the width is exponentially decreasing in the landward direction (Pillsbury, 1956; Langbein, 1963; Savenije, 2006, 2015; Toffolon and Lanzoni, 2010), which also implies an along-channel converging cross-sectional area. However, when depth and friction are not constant along-channel, for example linearly decreasing in the landward direction, less convergence in width is required to maintain constant energy per unit of width.

Therefore, many natural estuaries are neither in equilibrium nor in a condition of constant tidal energy per unit width. They deviate from the ideal ones as a result of a varying degree of sediment supply, lack of time for adaptation to changing upstream conditions and sea-level rise (Townend, 2012; de Haas et al., 2017). Whether continued sedimentation would reform bar-built estuaries with irregular planforms into ideal estuaries remains an open question. While we expect a somewhat different degree of convergence such that the ideal state of constant energy per unit of width is approximately maintained, we do not study the deviation of this convergence length from that in ideal estuaries.

Ideally, we would assess the degree to which an estuary is in equilibrium from an aerial photograph. However, the only indicator derivable is channel width and thus deviation from a converging width profile. Therefore, in Leuven et al. (2018a), we defined the excess width, which is the local width of the estuary minus our approximation of the potential ideal estuary width. Here, the ideal estuary width is approximated as an exponential fit on the width of the mouth and the width of the landward river. While the empirical measure of "ideal width" should not be confused with the "ideal state" of an estuary, it is the only practical way to estimate deviation from an ideal estuary based on the estuary outline only. Moreover, it proved to be a good indicator of occurring bar patterns (Leuven et al., 2018a) and will therefore be applied in this paper to study hypsometries.

2.2 General hypsometric curve

In the past, multiple authors have proposed empirical relations for the hypsometric shape of terrestrial landscapes (Strahler, 1952) and (partially) submerged bodies (Boon and Byrne, 1981; Wang et al., 2002; Toffolon and Crosato, 2007) (see Townend, 2008, for a review). All equations, except for Wang et al. (2002), predict a fairly similar hypsometric curve based on the volume and height range of the landform (Townend, 2008). While it is of interest to use these empirical relations to predict the occurring altitude variation of a landform, the framework here is different. Here, we apply the general hypsometric curve to characterise the occurring cross-sectional hypsometry along-channel. This approach is similar to the approach of Toffolon and Crosato (2007), who fitted a power function to 15 zones along the Western Scheldt. However, the zoned approach smooths out all the differences between bar complex and channel-dominated zones, which are of interest for this study. For this purpose, it is less relevant for which environment the hypsometric relation was proposed, as long as it is capable of describing the range of occurring hypsometries. For the case of the estuarine environment (Fig. 3), the hypsometric curve should be able to describe variations in concavity and variations in the slope of the curve at the inflection point. Here we use the original Strahler (1952) formulation, which is capable of doing so, but in principle any equation that fits well could be used.

Strahler (1952) formulated the general hypsometric curve as

$$h_z = \left[\frac{r}{r-1}\right]^z \left[\frac{1}{(1-r)y+r} - 1\right]^z, \quad (1)$$

in which h_z is the value of the bed elevation, above which fraction y of the width profile occurs. In other words, h_z is the proportion of total section height and y the proportion of section width. r sets the slope of the curve at the inflection point in a range of 0.01–0.50, with sharper curves for lower values of r (Fig. 2a, b). z determines the concavity of the function in a range of 0.03–2, with lower values giving a more convex profile and higher values giving a more concave profile (Strahler, 1952) (Fig. 2). Our approach changes the original definition of r and z to make them fitting parameters. It is expected that z values depend on excess width because the fraction of the width occupied by bars becomes larger with excess width, resulting in a more convex hypsometric profile (Fig. 3c, d), or the presence of bars generates excess width.

Excess width is defined as the local width minus the ideal width, which is given by

$$W_{\text{ideal}}(x) = W_{\text{m}} \cdot e^{-x/L_W}, \quad (2)$$

in which x is the distance from the mouth, W_{m} the width of the mouth and L_W is the width convergence length (Davies and Woodroffe, 2010), which can be obtained conservatively from a fit on the width of the mouth and the landward river width (Leuven et al., 2018a):

$$L_W = -s \frac{1}{\ln\left(\frac{W_{\text{s}}}{W_{\text{m}}}\right)}, \quad (3)$$

in which W_{m} is the local width measured at the mouth of the estuary, W_{s} is the width measured at the landward side of the estuary and s is the distance between these locations measured along the centreline. This practical method makes the convergence length somewhat sensitive to the selected position of the seaward and landward limit.

The landward limit was selected at the location where the width ceases to converge on an image at the resolution of the full estuary scale and the landward width was measured between the vegetated banks. The seaward limit was selected as the location with the minimum width in the case that bedrock geology, human engineering or a higher elevated spit confined the mouth because in these cases the minimum width limits the inflow of tidal prism. In other cases, the mouth was chosen at the point at which the first tidal flats were observed in the estuary or the sandy beach ends at the mouth of the estuary. However, when the mouth is chosen at a location where sand bars are present, the ideal width will be overestimated and the width of intertidal area underestimated. It is therefore recommended to either choose the mouth at a location where bars are absent or subtract the width of bars from the measured width at the mouth to obtain the ideal width profile.

2.3   Data availability and classification

Detailed bathymetries were available for four systems: the Western Scheldt esturary (NL), the Dovey estuary (Wales), the Eems–Dollard estuary (NL) and the Columbia River Estuary (USA) (Fig. 1, Table 1). Data for the Western Scheldt and Eems were obtained from Rijkswaterstaat (NL), for the Dovey estuary from Natural Resources Wales and for the Columbia River Estuary from the Lower Columbia Estuary Partnership. Bed elevations were extracted from these bathymetries as follows. First, the estuary outline was digitised, excluding fully developed salt marshes, and subsequently a centreline was determined within this polygon (following the approach of Leuven et al., 2018a). Bed elevations were collected on equally spaced transects perpendicular to the centreline of the estuary. The bed levels extracted at each transect were subsequently sorted by bed-level value and made dimensionless to obtain hypsometric profiles (see Fig. 3 for examples).

We classified the transects by morphological characteristics and potential susceptibility to errors. The following morphological classes were used: mouth, bar junction, bar complex, narrow bar, point bar, channel, pioneer marsh. The mouth is the location where the estuary transitions into the sea. A bar junction is the most seaward or most landward tip of tidal bars. A bar complex, also called a compound bar, is a location where a large bar is dissected by barb channels (Leuven et al., 2016) or multiple smaller bars are present. Narrow bar is used when the bars present were narrow along their entire length and often also relatively flat on their top. Point bar is a bar in the inner bend of a large meander. Channel was assigned when bars were largely absent. Pioneer marsh was assigned when aerial photographs or bathymetry gave visual indications of initial marsh formation, such as the presence of small tidal creeks and pioneering vegetation. Fully developed marsh is excluded from the outline.

The following classes were used to indicate possible errors: the presence of harbours, major dredging locations, the presence of a sand spit, the presence of drainage channels for agriculture, constraints by hard layers, human engineering works. Either a locally deep channel or scour occurred at one of the sides of these transects or they lacked a natural transition from channel to estuary bank, thus ending in their deepest part on one side of the transect. Major dredging locations have unnaturally deeper channels and shallower bars, resulting in a hypsometric shape that is relatively flat in the highest and lowest part and is steep in between (Fig. 3e, f). Furthermore, in a few cases side channels were perpendicular to the orientation of the main channel of the estuary. This resulted in transects being along-channel of these side channels, which biases transect data towards larger depth and creates a flat hypsometric profile at the depth of the side channel.

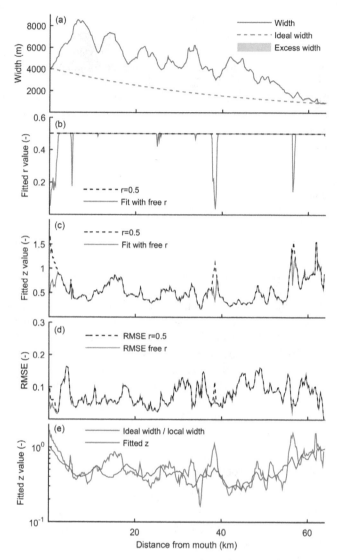

Figure 4. (a) Width along the Western Scheldt, with the maximum ideal converging width profile indicated. The green area is defined as the excess width cf. Leuven et al. (2018a). For each along-channel transect of the estuary, the optimal fit of z and r in the Strahler (1952) function (Fig. 2, Eq. 1) was determined. (b, c) Results for the Western Scheldt when both z and r are freely fitted (solid line) and the results when r is fixed to a constant value of 0.5 (dashed line). (d) The quality of the fits remains about the same when r is set to a fixed value of 0.5 as indicated by the root mean square error (RMSE). (e) Fitted z values show similar trends as the ideal width divided by local width.

2.4   Data processing

Least-squares fits resulted in optimal values of z and r in Eq. (1) (Fig. 2) for each transect, using three different approaches. First, a regular least-squares curve fitting was used for r and z, which resulted in along-channel varying values for z, but an almost entirely constant along-channel value for r of 0.5 (Fig. 4b, c, solid lines). In the second approach we

Table 1. Characteristics of the estuaries used in this study. h_m is the depth at the mouth, h_r is the depth at the landward river, W_m is the width at the mouth, W_r is the width at the landward river, a is the tidal amplitude, "Area" is the surface area, "% intertidal" is the percentage of intertidal area and Q_r is the river discharge.

	h_m (m)	h_r (m)	W_m (m)	W_r (m)	$2a$ (m)	Area (km^2)	% intertidal	Q_r (m^3 s^{-1})
Western Scheldt	25	15	4500	350	5	300	20	100
Columbia River	40	20	4000	800	2.5	900	30	7000
Dovey	10	2	450	50	3	12	75	30
Eems	25	8	3500	350	3.5	260	30	80

set r to a constant value of 0.5 and only fitted to obtain z (Fig. 4c, dashed line). We found that the quality of the fit was the same, as indicated by the root mean square error (RMSE) (Fig. 4d), and therefore apply this second approach in the remainder of this paper.

Locations where the RMSE was relatively large correspond to locations where major dredging occurred in the past century. This possibly resulted in a hypsometry characterised by a larger fraction of the width occupied by high tidal flats, a larger fraction of the width occupied by deep channels and a smaller fraction of the width occupied by the zone between channels and bars (Fig. 3e, f). Because the hypsometric curve at these locations deviated from the original Strahler function (Eq. 1), our third approach was to apply a modified function to find optimal values for z and r. To do so, the original formulation of Strahler (1952) was inverted to allow for hypsometries that describe steep transitions from bar top to channel bottom because the original does not fit nearly as well:

$$h_{z,\text{inv}} = \frac{\left[\frac{y^{1/z}(1-r)}{r} + 1 \right]^{-1} - r}{1 - r}. \qquad (4)$$

Applying this modified function resulted in better fits, but only at locations that were classified to be excluded because of possible errors. Therefore, results from this approach are not shown here and it is suggested to study the effect of dredging and dumping on hypsometry in more detail in future studies.

In principle, both the bed elevation (h_z) and the width fraction y in Eq. (1) are dimensionless. To compare the resulting predictions with measured values, the prediction needs to be dimensionalised. Values for y are scaled with the local estuary width. We test three options to scale h_z. The first option is to scale h_z between the highest bed elevation and lowest bed elevation in the given cross section, which is sensitive to the precise cut-off of the bathymetry. The second option is to scale h_z between the local high water level (HWL) and the maximum estuary depth in that cross section, which is sensitive to bathymetric information that is usually not available in unmapped estuaries.

The third option requires a prediction of depth at the upstream or downstream boundary. Width-averaged depth profiles along estuaries are often (near) linear (Savenije, 2015;

Leuven et al., 2018a), which includes horizontal profiles with constant depth. Therefore, only the channel depth at the mouth of the estuary and at the upstream river have to be estimated, and subsequently a linear regression can be made. Channel depth at an upstream river (h_r) is estimated with hydraulic geometry relations (e.g. Leopold and Maddock Jr, 1953; Hey and Thorne, 1986): $h_r = 0.12 W_r^{0.78}$. The depth at the mouth is estimated from relations between tidal prism and cross-sectional area (e.g. O'Brien, 1969; Eysink, 1990; Friedrichs, 1995; Lanzoni and D'Alpaos, 2015; Gisen and Savenije, 2015; Leuven et al., 2018a). Here we used

$$h_m = \frac{0.13 \times 10^{-3} P}{W_m}, \qquad (5)$$

in which P is the tidal prism, which can be estimated by multiplying the estuary surface area with the tidal range (Leuven et al., 2018a). This assumes a flat water surface elevation along the estuary and neglects portions of the estuary that might get dry during the tidal cycle (Boon, 1975). Locally, the maximum depth may be deeper or lower than predicted due to the presence of resistant layers in the subsurface or where banks are fixed or protected. While this may affect the accuracy of the locally predicted maximum channel depth, it has a minor effect on the calculations of subtidal and intertidal area. Moreover, the upper limit for dimensionalisation is chosen as the high water line, which implies that the supratidal area is not included in the predictions. We will show results with all methods, but only the third can be applied when information about depth is entirely lacking.

Statistical analyses in the remainder of the paper were approached as follows. In linear regressions, we minimised residuals in both the x and y directions. This results in regressions that are more robust than when residuals in only one direction are minimised. In the case that regressions are plotted, the legend will specify the multiplication factor that the confidence limits plot above or below the trend. R^2 values are given to indicate the variance around the regression. In cases in which the quality of the correlation between two along-channel profiles is assessed, we used the Pearson product-moment correlation coefficient (r).

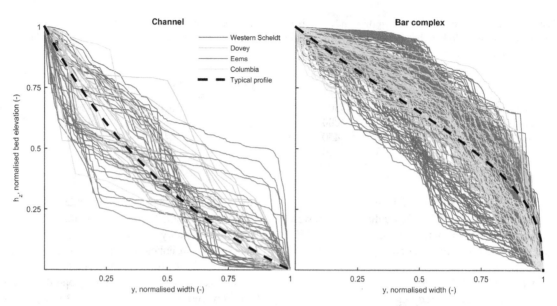

Figure 5. Hypsometric curves as extracted from bathymetry in the Western Scheldt, Dovey, Eems and Columbia River. Cross-sectional profiles were extracted along the centreline (Sect. 2.3) and were subsequently classified as channel when sand bars were (mostly) absent and as bar complex when one larger bar dissected by barb channels or multiple smaller bars was present. Channel-dominated morphology generally results in concave hypsometric profiles **(a)** and bar complexes in convex profiles **(b)**.

3   Results

We found a strong relation between along-channel variation in hypsometry and the degree to which an estuary deviates from its ideal shape. Below, we will first show how the hypsometry of typical channel morphology deviates from that of bar complexes. Subsequently, the data are presented per system and classified on their morphology and potential for errors due to human interference, method and other causes. Then we combine all data to derive an empirical relation to predict hypsometry. Last, we apply this relation to predict the along-channel variation in intertidal and subtidal width and validate the results with measurements from bathymetry.

3.1   Empirical relation between morphology and hypsometry

As hypothesised, it is indeed observed that channel-dominated morphology results in more concave hypsometry profiles (high z value), while bar complex morphology results in more convex hypsometry (low z value) (Fig. 5). Values for z in Eq. (1) range from 0.83 to 1.14 for channels, with an average value of 1.0. In contrast, z ranges from 0.36 to 0.41 for bar complexes, with an average of 0.39.

Clustering of morphological classes strongly suggests a relation between hypsometry and planform estuary shape (Fig. 6). Mouth and channel-dominated morphologies typically plot at the right-hand side of the plots in Fig. 6a, c, e and g, thus being locations close to ideal width. In the case of the Western Scheldt this results in the highest values of z. In the case of the Dovey and Columbia River, the mouth

region was respectively influenced by a spit and human engineering, which resulted in the formation of tidal flats on the side and thus led to a lower z value.

Bar complexes occur at the other end of the spectrum; these locations are generally much wider than the ideal shape and are characterised by hypsometries with a z value well below 1. Bar junctions, as well as narrow bars, are generally found at the transition from channel-dominated morphology to bar complex morphology and therefore also occur between these types in the plots. The point bar in the Western Scheldt (the Plaat van Ossenisse) shows hypsometry comparable to bar complexes (Fig. 6a), which reflects the complex history of formation by multiple bar amalgamations. Also, the locations in the Columbia River where pioneer marsh is present show the same trend as the locations where unvegetated bar complexes occur (Fig. 6g).

In a few cases, the transects used to extract bathymetry were not perpendicular to the main channel of the estuary. For example, landward and seaward of the point bar in the Western Scheldt (Plaat van Ossenisse) transects were inclined, covering a larger part of the channel than perpendicular transects, resulting in higher z values as a consequence of the apparent channel-dominated morphology. Immediately landward of the spit in the Dovey, transects are almost parallel to the shallow side channel. Fitting hypsometry at these locations resulted in relatively low z values because it is a relatively shallow side channel.

For the Western Scheldt and Eems it is known at which locations major dredging and dumping takes place (e.g. Swinkels et al., 2009; Jeuken and Wang, 2010; Bolle et al., 2010; Dam et al., 2015; Plancke and Vos, 2016). Even though

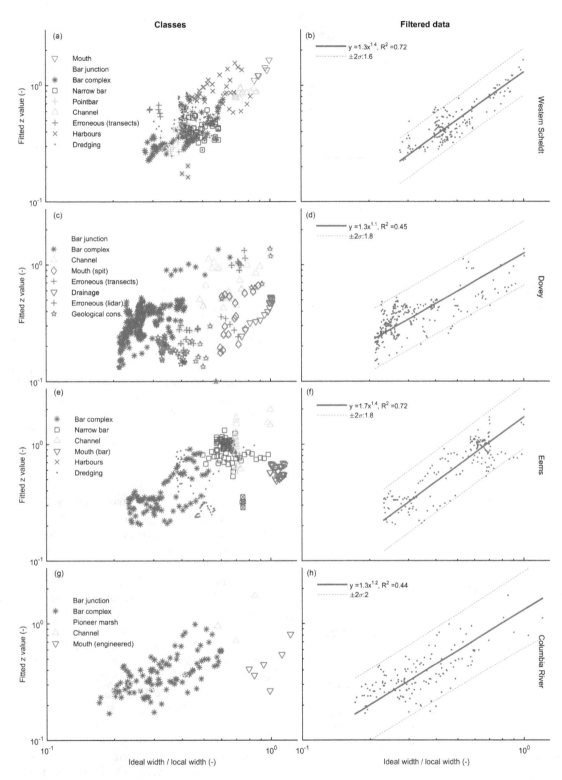

Figure 6. Results of hypsometry fitting in which clustering indicates a relation to planform geometry. **(a, c, e, g)** z values were fitted for cross-channel transects in the bathymetry of four estuaries and plotted by morphological classification. **(b, d, f, h)** Regressions for z value as a function of ideal width divided by local width. Data points that were influenced by human interference, bedrock geology, or errors in methodology or data, indicated in red **(a, c, e, g)**, were excluded. Confidence limits are plotted at 2 standard deviations above and below the regression and their multiplication factor compared to the trend is given in the legend.

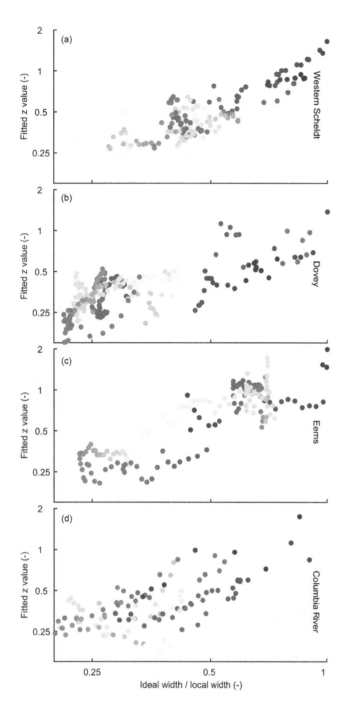

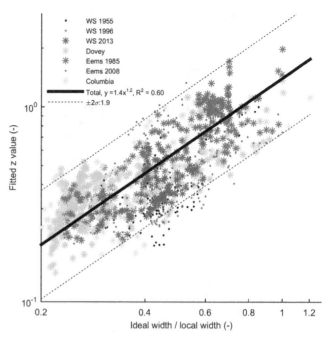

Figure 8. Fitted z values of filtered data increase with the fraction of ideal width and local width, which indicates that hypsometric shapes become progressively more concave when the local width approaches the ideal width and become more convex when the local width becomes larger than the ideal width (i.e. the excess width increases). The data shown as asterisks are used for the regression. Confidence limits are plotted at 2 standard deviations above and below the regression and their multiplication factor compared to the trend is given in the legend.

The filtered data show quasi-cyclicity in along-channel hypsometry (Fig. 7). In general, the width at the mouth of the estuary and at the upstream estuary is close to ideal and the hypsometry is concave, except in systems with wide mouths and bars in the inlet. The part in between is characterised by variations in the local width and therefore gradual increases and decreases in the ratio between local width and ideal width. In some cases, quasi-cyclic loops are visible (e.g. Fig. 7c) caused by the asymmetry in bar complexes. In other cases, the points show more zigzag or clustered patterns, which indicate minor variation in the bar complexes or scatter in the fit applied to the bathymetry.

3.2 Hypsometry predictor

The relations between excess width and hypsometric function are similar for all estuaries, which suggests that a universal function is of value. Combining all the filtered data resulted in a regression between the extent to which an estuary deviates from the ideal shape and the predicted z value in the hypsometry formulation (Fig. 8). Data from the Columbia River, Eems in 1985, the Western Scheldt in 2013 and Dovey were used to obtain this relation. Other data are shown in Fig. 8 but not used in the regression.

Figure 7. Fitted z values as a function of deviation from the ideal width. Colours indicate the location along the estuary, with dark blue colours at the mouth transitioning into dark red colours at the landward end. Some zones show scatter in fitted z values, while some other zones (e.g. green to orange to red in **c**) show quasi-periodic behaviour.

the resulting z values at these locations do not cause major outliers, the quality of the fits is typically lower and the inverted Strahler function (Eq. 4) fitted better. These points were therefore excluded from further analysis.

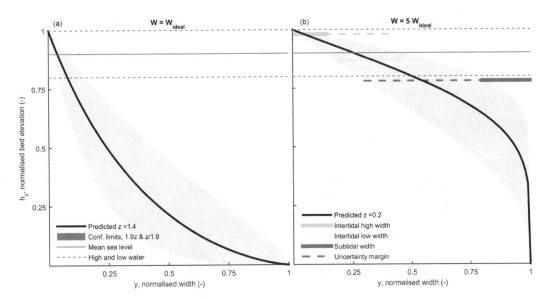

Figure 9. Illustration of uncertainty in the predicted hypsometry from Eq. (6) with uncertainty margins (Fig. 8). Resulting prediction for hypothetical location where **(a)** local width is equal to the ideal width and **(b)** local width is 5 times larger than the ideal width. Results are compared against a typical tidal range in order to show the uncertainty of predicted intertidal high width, intertidal low width and subtidal width as a fraction of the total estuary width.

These results mean that we found a predictive function for hypsometry, where r is set to a constant value of 0.5 and the $z(x)$ is calculated as

$$z(x) = 1.4 \left[\frac{W_{\text{ideal}}(x)}{W(x)} \right]^{1.2}, \qquad (6)$$

in which $W_{\text{ideal}}(x)$ is the ideal estuary width (Eq. 2) and $W(x)$ is the measured local width. The confidence limits of the regression plot a factor of 1.9 higher and lower than the regression, which indicates that the z value can be predicted within a factor of 2 (see Fig. 9 for an example of prediction with uncertainty). While not used in the regression, hypsometry from bathymetry in other years shows similar trends and scatter as the data used in the regression.

The predictor (Eq. 6) was applied to the Columbia River Estuary, Western Scheldt, Dovey and Eems to check the quality of the resulting along-channel predictions of intertidal high, intertidal low and subtidal width (Figs. 10 and 11). These zones can be derived after dimensionalising hypsometry and imposing a tidal range (Fig. 9b). For almost the entire along-channel profile, the predictions are within a factor of 2 of the measured value (Fig. 11) and the best agreement was obtained when the hypsometry was dimensionalised between the minimum and maximum measured bed level for each transect (Fig. 10).

4 Discussion

Results from this study illustrate that the bed-level distributions of channel and bar patterns in estuaries are topographically forced. The estuary outline that is observable from the

Table 2. Percentage of points predicted within a factor of 2 from the measured value.

Estuary	% for subtidal	% for intertidal
Western Scheldt	100	84
Columbia River	90	79
Dovey	54	71
Eems	91	59

surface translates into the three-dimensional patterns below the water surface. Bar-built estuaries typically have a quasi-periodic planform, in which major channel confluences occur at locations where the estuary is close to its ideal shape (Leuven et al., 2018a, b). The parts between the confluences are typically filled with intertidal bar complexes. These findings are consistent with hypsometry zonations previously found for the Western Scheldt with more concave hypsometries for channel-dominated morphology and more convex hypsometries for bar complex morphology (Toffolon and Crosato, 2007). Our cross-sectional approach additionally revealed quasi-periodic behaviour within these zones.

In contrast to an empirical description, ideally, a physics-based determination of the hypsometry would be favourable. However, with the current state of the art of bar theory (Leuven et al., 2016) and relations for intertidal area, tidal prism, cross-sectional area and flow velocities (O'Brien, 1969; Friedrichs and Aubrey, 1988) it is not yet possible to derive a theoretical prediction of hypsometry. For example, bar theory (Seminara and Tubino, 2001; Schramkowski et al., 2002) could predict occurring bar patterns on top of an (ideal)

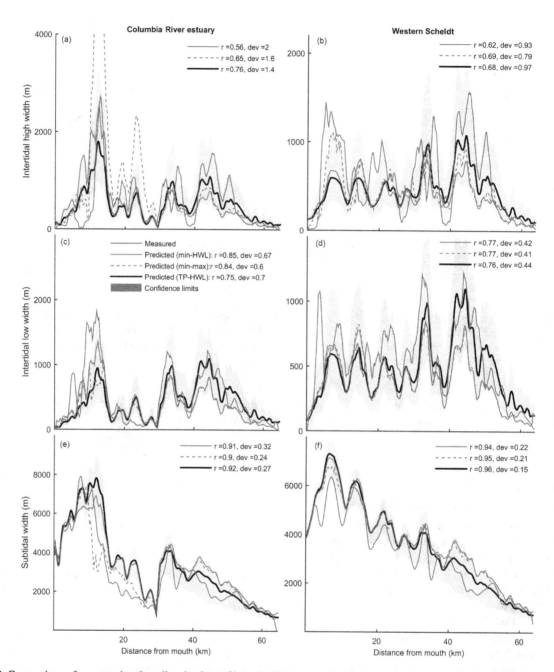

Figure 10. Comparison of measured and predicted values of intertidal high, intertidal low and subtidal width for the Columbia River Estuary (**a, c, e**) and the Western Scheldt (**b, d, f**). Predicted nondimensional hypsometry was dimensionalised for each cross section using three methods (explained in Methods) and uncertainty margins are given for one of the predictions (solid black line). In the legend, r indicates the Pearson product-moment correlation coefficient and "dev" the average factor of deviation between the predicted (TP-HWL) and measured lines.

estuary shape, but current theories overpredict their dimensions (Leuven et al., 2016) and it is still impossible to scale these to bed-level variations because the theories are linear. In addition to that, the resulting predictions would need to meet the requirement that the predicted bed levels and the intertidal area together lead to hydrodynamic conditions that fit the estuary as well.

Previously, hypsometry was used to summarise the geometry of entire tidal basins or estuaries (Boon and Byrne, 1981; Dieckmann et al., 1987; Townend, 2008). The whole system descriptions are consistent with the original Strahler (1952) concept of a basin hypsometry based on plan area, which is a valid description in a landform context. However, these descriptions oversimplify the along-channel variability in estuaries that are relatively long. These estuaries typi-

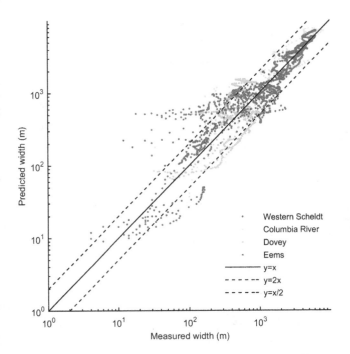

Figure 11. Comparison of measured and predicted width of intertidal and subtidal width. The (solid) line of equality indicates a perfect fit and dashed lines indicate a deviation of a factor of 2. The percentage of measurements within these margins is indicated in Table 2.

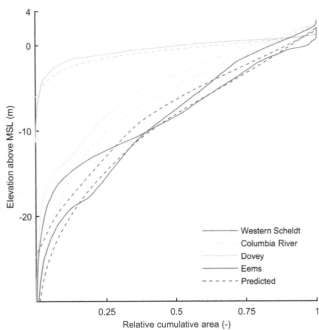

Figure 12. Hypsometry as summarised in a single curve for the entire estuaries. Solid lines are measured from bathymetry, and dashed lines are based on the predictions.

cally have a linear bed profile varying from an along-channel constant depth to strongly linear sloping (e.g. the Mersey in the UK). In the latter case, the elevation at which subtidal and intertidal area occurs varies significantly along-channel (Blott et al., 2006). Additionally, friction and convergence may cause the tidal range to either dampen or amplify, causing variation in tidal elevation, subtidal area and tidal prism (Savenije, 2006). Consequently, the along-channel cross section hypsometry should be assumed to be relative to an along-channel varying high water level or mean sea level rather than an along-channel fixed vertical datum. Interpreting these along-channel variations remains an open question because of the reasons outlined above. Nevertheless, if desired, along-channel varying hypsometry predictions can be converted into one single summarising curve (Fig. 12), which shows that the basin hypsometry can also be predicted when limited data are available.

Our results show that hypsometry is not only a tool to predict morphology when limited data are available, but that hypsometry can also be used to reduce a large dataset of bathymetry and to study the evolution of bathymetry over time. In the case of Strahler (1952), hypsometry fits result in along-channel profiles of z and r values, but in practice any function or shape could be fitted. For example, the locations along the Western Scheldt where major dredging and dumping took place showed a weaker correlation with the original Strahler (1952) shape (Fig. 13). In these cases, fits with

higher quality (lower RMSE) were obtained when we used the inverted hypsometric function (Eq. 4) (Fig. 13b, e). So in practice, one could fit a range of different hypsometry shapes and subsequently find out which of these shapes fit best on the dataset used. It can indicate that certain parts along the estuary require a separate hypsometric description. The fitting parameters are a method to describe the along-channel variation (Fig. 13b, d). Hypsometry can be fitted to compare data from nature, physical experiments and numerical modelling and subsequently study, for example, the effect of vegetation, cohesive mud and the influence of management on these systems.

4.1 Implications for management of estuaries

In many estuaries from around the world, subtidal channels are used as shipping fairways, while the intertidal bars (or shoals) form valuable ecological habitats (e.g. Bouma et al., 2005). For example in the Western Scheldt, the shipping fairway is now maintained at a depth of 14.5 m below the lowest astronomical tide (de Vriend et al., 2011; Depreiter et al., 2012). With empirical hypsometry predictions, we can estimate the width below a certain depth required for shipping, which gives estimates of what volume to dredge and at what locations along the estuary, which is relevant for the construction of future shipping fairways in estuaries for which we may have limited data.

In contrast, low-dynamic intertidal areas are valuable ecological habitats; for example, for the Western Scheldt there is

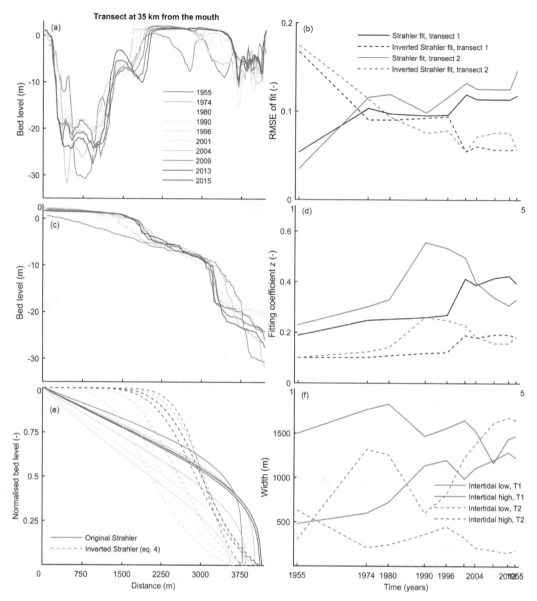

Figure 13. (a) Evolution of a cross section in the Western Scheldt where it has been significantly dredged and dumped (the Drempel van Hansweert). **(c)** Measured hypsometric profiles of the same time steps at the same cross section. **(e)** Best-fit hypsometries using the original Strahler equation (solid) and inverted Strahler function (Eq. 4) (dashed). **(b)** The quality of the two types of fits shows that the shape of the best-fitting hypsometric curve changes from the original Shrahler to the inverted equation in the 1970s–1980s. An additional transect is added in the panels on the right. **(d)** Fitting coefficients for z increase over time for both hypsometry types and both transects. **(f)** Intertidal high area increased over this period, while intertidal low area remained constant (transect 1) or decreased (transect 2).

an obligation to maintain a certain amount of intertidal area (Depreiter et al., 2012). Previously, Townend (2008) showed that basin hypsometry can be a tool to design breaches in managed realignment sites and can provide an indication of habitat composition. Hypsometry analysis per cross section shows that estuary outline translates into intertidal area, which implies that locations where the estuary is relatively wide have a relatively wide intertidal area. The ecological value is determined by the area of low-dynamical shallow water and intertidal areas (for settling and feeding) (Depre-

iter et al., 2012). This means that the edges should neither become steeper nor higher (leading to permanent dry fall) or deeper. Hypsometry fits (in the case of available data) or predictions (in the case of limited data) can indicate which locations along the estuary have a risk to transform away from low-dynamic area or have the potential to become low-dynamic area by the suppletion of dredged sediment.

The occurrence of vegetation species depends on bed elevation, salinity, maximum flow velocity and sediment type (de Jong, 1999; Gurnell et al., 2012). Even though predicted

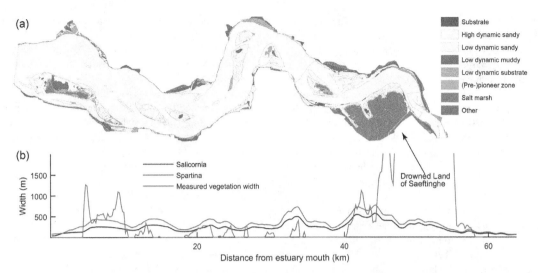

Figure 14. (a) Ecotope map of the Western Scheldt (2012) obtained from Rijkswaterstaat. **(b)** Prediction of the width in which *Salicornia* (black) and *Spartina* (blue) can occur when assuming that *Spartina* occurs between MSL and 1.5 m above and *Salicornia* occurs between 1.0 and 2.5 m, while ignoring velocity, salinity and sediment type constraints. Red line indicates measured width of vegetation based on ecotope map. The Drowned Land of Saeftinghe is excluded in the predictions because the high water line was the boundary of the analysed bathymetry, while it is included in the measured data.

hypsometry only gives bed elevations, a comparison of the height interval in which *Salicornia* and *Spartina* can occur (Mckee and Patrick, 1988; Davy et al., 2001; van Braeckel et al., 2008) showed similar trends and the same order of magnitude as the measured vegetation from ecotope maps of the Western Scheldt in 2012 (Fig. 14). Some underpredictions arise in parts along the estuary where bed elevations above the high water level occur, such as at the Drowned Land of Saeftinghe. However, in general, the vegetation width is overpredicted because (1) hypsometry is stretched between the high water line and channel depth and (2) other constraining biotic and abiotic factors were excluded.

5   Conclusions

We studied the relation between the along-channel planform geometry of sandy estuaries and their hypsometry, which characterises the distribution of along-channel and cross-channel bed levels. The vertical dimensions were found to relate to the horizontal dimensions. In other words, the degree to which the estuary width deviates from an ideal converging estuary shape is reflected in the occurring hypsometry. At locations where the width is much larger than ideal, convex hypsometric shapes are observed, contrary to the locations where the estuary width is close to ideal, where concave hypsometric shapes are observed. In between these extreme endmembers, a gradual transition with quasi-periodic variation was observed. This implies that it is possible to predict the along-channel varying hypsometry of estuaries, which is relevant for estuaries for which limited data are available. To obtain broad-brush estimates of the occurring bed levels, only the estuary outline and a typical tidal amplitude are re-

quired. The predictions can be used to study the presence and evolution of intertidal area, which forms valuable ecological habitats, and to get estimates of typical volumes that might need to be dredged when constructing shipping fairways.

Author contributions. The authors contributed in the following proportions to conception and design, data collection and processing, analysis and conclusions, and paper preparation: JRFWL (70, 65, 75, 70 %), SS (10, 30, 15, 15 %), MGK (20, 5, 10, 15 %).

Competing interests. The authors declare that they have no conflict of interest.

Acknowledgements. This research was supported by Future Deltas, Utrecht University (grant to JRFWL) and by the Dutch Technology Foundation TTW (grant Vici 016.140.316/13710 to MGK), which is part of the Netherlands Organisation for Scientific Research (NWO) and is partly funded by the Ministry of Economic Affairs. This work is part of the PhD research of JRFWL. We acknowledge the data processing contribution by Andy Bruijns as part of his MSc guided research. Reviews by Ian Townend and one anonymous reviewer helped to improve the paper.

References

Blott, S. J., Pye, K., van der Wal, D., and Neal, A.: Long-term morphological change and its causes in the Mersey Estuary, NW England, Geomorphology, 81, 185–206, 2006.

Bolle, A., Wang, Z. B., Amos, C., and de Ronde, J.: The influence of changes in tidal asymmetry on residual sediment transport in the Western Scheldt, Cont. Shelf Res., 30, 871–882, 2010.

Boon, J. D.: Tidal discharge asymmetry in a salt marsh drainage system, Limnol. Oceanogr., 20, 71–80, 1975.

Boon, J. D. and Byrne, R. J.: On basin hyposmetry and the morphodynamic response of coastal inlet systems, Marine Geol., 40, 27–48, https://doi.org/10.1016/0025-3227(81)90041-4, 1981.

Bouma, H., de Jong, D., Twisk, F., and Wolfstein, K.: A Dutch Ecotope system for coastal waters (ZES. 1), To map the potential occurence of ecological communities in Dutch coastal and transitional waters, Tech. rep., Rijkswaterstaat, Report RIKZ/2005.024, 2005.

Braat, L., van Kessel, T., Leuven, J. R. F. W., and Kleinhans, M. G.: Effects of mud supply on large-scale estuary morphology and development over centuries to millennia, Earth Surf. Dynam., 5, 617–652, https://doi.org/10.5194/esurf-5-617-2017, 2017.

Dam, G., van der Wegen, M., Roelvink, D., Labeur, R., and Bliek, B.: Simulation of long-term morphodynamics of the Western Scheldt, in: 36st IAHR congress, The Hague, Netherlands, 2015.

Davies, G. and Woodroffe, C. D.: Tidal estuary width convergence: Theory and form in North Australian estuaries, Earth Surf. Proc. Land., 35, 737–749, https://doi.org/10.1002/esp.1864, 2010.

Davy, A. J., Bishop, G. F., and Costa, C. S. B.: Salicornia L.(Salicornia pusilla J. woods, S. ramosissima J. woods, S. europaea L., S. obscura PW ball & tutin, S. nitens PW ball & tutin, S. fragilis PW ball & tutin and S. dolichostachya moss), J. Ecol., 89, 681–707, 2001.

de Haas, T., Pierik, H., van der Spek, A., Cohen, K., van Maanen, B., and Kleinhans, M.: Holocene evolution of tidal systems in The Netherlands: Effects of rivers, coastal boundary conditions, eco-engineering species, inherited relief and human interference, Earth-Sci. Rev., 177, 139–163, https://doi.org/10.1016/j.earscirev.2017.10.006, 2017.

de Jong, D.: Ecotopes in the Dutch marine tidal waters: a proposal for a classification of ecotopes and a method to map them, Tech. rep., Rijkswaterstaat, Report RIKZ/99.017, 1999.

de Vet, P., van Prooijen, B., and Wang, Z.: The differences in morphological development between the intertidal flats of the Eastern and Western Scheldt, Geomorphology, 281, 31–42, 2017.

de Vriend, H. J., Wang, Z. B., Ysebaert, T., Herman, P. M., and Ding, P.: Eco-morphological problems in the Yangtze Estuary and the Western Scheldt, Wetlands, 31, 1033–1042, https://doi.org/10.1007/s13157-011-0239-7, 2011.

Depreiter, D., Sas, M., Beirinckx, K., and Liek, G.-J.: Flexible Disposal Strategy: monitoring as a key to understanding and steering environmental responses to dredging and disposal in the Scheldt Estuary, Tech. rep., International Marine & Dredging Consultants, 2012.

Dieckmann, R., Osterthun, M., and Partenscky, H.-W.: Influence of water-level elevation and tidal range on the sedimentation in a German tidal flat area, Prog. Oceanogr., 18, 151–166, 1987.

Dronkers, J.: Convergence of estuarine channels, Cont. Shelf Res., 144, 120–133, https://doi.org/10.1016/j.csr.2017.06.012, 2017.

Eysink, W.: Morphologic response of tidal basins to changes, Coast. Eng. Proc., 1, 1948–1961, https://doi.org/10.1061/9780872627765.149, 1990.

Friedrichs, C. T.: Stability shear stress and equilibrium cross-sectional geometry of sheltered tidal channels, J. Coast. Res., 11, 1062–1074, 1995.

Friedrichs, C. T.: Barotropic tides in channelized estuaries, Cont. Iss. Est. Phys., 27–61, 2010.

Friedrichs, C. T. and Aubrey, D. G.: Non-linear tidal distortion in shallow well-mixed estuaries: a synthesis, Estuar. Coast. Shelf Sci., 27, 521–545, 1988.

Gardiner, S., Nicholls, R., and Tanton, T.: Management Implications of Flood/Ebb tidal dominance: its influence on saltmarsh and intertidal habitat stability in Poole Harbour, in: Littoral 2010–Adapting to Global Change at the Coast: Leadership, Innovation, and Investment, p. 06004, EDP Sciences, 2011.

Gisen, J. I. A. and Savenije, H. H.: Estimating bankfull discharge and depth in ungauged estuaries, Water Resour. Res., 51, 2298–2316, https://doi.org/10.1002/2014WR016227, 2015.

Gurnell, A. M., Bertoldi, W., and Corenblit, D.: Changing river channels: The roles of hydrological processes, plants and pioneer fluvial landforms in humid temperate, mixed load, gravel bed rivers, Earth-Sci. Rev., 111, 129–141, https://doi.org/10.1016/j.earscirev.2011.11.005, 2012.

Hey, R. D. and Thorne, C. R.: Stable channels with mobile gravel beds, J. Hydr. Eng., 112, 671–689, https://doi.org/10.1061/(ASCE)0733-9429(1986)112:8(671), 1986.

Hu, Z., Belzen, J., Wal, D., Balke, T., Wang, Z. B., Stive, M., and Bouma, T. J.: Windows of opportunity for salt marsh vegetation establishment on bare tidal flats: The importance of temporal and spatial variability in hydrodynamic forcing, J. Geophys. Res.-Biogeosci., 120, 1450–1469, 2015.

Jeuken, M. and Wang, Z.: Impact of dredging and dumping on the stability of ebb–flood channel systems, Coast. Eng., 57, 553–566, 2010.

Kirby, R.: Practical implications of tidal flat shape, Cont. Shelf Res., 20, 1061–1077, 2000.

Kleinhans, M. G. and van den Berg, J. H.: River channel and bar patterns explained and predicted by an empirical and a physics-based method, Earth Surf. Proc. Land., 36, 721–738, https://doi.org/10.1002/esp.2090, 2011.

Langbein, W.: The hydraulic geometry of a shallow estuary, Hydrol. Sci. J., 8, 84–94, https://doi.org/10.1080/02626666309493340, 1963.

Lanzoni, S. and D'Alpaos, A.: On funneling of tidal channels, J. Geophys. Res.-Earth, 120, 433–452, https://doi.org/10.1002/2014JF003203, 2015.

Leopold, L. B. and Maddock Jr, T.: The Hydraulic Geometry of Stream Channels and Some Physiographic Implications, Tech. rep., U.S. Geological Survey, Professional Paper 252, 1953.

Lesourd, S., Lesueur, P., Brun-Cottan, J.-C., Auffret, J.-P., Poupinet, N., and Laignel, B.: Morphosedimentary evolution of the macrotidal Seine estuary subjected to human impact, Estuar. Coast., 24, 940–949, 2001.

Leuven, J. R. F. W., Kleinhans, M. G., Weisscher, S. A. H., and van der Vegt, M.: Tidal sand bar dimensions and shapes in estuaries, Earth-Sci. Rev., 161, 204–233, https://doi.org/10.1016/j.earscirev.2016.08.004, 2016.

Leuven, J. R. F. W., Haas, T., Braat, L., and Kleinhans, M. G.: Topographic forcing of tidal sand bar patterns for irregular estuary planforms, Earth Surf. Proc. Land., 43, 172–186, https://doi.org/10.1002/esp.4166, 2018a.

Leuven, J. R. F. W., Braat, L., van Dijk, W. M., de Haas, T., and Kleinhans, M. G.: Growing forced bars determine non-ideal estuary planform Journal of Geophysical Research: Earth Surface (revision submitted, preprint: 10.31223/osf.io/hj27m), 2018b.

Mckee, K. L. and Patrick, W.: The relationship of smooth cordgrass (Spartina alterniflora) to tidal datums: a review, Estuaries, 11, 143–151, 1988.

Moore, R. D., Wolf, J., Souza, A. J., and Flint, S. S.: Morphological evolution of the Dee Estuary, Eastern Irish Sea, UK: a tidal asymmetry approach, Geomorphology, 103, 588–596, 2009.

O'Brien, M. P.: Equilibrium flow areas of inlets on sandy coasts, in: J. Geol. Soc. of the waterways and harbors division, Proceedings of the American Society of Civil Engineers, 43–52, ASCE, https://doi.org/10.1061/9780872620087.039, 1969.

O'Connor, B.: Short and long term changes in estuary capacity, J. Geol. Soc., 144, 187–195, 1987.

Pillsbury, G.: Tidal Hydraulics. Revised Edition, Corps of Engineers, US Army, May, 1956.

Plancke, Y. and Vos, G. R.: Sediment transport in the Scheldt-estuary: a comparison between measurements, transport formula and numerical models, in: Proceedings of the 4th IAHR Europe Congress, Liege, Belgium, 27–29 July 2016: Sustainable Hydraulics in the Era of Global Change, 498–503, 2016.

Savenije, H. H.: Salinity and tides in alluvial estuaries, Elsevier, 2006.

Savenije, H. H.: Prediction in ungauged estuaries: an integrated theory, Water Resour. Res., 51, 2464–2476, https://doi.org/10.1002/2015WR016936, 2015.

Schramkowski, G., Schuttelaars, H., and De Swart, H.: The effect of geometry and bottom friction on local bed forms in a tidal embayment, Cont. Shelf Res., 22, 1821–1833, https://doi.org/10.1016/S0278-4343(02)00040-7, 2002.

Seminara, G. and Tubino, M.: Sand bars in tidal channels. Part 1. Free bars, J. Fluid Mech., 440, 49–74, https://doi.org/10.1017/S0022112001004748, 2001.

Strahler, A. N.: Hypsometric (area-altitude) analysis of erosional topography, Geol. Soc. Am. Bull., 63, 1117–1142, https://doi.org/10.1130/0016-7606(1952)63[1117:HAAOET]2.0.CO;2, 1952.

Swinkels, C. M., Jeuken, C. M., Wang, Z. B., and Nicholls, R. J.: Presence of connecting channels in the Western Scheldt Estuary, J. Coast. Res., 627–640, 2009.

Toffolon, M. and Crosato, A.: Developing macroscale indicators for estuarine morphology: The case of the Scheldt Estuary, J. Coast. Res., 23, 195–212, 2007.

Toffolon, M. and Lanzoni, S.: Morphological equilibrium of short channels dissecting the tidal flats of coastal lagoons, J. Geophys.

Res.-Earth, 115, F04036, https://doi.org/10.1029/2010JF001673, 2010.

Townend, I.: An exploration of equilibrium in Venice Lagoon using an idealised form model, Cont. Shelf Res., 30, 984–999, 2010.

Townend, I.: The estimation of estuary dimensions using a simplified form model and the exogenous controls, Earth Surf. Proc. Land., 37, 1573–1583, https://doi.org/10.1002/esp.3256, 2012.

Townend, I. H.: Hypsometry of estuaries, creeks and breached sea wall sites, P. I. Civil Eng.-Mar. En., 161, 23–32, https://doi.org/10.1680/maen.2008.161.1.23, 2008.

van Braeckel, A., Vandevoorde, B., and van den Bergh, E.: Schorecotopen van de Schelde. Aanzet tot de ontwikkeling van één schorecotopenstelsel voor Vlaanderen en Nederland, Instituut voor Natuur-en Bosonderzoek, 2008.

van de Lageweg, W. I., Dijk, W. M., Box, D., and Kleinhans, M. G.: Archimetrics: a quantitative tool to predict three-dimensional meander belt sandbody heterogeneity, The Depositional Record, https://doi.org/10.1002/dep2.12, 2, 22–46, 2016.

van der Wegen, M. and Roelvink, J.: Long-term morphodynamic evolution of a tidal embayment using a two-dimensional, process-based model, J. Geophys. Res.-Oceans, 113, C03016, https://doi.org/10.1029/2006JC003983, 2008.

van der Wegen, M. and Roelvink, J.: Reproduction of estuarine bathymetry by means of a process-based model: Western Scheldt case study, the Netherlands, Geomorphology, 179, 152–167, https://doi.org/10.1016/j.geomorph.2012.08.007, 2012.

van Maren, D., Van Kessel, T., Cronin, K., and Sittoni, L.: The impact of channel deepening and dredging on estuarine sediment concentration, Cont. Shelf Res., 95, 1–14, https://doi.org/10.1016/j.csr.2014.12.010, 2015.

Wang, Z. and Winterwerp, J.: Impact of dredging and dumping on the stability of ebb-flood channel systems, in: Proceedings of the 2nd IAHR symposium on River, Coastal and Estuarine Morphodynamics, 515–524, 2001.

Wang, Z., Jeuken, M., Gerritsen, H., De Vriend, H., and Kornman, B.: Morphology and asymmetry of the vertical tide in the Westerschelde estuary, Cont. Shelf Res., 22, 2599–2609, 2002.

Wang, Z., Van Maren, D., Ding, P., Yang, S., Van Prooijen, B., De Vet, P., Winterwerp, J., De Vriend, H., Stive, M., and He, Q.: Human impacts on morphodynamic thresholds in estuarine systems, Cont. Shelf Res., 111, 174–183, 2015.

Wolanski, E. and Elliott, M.: Estuarine ecohydrology: an introduction, Elsevier, Amsterdam, 1–316, 2015.

5

Initial insights from a global database of rainfall-induced landslide inventories: the weak influence of slope and strong influence of total storm rainfall

Odin Marc[1], André Stumpf[1], Jean-Philippe Malet[1], Marielle Gosset[2], Taro Uchida[3], and Shou-Hao Chiang[4]

[1]École et Observatoire des Sciences de la Terre, Institut de Physique du Globe de Strasbourg, Centre National de la Recherche Scientifique UMR 7516, University of Strasbourg, 67084 Strasbourg CEDEX, France
[2]Géoscience Environnement Toulouse, Toulouse, France
[3]National Institute for Land and Infrastructure Management, Research Center for Disaster Risk Management, Tsukuba, Japan
[4]Center for Space and Remote Sensing Research, National Central University, Taoyuan City 32001, Taiwan

Correspondence: Odin Marc (odin.marc@unistra.fr)

Abstract. Rainfall-induced landslides are a common and significant source of damages and fatalities worldwide. Still, we have little understanding of the quantity and properties of landsliding that can be expected for a given storm and a given landscape, mostly because we have few inventories of rainfall-induced landslides caused by single storms. Here we present six new comprehensive landslide event inventories coincident with well identified rainfall events. Combining these datasets, with two previously published datasets, we study their statistical properties and their relations to topographic slope distribution and storm properties. Landslide metrics (such as total landsliding, peak landslide density, or landslide distribution area) vary across 2 to 3 orders of magnitude but strongly correlate with the storm total rainfall, varying over almost 2 orders of magnitude for these events. Applying a normalization on the landslide run-out distances increases these correlations and also reveals a positive influence of total rainfall on the proportion of large landslides. The nonlinear scaling of landslide density with total rainfall should be further constrained with additional cases and incorporation of landscape properties such as regolith depth, typical strength or permeability estimates. We also observe that rainfall-induced landslides do not occur preferentially on the steepest slopes of the landscape, contrary to observations from earthquake-induced landslides. This may be due to the preferential failures of larger drainage area patches with intermediate slopes or due to the lower pore-water pressure accumulation in fast-draining steep slopes. The database could be used for further comparison with spatially resolved rainfall estimates and with empirical or mechanistic landslide event modeling.

1 Introduction

Landslides associated with heavy rainfall cause significant economic losses and may injure several thousand people a year worldwide (Petley, 2012). In addition, the frequency of landsliding increases with the frequency of extreme rainfall events (Kirschbaum et al., 2012), which is expected to be enhanced by global climate change (Gariano and Guzzetti,

2016). Landslides are also recognized as a major geomorphic agent contributing to erosion and sediment yield in mountainous terrain (Hovius et al., 1997; Blodgett and Isacks, 2007). Yet, constraining quantitative relationships between landslides and rainfall metrics remains difficult.

There is limited theoretical understanding of how rainfall, through water infiltration in the ground, can increase pore-water pressures and trigger failures (Van Asch et al.,

1999; Iverson, 2000). Therefore, a variety of mechanistic models have been developed, usually by coupling a shallow hydrological model to a slope failure criterium (e.g., Montgomery and Dietrich, 1994; Baum et al., 2010; Arnone et al., 2011; Lehmann and Or, 2012; von Ruette et al., 2013). However, such deterministic approaches require not only appropriate physical laws but also an accurate and fine-scale quantification of several input parameters such as topography, cohesion, permeability, and rainfall pattern (Uchida et al., 2011). In most places, such a level of detailed information is currently unavailable, rendering deterministic approaches hardly applicable.

Data-driven studies have mostly focused on using precise information on individual landslide location and timing to decipher thresholds, typically based on preceding rainfall intensity and duration, at which a landslide would initiate (Caine, 1980; Guzzetti et al., 2008, and references therein). Although useful for hazard and early-warning purposes (e.g., Keefer et al., 1987), these approaches do not address the quantity and properties of landslides that can be triggered by a rainfall event. In order to understand the importance of rainfall on erosion rates or to anticipate landslide hazard associated with emerging cyclones and heavy rainstorms, it is highly desirable to quantitatively relate the properties of a landslide event L (total area, volume, size distribution) to the combination of site susceptibility, s, and rainfall forcing, f, properties, or equivalently to develop scaling relations of the form of

$$L = g(s(\text{slope, soil thickness, strength, permeability}, \ldots),$$
$$f(\text{total rainfall, intensity, antecedent rainfall}, \ldots)). \quad (1)$$

Note that variables in such an equation may be a statistical description at the catchment or landscape scale (being a simple mean or other moments of the distribution), and thus may not describe the fine-scale variability required by mechanistic models. Although being simplified versions of mechanistic models, such scaling laws can be useful to describe average properties of the phenomena, i.e., a population of landslides associated with a constrained trigger. The advantage of statistical or semi-deterministic approaches is that they are able to accurately predict global properties, while circumventing the difficulties of predicting specific local properties of individual landslides. Indeed, such scaling laws would allow prediction in data-scarce regions and possibly at various scales (hillslope scale, catchment scale, region scale, etc). This approach has driven important progress for both the understanding and hazard management of earthquake-induced landslides, thanks to the introduction of purely empirical, physically inspired, or mixed functional relations in the form of Eq. (1) (e.g., Jibson et al., 2000; Meunier et al., 2007, 2013; Nowicki et al., 2014; Marc et al., 2016, 2017). This progress has been possible thanks to detailed investigation of individual case studies with comprehensive landslide event inventories (e.g., Harp and Jibson, 1996; Liao and Lee,

2000; Yagi et al., 2009) and through their combined analysis as aggregated databases (Marc et al., 2016, 2017; Tanya et al., 2017). By comprehensive event-inventories we mean that all landslides larger than a given size were mapped, and that the spatial extent of the imagery allowed us to observe the landslide density fading away in all direction, tracking the reduction of the forcing intensity of the triggering event, whether shaking or rainfall.

In contrast, few studies on rainfall-induced landslides are based on comprehensive event inventories. Some studies are based on individual landslide information. For example, Saito et al. (2014) studied 4744 landslides in Japan, that occurred between 2001 and 2011, to better understand which rainfall properties control landslide size. This dataset, aggregating a small subset of the landslides triggered by rainfall events, misses the vast majority of landslides. For example, in Japan, Tropical Storm Talas alone caused a similar amount of landslides in a few days. It is, therefore, insufficient for a more advanced statistical analysis. At the global scale, Kirschbaum et al. (2009) presented a catalog containing information on 1130 landslide events worldwide, which occurred in 2003, 2007, and 2008. With this catalog, they underline the correlation between extreme rainfall and landsliding (Kirschbaum et al., 2012). However, such catalogs, mainly based on reports from various kinds, are rarely adequate to constrain the quantity and properties of landslides triggered by a rainfall event. Thus, we consider that neither studies, based on a small sample of individual landslides or on a global-scale analysis, will be able to effectively constrain Eq. (1), and that detailed storm-scale information is needed.

Few case studies rely on fragmentary event inventories (and are briefly reviewed in the next section) but they may contain too few landslides for statistical analyses or may be biased to specific locations (e.g., along roads or near settlements, within weak lithological units, or near rivers), thus complicating the deconvolution of forcing and site influences. However, in theory, satellite imagery allows for comprehensive mapping of landslides larger than the resolution limit, across all catchments affected by a large storm. In practice, obtaining useful images strictly constraining the landsliding caused by a single storm is not always possible, mainly because of cloud coverage, and detailed mapping across vast areas represents a significant work effort. As a result, landslide inventories triggered by rainfall during a whole season or a few years are used for testing mechanistic models (e.g., Baum et al., 2010; Arnone et al., 2011).

The purpose for this work is to present a compilation of new and past comprehensive rainfall-induced landslide (RIL) inventories, each containing the landslide population associated with an identified storm. They constitute the core of an expandable database, essential for further research. We first briefly review existing comprehensive and partially complete inventories associated with specific storms. Then we present six new inventories and analyze their statistical properties in terms of size (total area, landslide density), geometry (length, width, and depth), and relation to topographic slopes. We further analyze and discuss these properties with respect to rainfall observations in those cases and conclude on the various insights that can be derived from such an inventory compilation.

2 Data and methods

2.1 Review of pre-existing datasets

An in-depth literature review revealed that very few comprehensive, digital, RIL inventories have been published, such as the Colorado 1999 and Micronesia 2002 events detailed below. If we look for partial inventories, in which landslides have been mapped comprehensively in limited zones affected by a storm, a few more datasets exist.

For example, Hurricane Mitch hit Central America at the end of 1998 and triggered thousands of landslides across several countries. The rainfall was record-breaking in many places, with rain gauges recording up to 900, 1100, and 1500 mm in Honduras, Guatemala, and Nicaragua (Bucknam et al., 2001; Cannon et al., 2001; Crone et al., 2001; Harp et al., 2002). In the following weeks, the USGS performed a number of aerial surveys, identified the most affected areas in these three countries as well as in El Salvador (where the rainfall amount was less), and mapped a large number of the failures in selected zones. The resolution of their aerial photographs allowed them to distinguish failures down to a relatively small size ($< 100\,\text{m}^2$), but the mapping amalgamated multiple failures into single polygons, and combined very long debris flow paths and/or channel deposits to the source areas. Because of these limitations, we did not investigate this case in detail but note that these inventories may be corrected and used by later studies. Similarly in a number of studies, inventories of all the landslides caused by a given storm in a specific catchment or geographic zone can be found: in Liguria 2000 (Guzzetti et al., 2004), Umbria 2004 (Cardinali et al., 2006), Sicily 2009 (Ardizzone et al., 2012), Peru 2010 (Clark et al., 2016), Thailand 2011 (Ono et al., 2014), and Myanmar 2015 (Mondini, 2017), as well as in Taiwan for 10 typhoons between 2001 and 2009 (Chen et al., 2013). These inventories could not constrain the total landslide response to a storm, but may allow to constrain relationships between landslide properties and local rainfall properties, provided that enough landslides have been mapped for statistical analysis (e.g., > 50–100) and without any systematic sampling bias. However, a detailed assessment of these dataset properties and of their relation to rainfall is outside the scope of this study although it would probably interestingly complement our work in the future.

In this study, we analyzed two datasets previously published by the USGS. First, afternoon rain on 28 July 1999 that triggered numerous landslides and debris flows in the Colorado Front Range (Godt and Coe, 2007). Based on aerial

photograph interpretation and field inspection, landslides were mapped as polygons containing source areas, debris flow travel, and deposition zones. Initiation points were assumed to be the highest point upslope of each mapped landslide. In 57 out of 328 polygons, multiple initiation points (2 to > 15) were mapped for multi-headed polygons (Godt and Coe, 2007). These polygons are among the largest of the inventory and represent 61 % of the total landslide area. The surface of the source areas were often of similar width, suggesting equivalent contribution from each source to the transport and deposit areas, and rendering a manual splitting impractical. Thus, we instead conserve multi-headed polygons and we use the whole landslide area, A_l, perimeter, P_l, and number of source, N_s, for each multi-headed polygon to derive an equivalent area and perimeter associated with each source: $A_l^* = A_l/N_s$ and $P_s = P_l/N_s$. This first-order approach underestimates the perimeter of each component by one width, the segment that would be added for each new subpolygon; however, this underestimation decreases with the length/width ratio of the polygons, and is already below 10 % for $L/W > 4$. In any case, this assumption does not affect the total area affected, but it changes the landslide frequency-area and frequency-width distributions, and all terms derived from them.

The second dataset contains landslides caused by a summer typhoon in July 2002, mapped exhaustively with aerial photos on the islands of Micronesia (Harp et al., 2004). We digitized the original maps based on strong contrast between red polygons and the rest of the maps. A few artifacts due to this image processing were removed and a few amalgams were split. Again, scarps and deposits are not differentiated.

2.2 New comprehensive inventories of rainfall-induced landslides

We present the mapping methodology and imagery (Table S1 in the Supplement) used to produce six additional inventories. Here we consider landslides as a rapid downslope transport of material, disturbing vegetation outside of the fluvial domain, which we define by visible water flow in the imagery. We also consider individual landslides with a single source or scar areas to avoid amalgamation, and split polygons when necessary. Although the transition between hillslopes and channel may be blurry and in part subjective, the width estimation (cf. Sect. 2.4) will mitigate variations in the transport length, as long as large alluviated or flooded areas are not mapped as landslide deposits. Still the limit between scar, transport, and deposit areas could rarely be detected with the available imagery, and all polygons consider the whole disturbed areas on the hillslopes. Subsets of the inventories in Taiwan 2009 and Brazil 2011 were produced with an automatic algorithm, and then edited and corrected manually, while all others were manually mapped.

In 2008 around the Brazilian town of Blumenau, several days of intense rainfall at the end of a very wet fall triggered widespread landsliding and flooding, with some partial inventories published in the Brazilian literature (e.g., Pozzobon, 2013; Camargo, 2015), which were not reported in the international literature. The detection and manual mapping of landslides as georeferenced polygons was primarily done with a pair of Landsat 5 cloud-free images (1 February 2009 and 2 March 2008). The coarse resolution (30 m) of the images allowed us to only locate vegetation disturbances and accurate landslide delineation was only possible for the largest events. Therefore, we used extensive high-resolution imagery available in Google Earth (over > 90 % of the area of interest, AOI) acquired in May–June 2009 in most areas, and in 2010–2012 elsewhere, where scars were still visible. To avoid mapping post-event landslides, we only mapped the ones corresponding to vegetation radiometric index (e.g., NDVI) reduction for the pair of Landsat 5 images, present even for subpixel landslides (e.g., 10 m×5 m). Thus, the landslide mapping could be confirmed for ~ 90 % of the mapped polygons, and industrial digging or deforestation occurring on steep slopes could be avoided. This approach avoids amalgamating groups of neighboring landslides and allows for the mapping of very small landslides (~ 1 pixel in Landsat 5 images). However, some detailed field mapping in the surrounding of Blumenau reports up to twice the number of landslides that we observed (Pozzobon, 2013), indicating that we still miss a substantial number of small events. Nevertheless, these landslides must be quite small (not visible in ~ 1 m resolution imagery) and likely do not affect any of our statistics (area, volume, slope) apart from the total number of landslides.

The same approach was used to map the intense landsliding caused by a few days of intense rainfall between 10 and 12 January 2011 (Netto et al., 2013), in the mountains northeast of Rio de Janeiro. Near Teresópolis, first we used a pan-sharpened (10 m) EO-ALI and 30 m Landsat 7 images from February 2011 for co-registration and ortho-rectification.

> 95 % of the slides were cross-checked in Google Earth based on images from the 20 and 24 January 2011 (Fig. 1); and where clouds or no images where available we mapped landslides directly from Google Earth (available over >90 % of the AOI), even if poor ortho-rectification may create geometric distortions. Closer to Nova Friburgo, we used a pair of very-high-resolution GeoEye-1 images (2/0.5 m resolution in multispectral/panchromatic) from 26 May 2010 and 20 January 2011. On these images we applied the methods presented by Stumpf et al. (2014) to classify the whole image, detecting > 90 % of the landslides we could manually observe , but also including false positive. Thus, we manually screened the image to remove agricultural fields, inundated areas, channel deposits that were included, and split the amalgamated landslides frequently given the important clusters of landslides in many parts of the image. This correction seems sufficient given that the landslide size distribution for the three subparts of the inventory are consistent (Fig. S1 in the Supplement). Polygons from the automatic classification

display a slightly larger equivalent length/width ratio, maybe because some amalgamated polygons have been missed or simply because the classification allows for hollow polygons, biasing upward the length/width estimate based on a perimeter/area ratio (cf. Sect. 2.4).

From 1 to 4 September 2011, Tropical Storm Talas poured heavy rainfall on the Kii Peninsula, in Japan, resulting in several thousands of landslides. For disaster emergency response, the National Institute for Land and Infrastructure Management of Japan (NILIM) mapped landslides across most of the affected areas based mainly on post-typhoon aerial photographs and occasionally on Google Earth imagery (Uchida et al., 2012). Screening antecedent imagery (2010–2011) from Google Earth and Landsat 5, we identified and removed a few hundred pre-Talas polygons, mostly within 5 km of 136.25° E/34.29° N and 135.9° E/34.20°. With Google Earth we could validate NILIM mapping over about 85 % of the AOI and we added almost 200 polygons in areas were aerial photographs were not taken and split many large or multi-headed polygons that were amalgamated. Some polygons had distorted geometry or exaggerated width, most likely due to poor ortho-rectification of the aerial imagery and/or time constraints for the mapping. We could not systematically check all polygons, but we checked and corrected all polygons larger than 30 000 m^2 (3 % of the catalog but representing 45 % of the total area). We consider that the remaining distortions for some of the smaller polygons have minor impacts on the statistics discussed in the next sections.

In Taiwan, we collected landslide datasets associated with the 2008 Kalmaegi (16–18 July) and 2009 Morakot (6–10 August) typhoons, partially described by Chen et al. (2013). For 2008, we compared multispectral composite images and NDVI changes between (30 m) Landsat 5 images taken on the 21 June, 7 July, and 23 July. The image from 7 July is covered by clouds and light fog in many parts but allows us to identify that most places affected by landslides in the last images were still vegetated at this time. Thus all new landslides are attributed to the rainfall from typhoon Kalmaegi. For 2009, landslides were mapped with pre- and post-event FORMOSAT-2 satellite images (2 m panchromatic and 8 m multi-spectral; Chang et al., 2014). To cover most of the island, we mosaicked multiple mostly cloud-free pre-event (14 January, 8 May, 9 May, 10 May, 6 June 2009) and post-event (17 August, 19 August, 21 August, 28 August, 30 August, 6 September 2009) images. For a subsets of the inventory, especially to the east of the main divide, landslides were significantly amalgamated and bundled with river channel alluviation. We thus manually split the polygons and removed the channel areas.

In a few areas with clouds (< 5 % of the AOI) in the post-event mosaic, we mapped with Landsat 5 images (from 24 June and 12 September 2009), even if the spatial resolution limit may have censored the smallest landslides in these zones. Special attention was given to the separation of in-

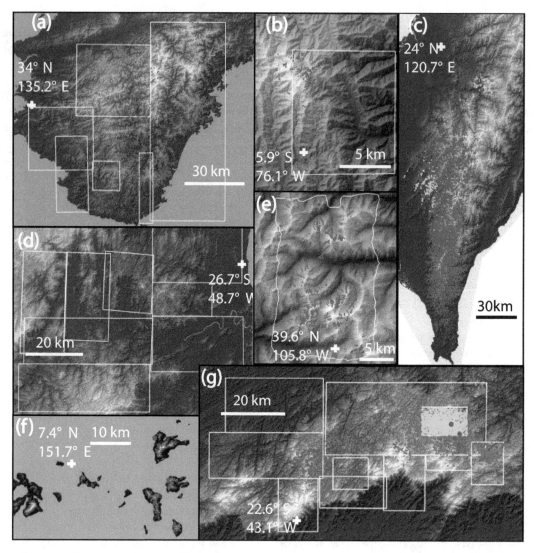

Figure 1. Landslides inventory superimposed on digital surface model for the events in Japan 2011 **(a)**, Colombia 2015 **(b)**, Taiwan 2008 and 2009 **(c)**, Brazil 2008 **(d)**, Colorado 1999 **(e)**, Micronesia 2002 **(f)** and Brazil 2011 **(g)**. Landslides are in purple, rain gauges used in this study are in red dots (and red crosses for Taiwan 2009), and the yellow frames show the availability of high-resolution imagery (Google Earth) used to check or perform the mapping. In **(c)** the green dots are landslides from 2008, while purple dots are from 2009. In **(g)** purple, red, and green dots are landslides mapped from EO-ALI, Google Earth, and automatic classification of GeoEye images, respectively.

dividual landslides by systematically checking and splitting polygons above 0.1 km² (2 % of the catalog but representing 30 % and 60 % of the total area and volume, respectively). However, it is clear that a number of smaller landslides are missed or merged with large ones; and, therefore, although total landsliding and landslide locations on slopes may be well represented, the size distribution of this catalog must be biased to some extent.

Between 15 and 17 May 2015, heavy rainfall in the mountains above the village of Salgar, Colombia, triggered catastrophic landslides and debris flow (> 80 deaths). Landslide mapping was carried out by comparing a (10 m) Sentinel-2 image from the 21 July 2016 and a pan-sharpened (15 m) Landsat 8 image from the 19 July and 26 December 2014.

These images were selected for their absence of clouds, good conditions of light, and similarity. High spatial resolution imagery from Google Earth, dated from 31 May 2015, shows fresh scars consistent with our mapping over most of the area (Fig. 1), and we assumed that the remaining landslides (< 15 % of the inventory) were also triggered by the same rainfall event.

2.3 Rainfall data

Rainfall data quality and amount are very variable for the different events, from none or one single gauge (for Micronesia or Colombia) to a dense gauge network and potentially weather radar coverage in Japan, Taiwan, and Brazil. Therefore, we selected a simple index that could be obtained for each case in order to discuss potential rainfall controls on the landslide properties. For each case we calculated an estimate of total rainfall, R_t, duration, D, and a peak rainfall intensity over 3 h, $I3$ (Table 1). Note that these variables do not represent an average value within the whole footprint of the storm, but rather a maximal forcing, usually colocated with the areas where landsliding was the most intense (Fig. 1) and derived mostly from one or a few rain gauges. Thus, these indexes may be taken as storm magnitude. A more detailed analysis of the spatiotemporal pattern of the rainfall and of its relations to the spatial pattern of landsliding is highly desirable, but challenging and is left for a future study.

The estimates from Taiwan and Japan are based on hourly gauge measurements from the Japan Meteorological Agency and Taiwan Institute for Flood and Typhoon research. In each case we took the three closest gauges within 5 to 15 km from the areas with the highest landslide densities (in 0.05 by 0.05° window, Fig. S2) and computed their average properties (Fig. 2). Minimum and maximum single gauge measurements give a coarse measure of the uncertainty. A single gauge is available in Micronesia, and we used the hourly rainfall from 1 to 3 July 2002 reported in Harp et al. (2004). For Colorado, we used the hourly rainfall from the rain gauge at Grizzly Peak, closest to the intense landsliding reported by Godt and Coe (2007). For this event, radar data indicate very localized, high-intensity precipitation located on the peaks where the debris flows occurred (Godt and Coe, 2007) and suggests that the single closest gauge is more representative than averaging with the other nearby ones. For the event in Brazil 2008 we considered the total daily rainfall from Luis Alves station (Fig. 1), where more than 130 mm day^{-1} were accumulated on the 21, 22, and 23 November and 250 mm on the 25, and intensity going up to 50 mm h^{-1} (Camargo, 2015). These days were also preceded by an abnormally wet period, with November 2008 accumulating $\sim$ 1000 mm, 7 times the long-term average for this month. In 2011 in Brazil, hourly rain data at Sitio Sao Paulista report 200 mm in 8 h before gauge failure, while there, and at nearby sites, the cumulative rainfall was $\sim$ 280 mm from the 10 to the morning of the 12 January (Netto et al., 2013). For these cases, rain gauges give a trustworthy estimate of the local rainfall, but are not constraining the large-scale rainfall pattern. Last, in Colombia, we could not find data from any nearby rain gauge and we thus use rainfall estimates from the GSMaP global satellite products (Kubota et al., 2006; Ushio et al., 2009) (Fig. S3). Here, the minimum, mean, and maximum rainfall are obtained by considering the triggering storm as the raining period at the time of debris flow occurrence, or the one from the previous day or merging both events, respectively (Fig. S3).

Defining storm duration accurately requires defining thresholds on rainfall intensity over given periods, to delimit the storm start and end. Given the variable quality of our data, we limit ourself to a first-order estimate of the continuous pe-

riod when rainfall was sustained (i.e., $I3 > 3$ mm h^{-1}). We consider these durations accurate within 10 %–20 % for the events with overall hourly data. For the less constrained cases B08, B11 and C15 durations are more uncertain. In any case, for these eight storms, we note a strong correlation between D and R_t and $I3$ and R_t (for power-law scalings, $R^2 = 0.9$ and $R^2 = 0.8$, respectively) (Fig. S4). Thus, given that spatial and temporal length scales are often linked in meteorology, the long events causing larger rainfall may also have larger footprints.

2.4 Landslide area, width and volume

Landslide plan view area and perimeter are directly obtained from each polygon. However, these values represent the total area disturbed, that is the scar, deposit and run-out areas, because a systematic delineation of the scar was not possible from most of the imagery. This means that landslide size statistics are resulting from processes affecting both landslide triggering and run out. Landslide volume, estimated based on area, may also be overestimated for long run-out slides. Therefore, we propose here a simple way to normalize for landslide run out and obtain an estimate of the scar area.

Following Marc and Hovius (2015), we computed an equivalent ellipse aspect ratio, K, using the area and perimeter of each polygons. For polygons with simple geometries, K is close to the actual length/width ratio, but this is a measure that also increases with polygon roughness or branching, and therefore with amalgamation (Marc and Hovius, 2015). Assuming an elliptic shape, polygon area can be approximated by $\pi LW/4$ with L and W being the polygon full length and width, respectively. This allows us to estimate $W \simeq \sqrt{4A/\pi K}$. To validate this geometric method to retrieve landslide width, we systematically measured the width of 418 randomly selected landslides across a wide range of polygon areas and aspect ratios, belonging to four inventories: J11, TW8, B11, and C15. For each polygon, we focused on the upper part of the landslide only, the likely scar, and averaged four width (i.e., length perpendicular to flow) measurements made in arcGIS. The width estimated based on P and A are within 30 % and 50 % of the measured width for 72 % and 92 % of the polygons, respectively (Fig. S5). We do not observe a trend in bias with area nor aspect ratio, except perhaps for the automatically mapped landslide in B11, where high aspect ratio correlates with underestimated width. Thus, for correctly mapped polygons, we can use P and A to derive W and a proxy of landslide scar area, $A_s \sim 1.5W^2$. We assume landslide scars have an aspect ratio of 1.5, as it was found to be the mean aspect ratio found across a wide range of landslide size within a global database of 277 measured landslide geometries (Domej et al., 2017). Even if this equivalent scar area may not exactly correspond to the real landslide scar, it effectively removes the contribution of the landslide run out to the landslide size and allows

Table 1. Rainfall data summary, containing the total rainfall, duration and maximum 3 h intensity for each storms. For TW8, TW9 and J11, we indicate the range for the three indexes that could be estimated from three gauges near the zone of maximal landsliding. We cannot perform this analysis for MI2 and C99, and can only assess a range of R_t for B08 and B11. For B08 the question mark indicates that the upper bound of R_t is under-constrained and taken as 115 % of the best estimate. For C15, we indicate by a star that we could only access satellite based rainfall estimates (GSMaP_MVK V04 ungauged products; Kubota et al., 2006; Ushio et al., 2009; see Fig. S3). Reference are as follows, 1: Godt and Coe (2007), 2: Harp et al. (2004), 3: Camargo (2015), 4: Netto et al. (2013) .

Event	C99	MI2	B08	TW8	TW9	B11	J11	C15
R_t, mm	45	500	695	670	2500	280	1300	65
			[680–800?]	[600–740]	[2100–2800]	[200–320]	[1000–1500]	[10–75]*
D, hours	4	20	100	24	105	36	62	10*
$I3$, mm h^{-1}	13	65	30	92	85	55	58	8*
				[78–116]	[83–87]		[38–88]	
Ref.	1	2	3	Us	Us	4	Us	GSMaP

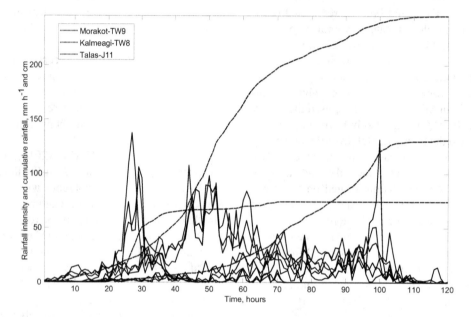

Figure 2. Rainfall history for typhoons Kalmaegi (TW8), Morakot (TW9), Talas (J11). For each event, hourly intensity is shown with solid curves for three gauges nearby the area with most intense landsliding (see Fig. 1 for locations). Dashed lines represent the mean cumulative rainfall from the three gauges.

us to compare different size distributions while reducing the impact of variable run-out distances.

We also assessed how using A_s affects estimates of landslide volumes and erosion, by computing landslide volume with the total landslide area and with A_s only. In both cases, we used $V = \alpha A^\gamma$, with α and γ and their 1σ globally derived by Larsen et al. (2010). Given that soil and bedrock slides have different shape and that soil slides are rarely larger than 10^5 m^2 (10^4 m^2 for soil scars; Larsen et al., 2010), we used the "all landslide" parameters ($\gamma = 1.332 \pm 0.005$; $\log_{10}(\alpha) = -0.836 \pm 0.015$) when $A < 10^5$, and the "bedrock" parameters ($\gamma = 1.35 \pm 0.01$; $\log_{10}(\alpha) = -0.73 \pm 0.06$) for larger landslides. Similarly, we used the soil scars ($\gamma = 1.262 \pm 0.009$; $\log_{10}(\alpha) = -0.649 \pm 0.021$) and bedrock scars ($\gamma = 1.41 \pm 0.02$; $\log_{10}(\alpha) = -0.63 \pm 0.06$) for $A_s <$

10^4 m^2 and $A_s >= 10^4$ m^2, respectively (Larsen et al., 2010). Marc et al. (2016) proposed a rudimentary version of such a run-out correction, where they effectively reduced landslide area by a factor of 2 for mixed landslides and of 3 for bedrock landslides, noting that volumes derived in this way were closer to field estimates for large landslides than without correction.

Uncertainties in this approach include the 1σ variability in the coefficient and exponent of the landslide area–volume relations given above, and an assumed standard deviation of 20 % of the mapped area. These uncertainties were propagated into the volume estimates using a Gaussian distribution. The standard deviation on the total landslide volumes, for the whole catalogs or for local subsets, were calculated assuming that the volume of each individual landslide was unrelated to that of any other, thus ignoring possible covariance. Although estimated 2σ for single landslides is typically from 60 % to 100 % of the individual volume, the 2σ for the total volume of the whole catalog is below 10 % for the eight datasets. However, for subsets with fewer landslides and with volume dominated by large ones, typical when we compute the total landslide volume density in a small area (e.g., 0.05°), 2σ uncertainty reaches 40 %–60 %. We note, however, that these uncertainty estimates do not consider potential errors in the identification of landslides, either missed because of occasional shadows or clouds, or erroneously attributed to the storm. Such uncertainty is hard to quantify but must scale with the area obscured in pre- and post-imagery. In most cases multiple pre- and post-event images mean that obscured areas typically represent less than 10 % of the affected area, and such errors may be between a few to ~ 20 % of the total area or volume, depending on whether obscured areas contain a landslide density higher or lower than the average observed throughout the affected area. Last, resolution may not allow us to detect the small landslides and in some cases the landslide number may be significantly underestimates, but not the total area and volume dominated by the larger landslides.

For each inventory, we also estimated the landslide distribution area, that is the size of the region within which landslides are distributed. Based on the landslide inventories we could delineate an envelope containing the overall landsliding. As discussed by Marc et al. (2017), such delineation is prone to high uncertainties as it is very dependent on individual isolated landslides. Thus for all cases, we give a range of distribution area, where the upper bound is a convex hull encompassing all the mapped landslides, while the lower bound is an envelope ignoring isolated and remote landslides (i.e., single or small cluster of landslides without other landslides within 5–10 km), if any. Although the spread can be large in absolute value, both approaches yield the same order of magnitude.

3 Results

The inventories contain from ~ 200 to $> 15\,000$ landslide polygons, representing total areas and total volumes (from scars) from 0.2 to 200 km^2 and 0.3 to 1000 Mm^3, respectively. The triggering rainfalls are characterized by a total precipitation of ~ 50 to 2500 mm in periods ranging from 4 h to 4 days, and caused landslides within areas ranging from ~ 50 to 10 000 km^2. Although the dominant landslide types are soil and regolith slumps, a number of large deep-seated bedrock landslides are also present in the inventories associated to the Talas and Morakot typhoons (Saito and Matsuyama, 2012; Chen et al., 2013). A more detailed description of the landslide types and materials involved was not possible with the available imagery; thus, our analysis does not consider landslide types. In the next sections, we present results obtained from these inventories in terms of landslide size statistics, landslide spatial patterns, and relation to slope, before correlating these landslide properties to rainfall parameters.

3.1 Landslide properties

3.1.1 Landslide size statistics

Frequency size distributions of landslide inventories have typically been fit by power-law tailed distributions, above a certain modal size (Hovius et al., 1997; Malamud et al., 2004). The modes and the decay exponents of these distributions are mainly related to the lithology (mechanical strength) or topographic landscape properties (i.e., susceptibility related) (Stark and Guzzetti, 2009; Frattini and Crosta, 2013; Katz et al., 2014; Milledge et al., 2014). Some authors suggested that this behavior could also be affected by the forcing processes. For example, analyzing earthquake-induced landslide catalogs, it was found that deeper earthquakes, thus with weaker strong-motions, have a smaller proportion of large landslides (Marc et al., 2016). Based on theoretical arguments, it has been proposed that short, high-intensity rainfall could cause pulses of high pore-water pressures at the soil–bedrock transition, initiating mainly small, shallow landslides, while long duration rainfall with high total precipitation could provoke significant elevation of the water table and trigger large, deep-seated landslides (Van Asch et al., 1999). To our knowledge, little empirical evidence has supported these assumptions, and we discuss next how our data compare to these ideas.

All landslide size distributions present a roll-over and then a steep decay (Fig. 3). The modal landslide area varies between ~ 3000 m^2 for TW8 and ~ 300 m^2 for B11, while the largest landslides are ~ 0.1 km^2 for most events and reach ~ 0.4 and 2.8 km^2 for J11 and TW9, respectively. The roll-over position certainly relates partly to the spatial resolution and acquisition parameters of the images (Stark and Hovius, 2001) (e.g., for TW8 and B08, where landslides were mostly mapped on a coarse spatial resolution image compared to aerial photographs for C99 and J11). However, mechanical parameters are also expected to influence the roll-over position (Stark and Guzzetti, 2009; Frattini and Crosta, 2013), as suggested by the fact that MI2, mapped with 1 m resolution aerial imagery, has larger modal area than C15, mapped with 10 m Sentinel-2 satellite imagery. Following Malamud et al. (2004), we use maximum likelihood estimation (MLE) to fit the whole distribution with an inverse gamma distribution (IGD) and obtain power-law decay exponents $\alpha + 1$ between ~ 2 and 3, consistent with the typical range found in the literature (Hovius et al., 1997; Malamud et al., 2004; Stark and Guzzetti, 2009; Frattini and Crosta, 2013). However, we note that at least three cases, B11, TW9 and C99, poorly follow an IGD, with a break in the distribution occurring in large areas, followed by a very steep decay.

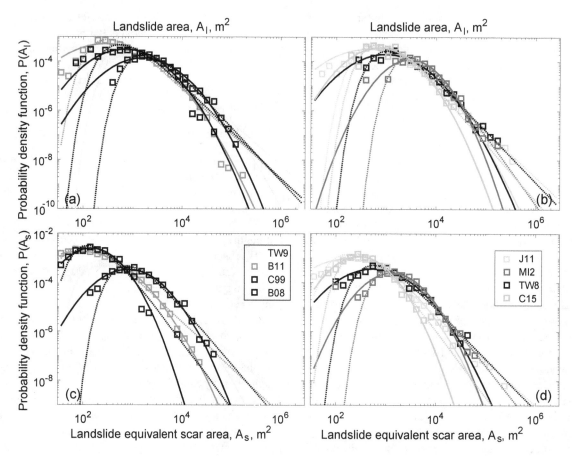

Figure 3. Probability density functions of landslide whole area (**a, b**) and estimated scar area (**c, d**). To improve visualization we split the 8 inventories in two groups. A Log-Normal and Inverse Gamma Distribution maximum-likelihood estimation for the whole distribution are shown by solid and dashed lines, respectively.

When considering landslide estimated scar sizes, that is essentially a correction to reduce landslide polygon aspect ratio to 1.5, we observe a reduction of the largest landslide size by 2 to 10 times, but a moderate reduction of the modal area. This is consistent with the fact that landslides with long runout distances are often over represented within the medium to large landslides (Fig. S6). We also note that after the runout distance variability is normalized, the distribution of C99 agrees better with an IGD. This is not the case, however, for B11 and TW9 that still feature a steepening of their distribution decay (and a divergence from IGD fit) beyond $\sim 10^3$ and $\sim 10^4$ m^2, respectively. Run out being normalized, this could be an artifact relating to residual amalgamation for TW9, but not for B11, where most landslides were mapped manually and amalgamation was avoided. In these two cases, for the whole landslide area or the landslide scar only, we note that a MLE fit of a log-normal distribution agrees better to the data (based on the result of both the Kolmogorov–Smirnov and the Anderson–Darling test). For other inventories, a log-normal fit is equivalent or worse than an IGD, but we note that the parameters describing the decay of both distributions are highly correlated (Fig. S7). Thus, we take $\alpha + 1$ as a rea-

sonable indicator of the relative proportion of large landslides within the different dataset and do not further explore the functional form of landslide size distribution and its implications, which we consider beyond the scope of this study.

3.1.2 Landslide and slope distribution

For all cases, we computed the frequency of slope angles above 5° based on the global 1 arcsec (~ 30 m) Shuttle Radar Topography Mission (SRTM) digital surface model (Farr and Kobrick 2000; Farr et al., 2007). In most cases, hillslopes have a distribution clearly independent from valley floors. However, for B08 and MI2, for example, the amount of plains in the study area do not allow for resolving the hillslope distribution. Therefore, for Micronesia we removed all slopes which are less than 10 m above sea level; and for Brazil, we extracted the slope cells in the landslide distribution area but with a mask excluding the wide valley bottoms, allowing us to obtain a hillslope distribution as an approximate Gaussian, with a mode significantly beyond our threshold of 5°. To focus on the scar area of each landslide polygon, we extracted only the slopes for the highest-elevation pixels representing a surface of $1.5\,W^2\,m^2$. Then, we computed the probability density function for the landslide-affected area and the whole topography (hereafter the "landslide" and "topographic" distributions) with a normal-kernel smoothing with an optimized bandwidth, as implemented in Matlab. We obtain topographic modal slopes, S_M, at 15.5 and 18.5° for the gentle landscape of Micronesia and Brazil, while in Japan and Taiwan we reach almost 30° (Fig. 4a). The landslide distributions are unimodal, except for C15 that seems to have secondary modes at $S_M - 5$ and $S_M + 25$, and are systematically shifted towards steeper slopes.

To further quantify the differences in slope sampling between these events, we computed the ratio of probability between the slope distribution of the whole topography and of the landslide-affected area only (P_L/P_T, Fig. 4). This ratio represents the tendency of landslide occurrence on a given slope to be more or less frequent than the expected occurrence of this given slope in the landscape. We refer to this as an oversampling or undersampling of the topographic slope distribution. To compare the events in different landscapes, we plot each event against $S - S_M$ (Fig. 4b). An important issue is to determine whether the landslide probability can be considered a random drawing from slopes of the topography or not. Given that landsliding affects less than 10 % of the landscape, the sampling of the topography by landslides can be approximated by a Bernoulli sampling. In this case, the central limit theorem gives the 95 % prediction interval as $P_T \pm 1.96\sqrt{P_T(1 - P_T)/N}$, with N the number of independent draws, here taken as the number of landslide scars. The convergence of N draws to P_T within the prediction interval is only valid if $N > 30$, $NP_T > 5$, and $N(1 - P_T) > 5$, implying that only very large samples can be interpreted towards the extremity of the topographic slope distribution, where P_T is small.

For all events, we observe that P_L is significantly different from a random drawing of the topography with oversampling of the slopes beyond S_M and undersampling below it (Fig. 4b). However, we note that for most events the undersampling and oversampling is smaller than a factor of 2. Some cases (C15, J11, and TW8) have stronger oversampling (> 4) for $S - S_M > 25$ but they may not be representative ratios given the limited number of landslides and of slopes this steep (i.e., $NP_T < 5$). The scars of C99 clearly depart from this behavior, with undersampling and oversampling of a factor of 10 and 6 at $S_M \pm 10°$, respectively. B08 has also strong undersampling below S_M but has a landslide distribution that rapidly converges to the topographic ones at high slopes.

3.2 Correlation between rainfall metrics and landslide properties

3.2.1 Total landsliding

For the eight inventories, we observe a nonlinear increase in all metrics of total landsliding with the storm total rain-

fall (Fig. 5). The increase is similar for the total area and volume, and best fit by exponential functions. We observe higher correlations with rainfall, when using the total scar area ($R^2 = 0.78$), estimated as W^2, instead of the total area ($R^2 = 0.72$). This is mainly due to the very large reductions of area for C99 and B11, where long run-out landslides were dominant (Fig. 5). Correlations are generally higher with volume and also increase when we derive total volume from scar estimates (from $R^2 = 0.81$ to $R^2 = 0.87$). Note also that with these scar metrics, the relation to rainfall becomes equally or better fit by a power-law function rather than an exponential function (Fig. 5). This is because when including landslide run out, the total landsliding of C99 and C15 is larger and creates an apparent asymptote, better fit by an exponential function. Last, we note that total volume values may change depending on which A-V scaling relations are used and with which assumptions, and their absolute values may be inaccurate but this should not affect the reported scaling form and exponents, considering that potential biases should be relatively uniform.

Total number of landslides also tends to increase with total rain but the scatter is much larger (Fig. 5). This is at least partly an artifact, given that for C99, MI2, and B11, high spatial resolution imagery allows us to delineate many more small landslides and to mitigate amalgamation, whereas for B08, TW8, and TW9, the limited spatial resolution, the density of landsliding, and our limited ability to split amalgamated landslides lead to an underestimation of the landslide number. Thus, even if landslide number may contain information, quantitative comparisons of the events are biased and we will not further interpret the total number of landslides in the following.

Last, we note that the landslide distribution areas (i.e., the regions within which landslides are distributed) also correlate strongly with the total rainfall. Only considering the eight inventories strongly suggest a power law form. However, based on the dataset reported for Hurricane Mitch, the distribution area was at least $100\,000\,km^2$ for maximum total rainfall of about 1500 mm (Cannon et al., 2001). Adding it to our fit, we found that power-law or exponential functions of the rainfall explained a similar amount of the variance, 72 % and 63 %, respectively.

In the next subsection, we compute landslide densities (in % of area), allowing us to study the intra-storm variability in landsliding.

3.2.2 Maximum and mean landslide density

Understanding what controls landslide density is a key objective to better constrain hazards and their consequences. For each storm, we compute the mean landslide density (in area and volume) by dividing total landsliding by the landslide distribution area (Fig. 6a). This density represents the whole affected area and hides important spatial variability (Fig. 1), thus we also compute the maximum landslide den-

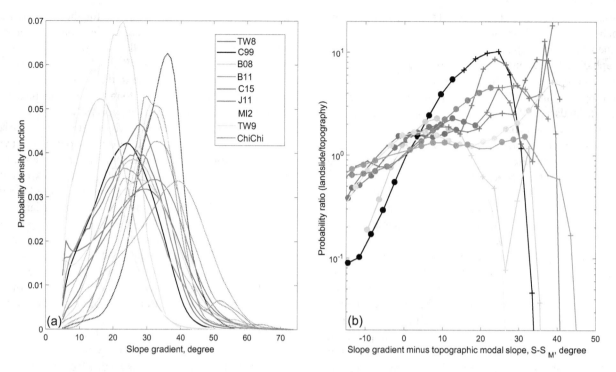

Figure 4. (a) Slope gradient probability distribution for the affected topography (solid) and the landslide scar areas only (dashed), for the eight rainfall events and as a comparison to the Chi-Chi earthquake. (b) Ratio of the two probability distributions against the difference between slope gradient and the modal topography. The ratios are estimated with the PDF averaged within 3° bins. Solid circles and dots represent ratios where the landslide probability is beyond or within, respectively, the 95 % prediction interval of the topography distribution. Crosses indicate bins where data are insufficient for the validity of the central limit theorem required to estimate prediction interval.

sity by computing the total landsliding (again in area and volume) within a moving window of 0.05° ($\sim$ 25 km^2), assigning landslides to a cell based on their centroid locations and selecting the maximal value (Fig. 6b). Given the better correlation obtained above with a run-out normalization, we focus on area and volume densities derived from scar estimates.

The mean landslide densities vary between 0.01 %–1 % and 100–10 000 m^3 km^{-2} but with poor correlation with total rainfall ($R^2 = 0.01$ and $R^2 = 0.46$ for area and volume density, respectively). Indeed, given that both total landsliding and distribution area increase strongly with total rainfall, their ratio is relatively independent. In contrast, the maximum landslide scar density and volume density range from 0.1 % to 5 % and 0.002 to 1.5 millions m^3 km^{-2}, respectively, and are strongly correlated with a power-law of total rainfall ($R^2 = 0.76$ and $R^2 = 0.95$). We found very similar correlations when computing the local density on a grid of 0.03 or 0.1°, but degraded correlations when using the whole landslide area to compute landslide density ($R^2 = 0.40$ and $R^2 = 0.69$). We also note that, as for the total landsliding, maximum landslide density and volume density are significantly correlated with peak rainfall intensity, $I3$ ($R^2 = 0.58$ and $R^2 = 0.67$, respectively), and duration, D ($R^2 = 0.70$ and $R^2 = 0.73$, respectively), although less strongly than with total rainfall.

3.2.3   Landslide size, run out, and position on slope

The decay exponents of the distribution of landslide area do not correlate significantly with any storm metrics (intensity, duration, or total rainfall; $|R| < 0.1$). However, after run-out normalization, the decay exponents of landslide scar area correlate with all metrics, although with significant scatter ($R^2 \sim 0.5$ Figs. 3, 7). The two largest storms (J11 and TW9) have the lowest exponents ($\alpha + 1 \sim 1.8$), and thus a large proportion of very large landslides, while the two smallest storms (C15 and C99) have a small proportion of large landslides and large exponents ($\alpha + 1 \sim 2.7$). However, intermediate cases are very scattered, as B11 and TW8 have similar total rainfall, peak intensity, and duration but very different distribution with $\alpha + 1 = 1.9$ and with $\alpha + 1 = 2.6$, respectively. Still, randomly removing one event (i.e., jackknife sampling) we obtained R^2 between 0.4 and 0.7, with a similar mean R^2 of about 0.5.

The decay exponents of the equivalent aspect ratio (Figs. 3, S6) do not correlate significantly with any storm metrics (intensity, duration, or total rainfall; $|R| < 0.2$). Indeed, long run-out landslides are abundant for the smallest storms, C99 and C15, as well as for the second largest storm, J11, but are relatively rare for other storms (e.g., MI2, TW8, B08), spanning the whole range of storm indexes. Similarly, the mean and modal aspect ratio are similar for all events

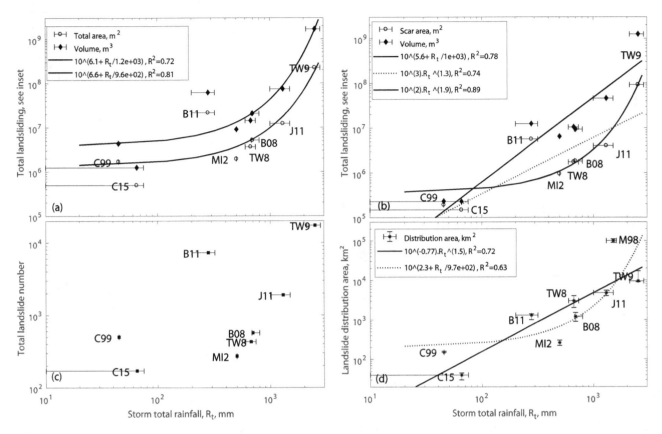

Figure 5. Total landsliding in area and volume derived from whole landslides (**a**) or from scar estimates only (**b**), total landslide number (**c**) and landslide distribution area (**d**) against storm total rainfall. M98 is for Mitch 1998. Horizontal error bars show a range of maximum storm rainfall when available (cf., Table 1). In (**a**) and (**b**) 1σ uncertainty in the total volumes and areas, ignoring potential landslide mis-detections (cf. methods), are smaller than the symbols ($< 10\%$). Vertical error bars are based on the range of affected areas in (**d**), while we could not obtain quantitative uncertainties in the total number (**c**).

across all storm metrics, except for C99 which is heavily dominated by debris flow and has a modal aspect ratio > 10.

We have observed that almost the eight events behave similarly with respect to the distribution of topographic slopes, not suggesting a strong link with the individual storm properties. The C99 event has a different behavior that may relate to the fact that it was the shortest storm with the smallest total, or that it was the only case occurring in high-elevation terrain with sparse vegetation. C15, the second shortest and smallest storm event may also have strong oversampling about $20°$ beyond S_M but the limited number of landslides does not allow us to confirm the significance of this oversampling.

4 Discussion

4.1 Scaling between rainfall and landsliding

We found that total landsliding, peak landslide density, and the distribution area of landsliding were all best described as increasing as a power-law or exponential function of the total storm rainfall, R_t. Our mechanistic understanding of lands-

liding predicts that, for a given site, the mechanism leading to failure is the reduction of the normal load and friction due to the increasing pore-water pressure (Iverson, 2000). This requires progressive saturation of the material above the failure plane and depends directly on the total amount of water poured on the slopes. However, we can envision that landscapes may rapidly reach an equilibrium in which all unstable slopes under rainfall conditions frequently occurring would have been removed. In this framework, the rainfall amount relative to the local climate would be more relevant than absolute rainfall, requiring an analysis in terms of deviation from the mean rainfall or in terms of rainfall percentiles (e.g., Guzzetti et al., 2008). Although we could not define rainfall percentiles in each area, we note that normalizing R_t by the mean monthly rainfall relevant for each storm, we still find a decent correlation with the peak landslide density, implying climate normalized rainfall variable may be driving landsliding (Fig. S8).

The antecedent rainfall is also expected to play a key role in controlling the saturation level before the triggering storm (e.g., Gabet et al., 2004; Godt et al., 2006). However, if the

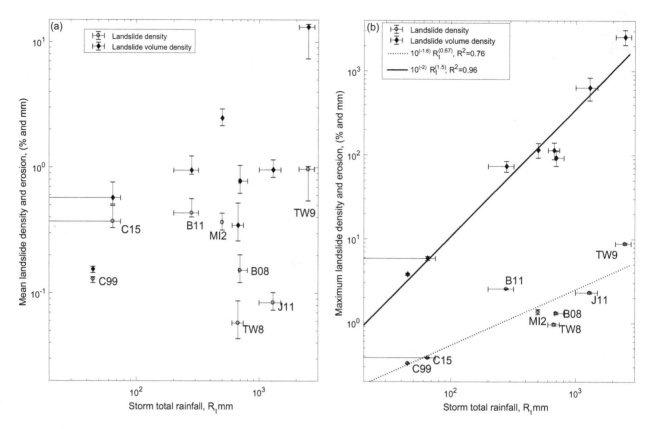

Figure 6. Mean landslide density **(a)** and peak landslide density in a 0.05° sliding window **(b)**, against storm total rainfall. Landslide area and volume are derived from scar estimates, i.e., removing run-out contribution. Horizontal error bars show a range of maximum storm rainfall when available (cf., Table 1). Vertical error bars are based on the range of affected area in **(a)** (the most uncertain term), and represent 1σ uncertainty in the total volume and area density in **(b)**, ignoring potential landslide mis-detections (cf. methods).

regolith is already close to the field capacity, significant parts of antecedent rainfall may be drained from the regolith within some hours or days (Wilson and Wieczorek, 1995), and as a result, the contribution of past storms may be negligible compared to heavy rainfalls over relatively short time intervals (1–4 days). However, for moderate storms, like C15 or C99, and especially during dry periods when the slopes are saturated below field capacity, the role of antecedent rainfall may be more substantial. Thus, we expect that moderate storms happening after prolonged dry or wet periods may deviate downward or upward from the scaling, respectively. We also note that the abundance of larger and deeper landslides, strongly influencing the total volume or erosion, may depend on deeper water level rather than regolith saturation and thus may be most sensitive to water accumulation over several days rather than a few hours (Van Asch et al., 1999; Uchida et al., 2013). Therefore, although we obtained a good correlation without considering antecedent rainfall, its role should be assessed in future refined scalings. Last, the scaling reported here is based on events where all landslides occurred within a short time frame (few hours to few days), and would not apply to a monsoon setting where landslides occur more or less continuously during several weeks (Gabet et al., 2004;

Dahal and Hasegawa, 2008), driven by continuous, heavy but unexceptional, rainfall. Indeed, in a long period with fluctuating rainfall such as the monsoon, drainage and storage of water will certainly not be negligible and the derivation of a soil water content proxy will be necessary (e.g., Gabet et al., 2004).

The strong correlation between R_t and A_d suggests that storms able to generate greater amounts of rainfall also tend to deliver a sufficient amount of rain over broader areas. For tropical storms and hurricanes (5 out of 9 cases in Fig. 5d) a number of studies (cf., Jiang et al., 2008, and references therein) found that the maximum inland storm total rainfall (i.e., R_t for us) correlated well ($R > 0.7$) with a rainfall potential defined as the product of storm diameter and storm mean rainfall rate within this diameter over storm velocity, each term measured 1–3 days before the storm made landfall. It was also generally observed that rainfall intensity is higher closer of the storm core, thus potentially tightening the link between R_t and a given storm radius with intense rainfall and high landslide probability. These observations would imply linear proportionality between R_t and A_d and could be consistent with the observed power-law trend (exponent 1.5; Fig. 5), especially if some further links between

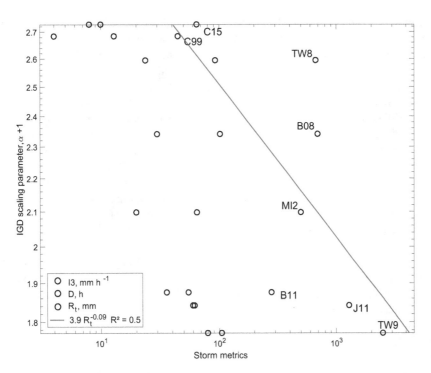

Figure 7. Landslide scar size distribution decay exponents against storm total rainfall (red), storm duration (blue) and storm peak intensity (black). The best least-squares fit is shown in red. The reduction of the decay exponents with increasing storm magnitude indicates an increase in the proportion of large landslides relative to small landslides.

R_t and mean storm intensity or velocity exist. Potential links between R_t and A_d for smaller-scale storms (C99, C15, B08, and B11) are harder to interpret, and we cannot exclude that it is a coincidence allowed for by our small number of events.

In any case, the broader zone is not likely to receive homogeneous rainfall amount, decoupling mean landslide density from storm maximum strength (Fig. 6a). The variability in rainfall within these extended zones is likely a main control on the spatial variability in landslide density, although lithological properties or slope distribution may also matter. Indeed lithological boundaries or a lack of steep slopes can sometimes explain spatial variability in landsliding, but not all of it (e.g., Fig. S9). In any case, it seems clear that to predict the spatial variability of landsliding, the rainfall spatiotemporal pattern is a primary requirement. The good correlation between storm total rainfall and peak landslide density is encouraging and suggests that, as most mountainous regions may have sparse instrumental coverage, the use of satellite measurements (Ushio et al., 2009; Huffman et al., 2007) or meso-scale meteorological models (e.g., Lafore et al., 1997) may be required to understand the spatial pattern of rainfall-induced landsliding.

A few nonlinear scalings between total landsliding and total rainfall have been reported at the catchment scale, but were derived from datasets not easily comparable to the one presented in this study (Reid, 1998; Chen et al., 2013; Marc et al., 2015). The details of this scaling are of importance in

order to understand the impact of extreme rainfall events and more generally which type of rainfall event contributes most to sediment transfer over long timescales (Reid and Page, 2003; Chen et al., 2015). We also found nonlinear scaling between R_t and total landslide area, but without a strong statistical difference between exponential or power-law functions. Exponential functions yield a minimum landsliding amount at low rainfall, that is not physically justified. This apparent contradiction may, however, be resolved by considering a rainfall threshold below which landsliding is null. The higher correlation between R_t and total volume is likely due to the fact that R_t correlates well with maximum landslide size ($R^2 = 0.8$ with whole landslides, $R^2 = 0.9$ and almost linear correlation with scar estimates, Fig. 10), with large landslides contributing most of the total volume and erosion. A correlation between R_t and large landslides may arise because landslide stability is determined by the ratio between pore pressure and the total normal stress on the slip plane, meaning that larger landslides that usually have deeper failure planes (Larsen et al., 2010) may only fail with greater precipitation amount. However, given that the trend between total rainfall and the landslide size distribution is much weaker, this correlation may also partly result from a sampling bias as the probability to draw large landslides increases with the total number of landslides. For now, our unreliable estimates of total landslide number do not allow for quantifying this effect.

In any case, several caveats should be taken with the preliminary scalings between total storm rainfall and total landsliding. First, the definition and limit of a single "storm" is not generally agreed in the meteorological community, because the atmospheric fluids suffer perturbations with scale interactions, and therefore with events not independent from each other. Ideally, future studies could categorize storms according to some space-time filtering and analyze the scalings with total landsliding for each storm category. Currently, our database is not sufficient for this. Second, linking total rainfall in a limited area and the total landsliding within the storm footprint implicitly suggests that storm rainfall is somewhat structured with internal correlations between peak rainfall, storm size, and the spatial pattern of rainfall intensity within the storm. This seems to be the case for large tropical storms (Jiang et al., 2008), but should be explored for a broader range of storm types. Orographic effects (e.g., Houze, 2012; Taniguchi et al., 2013), focussing high-intensity rainfall on topographic barriers, may also enhance such a correlation between local total rainfall and the broader pattern of rainfall and landsliding. Last, the scaling with rainfall may also be obscured by outliers due to processes not controlled by rainfall. For example, the inclusion of the very long run-out components in several inventories led to larger scattering for both power-law and exponential models and to favor the latter. Therefore, the proposed run-out correction seems essential for future studies. Another issue concerns the normalization of landscape parameters affecting the susceptibility to landsliding, such as hillslope steepness and mechanical strength (Schmidt and Montgomery, 1995; Parise and Jibson, 2000; Marc et al., 2016). Nevertheless, the proportion of flat or submerged land within the area of the most intense rainfall must limit the total landsliding, as it was certainly the case for MI2 or B08 (Fig. 1). Recent, widespread antecedent landsliding may also reduce subsequent susceptibility to rainfall triggered by removing the weak layer of soil or regolith on steep slopes. In the pre-event imagery, we did not see specific evidence of such a limitation, except maybe for J11, where abundant pre-event fresh landsliding were visible near 136,25° E/34,29° N and 135,9° E/34,20° and very few new landslides occurred. A more systematic evaluation of this effect may be important when quantitatively comparing the landslide and rainfall patterns. In any case, it is clear that further analysis of this database, possibly extended with additional landslide inventories, should be used by future studies to refine the scaling with rainfall and incorporate the effects of controlling parameters such as available topography, antecedent rainfall or regolith properties (e.g., strength and permeability).

4.2   Relation between rainfall and landslide properties

We found an increase in the proportion of large landslide scars with all storm metrics, but clearest with the total rainstorm (Fig. 7). This is consistent with the idea that large landslides require larger amounts of rainfall to be triggered (Van Asch et al., 1999), as discussed above and exemplified with the strong correlation between R_t and maximal landslide scar (Fig. S10). The large remaining scatter suggests that other differences between the inventories matter, such as differences in the mechanical properties of the substrate (e.g., Stark and Guzzetti, 2009). Indeed, broad lithological contrast exists between each event, and sometimes within an event (Fig. S9). The variability in extent and thickness of weak superficial layers (i.e., soils) between the different landscapes affected may also be important. Variations in slope distribution and relief are also wide between each case (Figs. 1, 4) and have also been reported to influence landslide size (Frattini and Crosta, 2013).

We note that the positive correlation between peak rainfall intensity and large landslide abundance is opposed to what could be expected, as more small landslides are expected for pulses of very intense rain leading to the occurrence of transient high pore-water pressure pulses at shallow depth (Iverson, 2000). Given that water retention and hydraulic conductivity may easily change by orders of magnitude between different environments, it may be needed to normalize intensity by the regolith hydraulic conductivity (Iverson, 2000) to understand its potential influence. For the moment, we consider that the correlation between D and $I3$ and the landslide size distribution exponents likely arises because of the correlation between these storm metrics and R_t (Fig. S4)

In any case, our results suggest that it is not only the landscape properties that set the landslide size distribution but also the trigger characteristics, as previously reported for earthquake-induced landslide size distributions (Marc et al., 2016). This means, for example, that the influence of forcing variability should be assessed and normalized before inverting landslide size distribution parameters to obtain regional variations of mechanical properties (e.g., Gallen et al., 2015).

In contrast, aspect ratio or run out did not correlate well with storm metrics and thus obscured any direct correlation between storm metrics and the decay exponents of whole landslide area. This underlines again the importance to isolate scar geometry to deconvolve processes driving landslide initiation and landslide run out. As for the landslide size distribution, landslide run out may likely be influenced by slope and relief distributions, as well as by hydrologic processes. The case of C99, with exceptional run out for most of its landslides is interpreted as the effect of low infiltration rates favoring large runoff generation (Godt and Coe, 2007). This may also explain the abundance of debris flow in other places (C15, J11) but cannot be verified without information on infiltration rate in these places to normalize the intensity variations. An alternative could be to study various storms occurring over the same region, and where infiltration rate or conductivity could be assumed constant; for example, with datasets from multiple typhoons in Taiwan (Chen et al., 2013).

Finally, we observed that most rainfall-induced landslide inventories sample the topographic slope distribution with a minor oversampling beyond the topographic modal slope (Fig. 4). This is in contrast with the case of earthquake-triggered landslides, where we systematically observe preferential landsliding on the steepest slopes (Parise and Jibson, 2000; Gorum et al., 2013, 2014; Fig. 4), quite similar to the case of C99. One-dimensional static force balance shows that the steepest slopes are the most unstable, and therefore the oversampling of steep slopes must be expected if the forcing (pore-water pressure or shaking) is randomly distributed across the whole topography. To obtain equal sampling or undersampling of steep slopes, the forcing intensity must be anti-correlated with slope gradient. Rainfall may be mostly independent of local slopes, but probably not the pore pressure rise that depends on the underground water circulation and thus topography. The pore pressure will thus crucially depend on vertical infiltration and drainage, but also on along-slope contributions. For example, under moderate intensity, but long rainfall, pore-water pressure will reach a higher level in concave, downslope areas (Montgomery and Dietrich, 1994) with a relatively large drainage area, and thus lower slope gradient (Montgomery, 2001). In such a view, landslide slope statistics would bear information on the type of rainfall, short and intense (relative to local permeability) for steep oversampling, while equal sampling and undersampling would relate to moderate and long rainfall. This framework might explain the preferential location on steep slopes observed for the very short duration C99 and possibly C15 (Fig. 4). However, the statistics of C15 are weak, and C99 strong oversampling may mainly relate to specific mass movement triggered by surface runoff such as rilling and the "fire hose effect" (cf., Godt and Coe, 2007). These processes also require high-intensity short-duration events, but also low surficial infiltration rate leading to overland flow able to mobilize relatively loose surface material. For other events, we analyzed the slope-gradient to drainage area relationship for a topography and landslide subset and did not find clear oversampling of high-drainage and gentle-gradient areas in the landslide distribution. A 30 m digital elevation model (DEM) may not be able to accurately resolve the fine-scale pattern of slope gradients and drainage areas on the hillsides, where landslides occur, but it may also suggest that the upslope drainage area is not the main explanation. For example, the subsurface drainage efficiency may also increase with slope gradient, thus making very steep areas less likely to develop large pore pressure and possibly explaining the preferential landsliding of slopes just above the modal slopes for almost all events, independent of rainfall properties. Hydro-mechanical modeling at the catchment scale (e.g., von Ruette et al., 2013), applied on several of our dataset may be the only way to test between these different hypotheses. Further constraints on the processes controlling rainfall-induced landslides may also be achieved through a discussion in terms of relative distance from ridge and river

(cf, Meunier et al., 2008), as intense and brief storms should yield uniformly distributed landsliding, in contrast to longer, less-intense storms favoring near-river slides.

5 Conclusions

We present landslide inventories (comprising from a few hundreds to more than 15 000 polygons) associated with eight triggering rain storms from Asia, South America, and North America. We hypothesize that these datasets constitute a global database of rainfall-induced landslides, which allows for studying a number of landslide metrics and their relations to rainfall and landscape properties. Indeed, although spanning a large range of landscape settings, whether in terms of topography, climate, or vegetation, the magnitude of landsliding scales nonlinearly with the magnitude of the storm, here quantified with estimates of the total rainfall. A preliminary analysis indicates that these correlations hold when total rainfall is normalized by long-term monthly rainfall, and how to normalize the storm rainfall by the regional climate should be further investigated. We also found that the correlation between landsliding and rainfall is higher when considering landslide scar estimates obtained through a normalization of landslide run out, as the run-out distribution does not clearly correlate to rainfall. Therefore, after removing the run-out contribution (i.e., focussing on scars) we also find that the landslide size distribution decay exponent seems to be partly controlled by total rainfall, with a greater proportion of large landslides for larger total precipitation. This implies that variations in landslide size distribution cannot be directly interpreted as variations in landscape properties. For total landsliding and maximum local landslide density, power-law scaling based on total rainfall explains 74 % (87 % for total volume) and 76 % (95 % for volume density) of the variance, respectively. Adding a number of other storm events as well as integrating other rainfall forcing parameters or landscape susceptibility properties has, therefore, the potential to yield robust prediction on the magnitude of rainfall-induced landsliding. Finally, we observe that compared to earthquakes, storms tend to trigger landslides that only slightly oversample the topographic slope distribution, possibly due to faster drainage on steep slopes or to underground water accumulation on high-drainage–low-gradient portions of the hillslope. This may bring new, although less straightforward, implications for the difference in resulting topography of bedrock landscape dominated by rainfall-induced or earthquake-induced landslides (Densmore and Hovius, 2000). Although preliminary these insights and scalings show the value of systematically mapping a large sample of the landslides that can be related to a single storm and we identified a number of recent storm events where such a type of inventory could be produced. Although not thoroughly investigated here, a landslide density spatial pattern is likely strongly related to the spatiotemporal pattern of rainfall, and constraining the quantitative links between the two is another challenge that may be addressed with some of the inventories presented here. More generally, the database presented here may also serve as a benchmark for developing and comparing rainfall-induced landslide models, whether empirical, semi- or fully deterministic. These future developments are important challenges in order to understand the natural hazards posed by rainfall-induced landslides as well as their specific implication for the erosion and topographic evolution of landscapes in different climatic settings.

Data availability. All imagery used to map the landslides are available in public repositories except the images from FORMOSAT-2, GeoEye-1, and aerial photos used for the events in Colorado, Micronesia, and Japan. For the two former, landslide maps are directly available from Godt and Coe (2003) and Harp et al. (2004). For Japan and Brazil, extensive high-resolution imagery is available over most of the areas of interest in Google Earth. The new digitized landslide inventories are available upon request. The hourly rainfall data used in Taiwan and Japan are available in the Supplement, while rainfall data for the other cases are available in the literature. The authors have retrieved the GSMaP_MVK V04 products (Kubota et al., 2006; Ushio et al., 2009) from the "JAXA Global Rainfall Watch" website (https://sharaku.eorc.jaxa.jp/GSMaP/, last access: 29 September 2018). The SRTM-30 m DEM data (NASA JPL, 2013) were retrieved from the online Global Data Explorer, courtesy of the NASA EOSDIS Land Processes Distributed Active Archive Center (LP DAAC), USGS/Earth Resources Observation and Science (EROS) Center, Sioux Falls, South Dakota, (https://gdex.cr.usgs.gov/gdex/, last access: 29 September 2018).

Author contributions. OM designed the study and performed all analyses with some input from AS, JPM, and MG. TU provided landslide and rainfall data for the J11 event. SHC provided imagery and landslide data for the TW9 event. OM wrote the manuscript with contributions from all co-authors.

Competing interests. The authors declare that they have no conflict of interest.

Acknowledgements. The authors are grateful to the thorough and constructive reviews from Dave Milledge and an anonymous reviewer that helped to improve and considerably clarify this manuscript. Odin Marc thanks Patrick Meunier for discussions on prediction interval and the statistical significance of probability distribution ratios. This work was carried with the support of the French space agency (CNES) through the project STREAM-LINE GLIDERS "SaTellite-based Rainfall Measurement and LandslIde detectioN for Global LandslIDE–Rainfall Scaling". Additional support by the open partial agreement "Major Hazards" of the Council of Europe through the project "Space data in Disaster Risk Reduction: using satellite precipitation data and hydrological information for leapfrogging landslide forecasting" was available. The authors acknowledge the Taiwan Typhoon and Flood Research Institute (National Applied Research Laboratories) and the Japan Meteorological Agency for providing rain gauge data.

References

Ardizzone, F., Basile, G., Cardinali, M., Casagli, N., Conte, S. D., Ventisette, C. D., Fiorucci, F., Garfagnoli, F., Gigli, G., Guzzetti, F., Iovine, G., Mondini, A. C., Moretti, S., Panebianco, M., Raspini, F., Reichenbach, P., Rossi, M., Tanteri, L., and Terranova, O.: Landslide inventory map for the Briga and the Giampilieri catchments, NE Sicily, Italy, J. Maps, 8, 176–180, https://doi.org/10.1080/17445647.2012.694271, 2012.

Arnone, E., Noto, L. V., Lepore, C., and Bras, R. L.: Physically-based and distributed approach to analyze rainfall-triggered landslides at watershed scale, Geomorphology, 133, 121–131, https://doi.org/10.1016/j.geomorph.2011.03.019, 2011.

Baum, R. L., Godt, J. W., and Savage, W. Z.: Estimating the timing and location of shallow rainfall-induced landslides using a model for transient, unsaturated infiltration, J. Geophys. Res.-Earth, 115, F03013, https://doi.org/10.1029/2009JF001321, 2010.

Blodgett, T. A. and Isacks, B. L.: Landslide Erosion Rate in the Eastern Cordillera of Northern Bolivia, Earth Interact., 11, 1–30, https://doi.org/10.1175/2007EI222.1, 2007.

Bucknam, R. C., Coe, J. A., Chavarria, M. M., Godt, J. W., Tarr, A. C., Bradley, L.-A., Rafferty, S. A., Hancock, D., Dart, R. L., and Johnson, M. L.: Landslides triggered by Hurricane Mitch in Guatemala – inventory and discussion, USGS Numbered Series 2001-443, https://doi.org/10.3133/ofr01443, 2001.

Caine, N.: The Rainfall Intensity: Duration Control of Shallow Landslides and Debris Flows, Geogr. Ann. A, 62, 23–27, https://doi.org/10.2307/520449, 1980.

Camargo, L. P.: Análise integrada no meio físico dos ribeirões Braço Serafim e Máximo com ênfase nas áreas de fragilidade estrutural, Luís Alves, (SC), PhD thesis, Universidade Federal de Santa Catarina, Florianopolis, available at: https://repositorio.ufsc.br/handle/123456789/157291 (last access: 2 October 2018), 2015.

Cannon, S. H., Haller, K. M., Ekstrom, I., Schweig, E. S., Devoli, G., Moore, D. W., Rafferty, S. A., and Tarr, A. C.: Landslide response to Hurricane Mitch rainfall in seven study areas in Nicaragua, USGS Numbered Series 2001-412-A, https://doi.org/10.3133/ofr01412A, 2001.

Cardinali, M., Galli, M., Guzzetti, F., Ardizzone, F., Reichenbach, P., and Bartoccini, P.: Rainfall induced landslides in December 2004 in south-western Umbria, central Italy: types, extent, damage and risk assessment, Nat. Hazards Earth Syst. Sci., 6, 237–260, https://doi.org/10.5194/nhess-6-237-2006, 2006.

Chang, K.-T., Chiang, S.-H., Chen, Y.-C., and Mondini, A. C.: Modeling the spatial occurrence of shallow landslides triggered by typhoons, Geomorphology, 208, 137–148, https://doi.org/10.1016/j.geomorph.2013.11.020, 2014.

Chen, Y.-C., Chang, K.-T., Chiu, Y.-J., Lau, S.-M., and Lee, H.-Y.: Quantifying rainfall controls on catchment-scale landslide erosion in Taiwan, Earth Surf. Proc. Land., 38, 372–382, https://doi.org/10.1002/esp.3284, 2013.

Chen, Y.-C., Chang, K.-T., Lee, H.-Y., and Chiang, S.-H.: Average landslide erosion rate at the watershed scale in southern Taiwan estimated from magnitude and frequency of rainfall, Geomorphology, 228, 756–764, https://doi.org/10.1016/j.geomorph.2014.07.022, 2015.

Clark, K. E., West, A. J., Hilton, R. G., Asner, G. P., Quesada, C. A., Silman, M. R., Saatchi, S. S., Farfan-Rios, W., Martin, R. E., Horwath, A. B., Halladay, K., New, M., and Malhi, Y.: Storm-triggered landslides in the Peruvian Andes and implications for topography, carbon cycles, and biodiversity, Earth Surf. Dynam., 4, 47–70, https://doi.org/10.5194/esurf-4-47-2016, 2016.

Crone, A. J., Baum, R. L., Lidke, D. J., Sather, D. N., Bradley, L.-A., and Tarr, A. C.: Landslides induced by Hurricane Mitch in El Salvador – an inventory and descriptions of selected features, USGS Numbered Series 2001-444, https://doi.org/10.3133/ofr01444, 2001.

Dahal, R. K. and Hasegawa, S.: Representative rainfall thresholds for landslides in the Nepal Himalaya, Geomorphology, 100, 429–443, https://doi.org/10.1016/j.geomorph.2008.01.014, 2008.

Densmore, A. L. and Hovius, N.: Topographic fingerprints of bedrock landslides, Geology, 28, 371–374, https://doi.org/10.1130/0091-7613(2000)28<371:TFOBL>;2.0.CO;2, 2000.

Domej, G., Bourdeau, C., and Lenti, L.: Mean Landslide Geometries Inferred from a Global Database of Earthquake- and Non-Earthquake-Triggered Landslides, Italian Journal of Engineering Geology and Environment, 87–107, https://doi.org/10.4408/IJEGE.2017-02.O-05, 2017.

Farr, T. G. and Kobrick, M.: Shuttle radar topography mission produces a wealth of data, Eos, 81, 583–585, https://doi.org/10.1029/EO081i048p00583, 2000.

Farr, T. G., Rosen, P. A., Caro, E., Crippen, R., Duren, R., Hensley, S., Kobrick, M., Paller, M., Rodriguez, E., Roth, L., Seal, D., Shaffer, S., Shimada, J., Umland, J., Werner, M., Oskin, M., Burbank, D., and Alsdorf, D. E.: The shuttle radar topography mission, Rev. Geophys., 45, n RG2004, https://doi.org/10.1029/2005RG000183, 2007.

Frattini, P. and Crosta, G. B.: The role of material properties and landscape morphology on landslide size distributions, Earth Planet. Sc. Lett., 361, 310–319, https://doi.org/10.1016/j.epsl.2012.10.029, 2013.

Gabet, E. J., Burbank, D. W., Putkonen, J. K., Pratt-Sitaula, B. A., and Ojha, T.: Rainfall thresholds for landsliding in the Himalayas of Nepal, Geomorphology, 63, 131–143, https://doi.org/10.1016/j.geomorph.2004.03.011, 2004.

Gallen, S. F., Clark, M. K., and Godt, J. W.: Coseismic landslides reveal near-surface rock strength in a high-relief, tectonically active setting, Geology, 43, 11–14, https://doi.org/10.1130/G36080.1, 2015.

Gariano, S. L. and Guzzetti, F.: Landslides in a changing climate, Earth-Sci. Rev., 162, 227–252, https://doi.org/10.1016/j.earscirev.2016.08.011, 2016.

Godt, J. W. and Coe, J. A.: Map showing alpine debris flows triggered by a July 28, 1999 thunderstorm in the central Front Range of Colorado, USGS Open-File Report, https://doi.org/10.3133/ofr0350, 2003.

Godt, J. W. and Coe, J. A.: Alpine debris flows triggered by a 28 July 1999 thunderstorm in the central Front Range, Colorado, Geomorphology, 84, 80–97, https://doi.org/10.1016/j.geomorph.2006.07.009, 2007.

Godt, J. W., Baum, R. L., and Chleborad, A. F.: Rainfall characteristics for shallow landsliding in Seattle, Washington, USA, Earth Surf. Proc. Land., 31, 97–110, https://doi.org/10.1002/esp.1237, 2006.

Gorum, T., van Westen, C. J., Korup, O., van der Meijde, M., Fan, X., and van der Meer, F. D.: Complex rupture mechanism and topography control symmetry of mass-wasting pattern, 2010 Haiti earthquake, Geomorphology, 184, 127–138, https://doi.org/10.1016/j.geomorph.2012.11.027, 2013.

Gorum, T., Korup, O., van Westen, C. J., van der Meijde, M., Xu, C., and van der Meer, F. D.: Why so few? Landslides triggered by the 2002 Denali earthquake, Alaska, Quaternary Sci. Rev., 95, 80–94, https://doi.org/10.1016/j.quascirev.2014.04.032, 2014.

Guzzetti, F., Cardinali, M., Reichenbach, P., Cipolla, F., Sebastiani, C., Galli, M., and Salvati, P.: Landslides triggered by the 23 November 2000 rainfall event in the Imperia Province, Western Liguria, Italy, Eng. Geol., 73, 229–245, https://doi.org/10.1016/j.enggeo.2004.01.006, 2004.

Guzzetti, F., Peruccacci, S., Rossi, M., and Stark, C. P.: The rainfall intensity-duration control of shallow landslides and debris flows: an update, Landslides, 5, 3–17, https://doi.org/10.1007/s10346-007-0112-1, 2008.

Harp, E. and Jibson, R.: Landslides triggered by the 1994 Northridge, California, earthquake, B. Seismol. Soc. Am., 86, S319–S332, 1996.

Harp, E. L., Hagaman, K. W., Held, M. D., and McKenna, J. P.: Digital inventory of landslides and related deposits in Honduras triggered by Hurricane Mitch, USGS Numbered Series 2002-61, U.S. Geological Survey, Reston, VA, https://doi.org/10.3133/ofr0261, 2002.

Harp, E. L., Reid, M. E., and Michael, J. A.: Hazard analysis of landslides triggered by Typhoon Chata'an on July 2, 2002, in Chuuk State, Federated States of Micronesia, USGS Numbered Series 2004-1348, https://doi.org/10.3133/ofr20041348, 2004.

Houze, R. A.: Orographic effects on precipitating clouds, Rev. Geophys., 50, RG1001, https://doi.org/10.1029/2011RG000365, 2012.

Hovius, N., Stark, C. P., and Allen, P. A.: Sediment flux from a mountain belt derived by landslide mapping, Geology, 25, 231–234, https://doi.org/10.1130/0091-7613(1997)025<0231:SFFAMB>2.3.CO;2, 1997.

Huffman, G. J., Bolvin, D. T., Nelkin, E. J., Wolff, D. B., Adler, R. F., Gu, G., Hong, Y., Bowman, K. P., and Stocker, E. F.: The TRMM Multisatellite Precipitation Analysis (TMPA): Quasi-Global, Multiyear, Combined-Sensor Precipitation Estimates at Fine Scales, J. Hydrometeorol., 8, 38–55, https://doi.org/10.1175/JHM560.1, 2007.

Iverson, R. M.: Landslide triggering by rain infiltration, Water Resour. Res., 36, 1897–1910, https://doi.org/10.1029/2000WR900090, 2000.

Jiang, H., Halverson, J. B., Simpson, J., and Zipser, E. J.: Hurricane "Rainfall Potential" Derived from Satellite Observations

Aids Overland Rainfall Prediction, J. Appl. Meteorol. Clim., 47, 944–959, https://doi.org/10.1175/2007JAMC1619.1, 2008.

Jibson, R. W., Harp, E. L., and Michael, J. A.: A method for producing digital probabilistic seismic landslide hazard maps, Eng. Geol., 58, 271–289, https://doi.org/10.1016/S0013-7952(00)00039-9, 2000.

Katz, O., Morgan, J. K., Aharonov, E., and Dugan, B.: Controls on the size and geometry of landslides: Insights from discrete element numerical simulations, Geomorphology, 220, 104–113, https://doi.org/10.1016/j.geomorph.2014.05.021, 2014.

Keefer, D. K., Wilson, R. C., Mark, R. K., Brabb, E. E., Brown, W. M., Ellen, S. D., Harp, E. L., Wieczorek, G. F., Alger, C. S., and Zatkin, R. S.: Real-Time Landslide Warning During Heavy Rainfall, Science, 238, 921–925, https://doi.org/10.1126/science.238.4829.921, 1987.

Kirschbaum, D., Adler, R., Adler, D., Peters-Lidard, C., and Huffman, G.: Global Distribution of Extreme Precipitation and High-Impact Landslides in 2010 Relative to Previous Years, J. Hydrometeorol., 13, 1536–1551, https://doi.org/10.1175/JHM-D-12-02.1, 2012.

Kirschbaum, D. B., Adler, R., Hong, Y., Hill, S., and Lerner-Lam, A.: A global landslide catalog for hazard applications: method, results, and limitations, Nat. Hazards, 52, 561–575, https://doi.org/10.1007/s11069-009-9401-4, 2009.

Kubota, T., Shige, S., Hashizume, H., Ushio, T., Aonashi, K., Kachi, M., and Okamoto, K.: Global Precipitation Map using Satelliteborne Microwave Radiometers by the GSMaP Project: Production and Validation, in: 2006 IEEE MicroRad, 290–295, https://doi.org/10.1109/MICRAD.2006.1677106, 2006.

Lafore, J. P., Stein, J., Asencio, N., Bougeault, P., Ducrocq, V., Duron, J., Fischer, C., Héreil, P., Mascart, P., Masson, V., Pinty, J. P., Redelsperger, J. L., Richard, E., and Arellano, J. V.-G. D.: The Meso-NH Atmospheric Simulation System. Part I: adiabatic formulation and control simulations, Ann. Geophys., 16, 90–109, https://doi.org/10.1007/s00585-997-0090-6, 1997.

Larsen, I., Montgomery, D., and Korup, O.: Landslide erosion controlled by hillslope material, Nat. Geosci., 3, 247–251, 2010.

Lehmann, P. and Or, D.: Hydromechanical triggering of landslides: From progressive local failures to mass release, Water Resour. Res., 48, W03535, https://doi.org/10.1029/2011WR010947, 2012.

Liao, H.-W. and Lee, C.: Landslides triggered by Chi-Chi earthquake, in: Proceedings of the 21st Asian conference on remote sensing, 1, 383–388, 2000.

Malamud, B. D., Turcotte, D. L., Guzzetti, F., and Reichenbach, P.: Landslide inventories and their statistical properties, Earth Surf. Proc. Land., 29, 687–711, https://doi.org/10.1002/esp.1064, 2004.

Marc, O. and Hovius, N.: Amalgamation in landslide maps: effects and automatic detection, Nat. Hazards Earth Syst. Sci., 15, 723–733, https://doi.org/10.5194/nhess-15-723-2015, 2015.

Marc, O., Hovius, N., Meunier, P., Uchida, T., and Hayashi, S.: Transient changes of landslide rates after earthquakes, Geology, 43, 883–886, https://doi.org/10.1130/G36961.1, 2015.

Marc, O., Hovius, N., Meunier, P., Gorum, T., and Uchida, T.: A seismologically consistent expression for the total area and volume of earthquake-triggered landsliding, J. Geophys. Res.-Earth, 121, 640–663, https://doi.org/10.1002/2015JF003732, 2016.

Marc, O., Meunier, P., and Hovius, N.: Prediction of the area affected by earthquake-induced landsliding based on seismological parameters, Nat. Hazards Earth Syst. Sci., 17, 1159–1175, https://doi.org/10.5194/nhess-17-1159-2017, 2017.

Meunier, P., Hovius, N., and Haines, A. J.: Regional patterns of earthquake-triggered landslides and their relation to ground motion, Geophys. Res. Lett., 34, L20408, https://doi.org/10.1029/2007GL031337, 2007.

Meunier, P., Hovius, N., and Haines, J. A.: Topographic site effects and the location of earthquake induced landslides, Earth Planet. Sc. Lett., 275, 221–232, https://doi.org/10.1016/j.epsl.2008.07.020, 2008.

Meunier, P., Uchida, T., and Hovius, N.: Landslide patterns reveal the sources of large earthquakes, Earth Planet. Sc. Lett., 363, 27–33, https://doi.org/10.1016/j.epsl.2012.12.018, 2013.

Milledge, D. G., Bellugi, D., McKean, J. A., Densmore, A. L., and Dietrich, W. E.: A multidimensional stability model for predicting shallow landslide size and shape across landscapes, J. Geophys. Res.-Earth, 119, 2014JF003135, https://doi.org/10.1002/2014JF003135, 2014.

Mondini, A. C.: Measures of Spatial Autocorrelation Changes in Multitemporal SAR Images for Event Landslides Detection, Remote Sensing, 9, 554, https://doi.org/10.3390/rs9060554, 2017.

Montgomery, D. R.: Slope Distributions, Threshold Hillslopes, and Steady-state Topography, Am. J. Sci., 301, 432–454, https://doi.org/10.2475/ajs.301.4-5.432, 2001.

Montgomery, D. R. and Dietrich, W. E.: A physically based model for the topographic control on shallow landsliding, Water Resour. Res., 30, 1153–1171, https://doi.org/10.1029/93WR02979, 1994.

NASA JPL: NASA Shuttle Radar Topography Mission Global 1 arc second [Data set], NASA EOSDIS Land Processes DAAC, https://doi.org/10.5067/MEaSUREs/SRTM/SRTMGL1.003, 2013.

Netto, A. L. C., Sato, A. M., Avelar, A. D. S., Vianna, L. G. G., Araújo, I. S., Ferreira, D. L. C., Lima, P. H., Silva, A. P. A., and Silva, R. P.: January 2011: The Extreme Landslide Disaster in Brazil, in: Landslide Science and Practice, Springer, Berlin, Heidelberg, 377–384, https://doi.org/10.1007/978-3-642-31319-6_51, 2013.

Nowicki, M. A., Wald, D. J., Hamburger, M. W., Hearne, M., and Thompson, E. M.: Development of a globally applicable model for near real-time prediction of seismically induced landslides, Eng. Geol., 173, 54–65, https://doi.org/10.1016/j.enggeo.2014.02.002, 2014.

Ono, K., Kazama, S., and Ekkawatpanit, C.: Assessment of rainfall-induced shallow landslides in Phetchabun and Krabi provinces, Thailand, Natural Hazards, 74, 2089–2107, https://doi.org/10.1007/s11069-014-1292-3, 2014.

Parise, M. and Jibson, R. W.: A seismic landslide susceptibility rating of geologic units based on analysis of characteristics of landslides triggered by the 17 January, 1994 Northridge, California earthquake, Eng. Geol., 58, 251–270, https://doi.org/10.1016/S0013-7952(00)00038-7, 2000.

Petley, D.: Global patterns of loss of life from landslides, Geology, 40, 927–930, https://doi.org/10.1130/G33217.1, 2012.

Pozzobon, M.: Análise da suscetibilidade a deslizamentos no município de blumenau/sc: uma abordagem probabilística atravées da Aplicação da Técnica pesos de evidência, PhD thesis, Universidade federal do paranã, Curitiba, available at: http://www.floresta.ufpr.br/pos-graduacao/defesas/pdf_dr/2013/t342_0370-D.pdf (last access: 2 October 2018), 2013.

Reid, L. M.: Calculation of average landslide frequency using climatic records, Water Resour. Res., 34, 869–877, https://doi.org/10.1029/97WR02682, 1998.

Reid, L. M. and Page, M. J.: Magnitude and frequency of landsliding in a large New Zealand catchment, Geomorphology, 49, 71–88, https://doi.org/10.1016/S0169-555X(02)00164-2, 2003.

Saito, H. and Matsuyama, H.: Catastrophic Landslide Disasters Triggered by Record-Breaking Rainfall in Japan: Their Accurate Detection with Normalized Soil Water Index in the Kii Peninsula for the Year 2011, Sola, 8, 81–84, https://doi.org/10.2151/sola.2012-021, 2012.

Saito, H., Korup, O., Uchida, T., Hayashi, S., and Oguchi, T.: Rainfall conditions, typhoon frequency, and contemporary landslide erosion in Japan, Geology, 42, 999–1002, https://doi.org/10.1130/G35680.1, 2014.

Schmidt, K. M. and Montgomery, D. R.: Limits to Relief, Science, 270, 617–620, https://doi.org/10.1126/science.270.5236.617, 1995.

Stark, C. P. and Guzzetti, F.: Landslide rupture and the probability distribution of mobilized debris volumes, J. Geophys. Res.-Earth, 114, F00A02, https://doi.org/10.1029/2008JF001008, 2009.

Stark, C. P. and Hovius, N.: The characterization of landslide size distributions, Geophys. Res. Lett., 28, 1091–1094, https://doi.org/10.1029/2000GL008527, 2001.

Stumpf, A., Lachiche, N., Malet, J.-P., Kerle, N., and Puissant, A.: Active Learning in the Spatial Domain for Remote Sensing Image Classification, IEEE T. Geosci. Remote, 52, 2492–2507, https://doi.org/10.1109/TGRS.2013.2262052, 2014.

Taniguchi, A., Shige, S., Yamamoto, M. K., Mega, T., Kida, S., Kubota, T., Kachi, M., Ushio, T., and Aonashi, K.: Improvement of High-Resolution Satellite Rainfall Product for Typhoon Morakot (2009) over Taiwan, J. Hydrometeorol., 14, 1859–1871, https://doi.org/10.1175/JHM-D-13-047.1, 2013.

Tanyaş, H., van Westen, C. J., Allstadt, K. E., Anna Nowicki Jessee, M., Görüm, T., Jibson, R. W., Godt, J. W., Sato, H. P., Schmitt, R. G., Marc, O., and Hovius, N.: Presentation and Analysis of a Worldwide Database of Earthquake-Induced Landslide Inventories, J. Geophys. Res.-Earth, 122, 2017JF004236, https://doi.org/10.1002/2017JF004236, 2017.

Ushio, T., Sasashige, K., Kubota, T., Shige, S., Okamoto, K., Aonashi, K., Inoue, T., Takahashi, N., Iguchi, T., Kachi, M., Oki, R., Morimoto, T., and Kawasaki, Z.-I.: A Kalman Filter Approach to the Global Satellite Mapping of Precipitation (GSMaP) from Combined Passive Microwave and Infrared Radiometric Data, J. Meteorol. Soc. Jpn., 87A, 137–151, https://doi.org/10.2151/jmsj.87A.137, 2009.

Uchida, T., Tamur, K., and Akiyama, K.: The role of grid cell size, flow routing algolithm and spatial variability of soil depth on shallow landslide prediction, in: 5th International Conference on Debris-Flow Hazards Mitigation: Mechanics, Prediction and Assessment, 149–157, Italian journal of engineering geology and environment, Padua, Italy, 2011.

Uchida, T., Sato, T., Mizuno, M., and Okamoto, A.: The role of rainfall magnitude on landslide characteristics triggered by Typhoon Tales, 2011, Civil Engineering Journal, 54, 10–13, 2012 (in Japanese).

Uchida, T., Okamoto, A., Kanbara, J. I., and Kuramoto, K.: RAINFALL THRESHOLDS FOR DEEP-SEATED RAPID LANDSLIDES, in: International Conference on Vajont 1963-2013/Proceedings – Thoughts and analyses after 50 years since the catastrophic landslide, ITAlian journal of engineering geology and environment, Padua, Italy, 211–217, 2013.

Van Asch, T. W. J., Buma, J., and Van Beek, L. P. H.: A view on some hydrological triggering systems in landslides, Geomorphology, 30, 25–32, https://doi.org/10.1016/S0169-555X(99)00042-2, 1999.

von Ruette, J., Lehmann, P., and Or, D.: Rainfall-triggered shallow landslides at catchment scale: Threshold mechanics-based modeling for abruptness and localization, Water Resour. Res., 49, 6266–6285, https://doi.org/10.1002/wrcr.20418, 2013.

Wilson, R. C. and Wieczorek, G. F.: Rainfall Thresholds for the Initiation of Debris Flows at La Honda, California, Environ. Eng. Geosci., I, 11–27, https://doi.org/10.2113/gseegeosci.I.1.11, 1995.

Yagi, H., Sato, G., Higaki, D., Yamamoto, M., and Yamasaki, T.: Distribution and characteristics of landslides induced by the Iwate-Miyagi Nairiku Earthquake in 2008 in Tohoku District, Northeast Japan, Landslides, 6, 335–344, 2009.

6

Estimates of late Cenozoic climate change relevant to Earth surface processes in tectonically active orogens

Sebastian G. Mutz[1], **Todd A. Ehlers**[1], **Martin Werner**[2], **Gerrit Lohmann**[2], **Christian Stepanek**[2], **and Jingmin Li**[1,a]

[1]Department of Geosciences, University Tübingen, 72074 Tübingen, Germany
[2]Department of Paleoclimate Dynamics, Alfred Wegener Institute, Helmholtz Centre for Polar and Marine Research, 27570 Bremerhaven, Germany
[a]now at: Institute for Geography and Geology, University of Würzburg, Würzburg, 97074 Germany

Correspondence: Sebastian G. Mutz (sebastian.mutz@uni-tuebingen.de)

Abstract. The denudation history of active orogens is often interpreted in the context of modern climate gradients. Here we address the validity of this approach and ask what are the spatial and temporal variations in palaeoclimate for a latitudinally diverse range of active orogens? We do this using high-resolution (T159, ca. 80×80 km at the Equator) palaeoclimate simulations from the ECHAM5 global atmospheric general circulation model and a statistical cluster analysis of climate over different orogens (Andes, Himalayas, SE Alaska, Pacific NW USA). Time periods and boundary conditions considered include the Pliocene (PLIO, ~ 3 Ma), the Last Glacial Maximum (LGM, ~ 21 ka), mid-Holocene (MH, ~ 6 ka), and pre-industrial (PI, reference year 1850). The regional simulated climates of each orogen are described by means of cluster analyses based on the variability in precipitation, 2 m air temperature, the intra-annual amplitude of these values, and monsoonal wind speeds where appropriate. Results indicate the largest differences in the PI climate existed for the LGM and PLIO climates in the form of widespread cooling and reduced precipitation in the LGM and warming and enhanced precipitation during the PLIO. The LGM climate shows the largest deviation in annual precipitation from the PI climate and shows enhanced precipitation in the temperate Andes and coastal regions for both SE Alaska and the US Pacific Northwest. Furthermore, LGM precipitation is reduced in the western Himalayas and enhanced in the eastern Himalayas, resulting in a shift of the wettest regional climates eastward along the orogen. The cluster-analysis results also suggest more climatic variability across latitudes east of the Andes in the PLIO climate than in other time slice experiments conducted here. Taken together, these results highlight significant changes in late Cenozoic regional climatology over the last ~ 3 Myr. Comparison of simulated climate with proxy-based reconstructions for the MH and LGM reveal satisfactory to good performance of the model in reproducing precipitation changes, although in some cases discrepancies between neighbouring proxy observations highlight contradictions between proxy observations themselves. Finally, we document regions where the largest magnitudes of late Cenozoic changes in precipitation and temperature occur and offer the highest potential for future observational studies that quantify the impact of climate change on denudation and weathering rates.

1 Introduction

Interpretation of orogen denudation histories in the context of climate and tectonic interactions is often hampered by a paucity of terrestrial palaeoclimate proxy data needed to reconstruct spatial variations in palaeoclimate. While it is self-evident that palaeoclimate changes could influence palaeodenudation rates, it is not always self-evident what the magnitude of climate change over different geologic timescales is, or what geographic locations offer the greatest potential to investigate palaeoclimate impacts on denudation. Palaeoclimate reconstructions are particularly beneficial when denudation rates are determined using geo- and thermo-chronology techniques that integrate over timescales of 10^3–10^{6+} years (e.g. cosmogenic radionuclides or low-temperature thermochronology; e.g. Kirchner et al., 2001; Schaller et al., 2002; Bookhagen et al., 2005; Moon et al., 2011; Thiede and Ehlers, 2013; Lease and Ehlers, 2013). However, few studies using denudation rate determination methods that integrate over longer timescales have access to information about past climate conditions that could influence these palaeodenudation rates. Palaeoclimate modelling offers an alternative approach to sparsely available proxy data for understanding the spatial and temporal variations in precipitation and temperature in response to changes in orography (e.g. Takahashi and Battisti, 2007a, b; Insel et al., 2010; Feng et al., 2013) and global climate change events (e.g. Salzmann et al., 2011; Jeffery et al., 2013). In this study, we characterise the climate at different times in the late Cenozoic and the magnitude of climate change for a range of active orogens. Our emphasis is on identifying changes in climate parameters relevant to weathering and catchment denudation to illustrate the potential importance of various global climate change events on surface processes.

Previous studies of orogen-scale climate change provide insight into how different tectonic or global climate change events influence regional climate change. For example, sensitivity experiments demonstrated significant changes in regional and global climate in response to landmass distribution and topography of the Andes, including changes in moisture transport, the north–south asymmetry of the Intertropical Convergence Zone (e.g. Takahashi and Battisti, 2007a; Insel et al., 2010), and (tropical) precipitation (Maroon et al., 2015, 2016). Another example is the regional and global climate changes induced by the Tibetan Plateau surface uplift due to its role as a physical obstacle to circulation (Raymo and Ruddiman, 1992; Kutzbach et al., 1993; Thomas, 1997; Bohner, 2006; Molnar et al., 2010; Boos and Kuang, 2010). The role of tectonic uplift in long-term regional and global climate change remains a focus of research and continues to be assessed with geologic datasets (e.g. Dettman et al., 2003; Caves, 2017; Kent-Corson et al., 2006; Lechler et al., 2013; Lechler and Niemi, 2011; Licht et al., 2017; Methner et al., 2016; Mulch et al., 2015, 2008; Pingel et al., 2016) and climate modelling (e.g. Kutzbach et al., 1989; Kutzbach

et al., 1993; Zhisheng, 2001; Bohner, 2006; Takahashi and Battisti, 2007a; Ehlers and Poulsen, 2009; Insel et al., 2010; Boos and Kuang, 2010). Conversely, climate influences tectonic processes through erosion (e.g. Molnar and England, 1990; Whipple et al., 1999; Montgomery et al., 2001; Willett et al., 2006; Whipple, 2009). Quaternary climate change between glacial and interglacial conditions (e.g. Braconnot et al., 2007; Harrison et al., 2013) resulted in not only the growth and decay of glaciers and glacial erosion (e.g. Yanites and Ehlers, 2012; Herman et al., 2013; Valla et al., 2011) but also global changes in precipitation and temperature (e.g. Otto-Bliesner et al., 2006; Li et al., 2017) that could influence catchment denudation in non-glaciated environments (e.g. Schaller and Ehlers, 2006; Glotzbach et al., 2013; Marshall et al., 2015). These dynamics highlight the importance of investigating how much climate has changed over orogens that are the focus of studies of climate–tectonic interactions and their impact on erosion.

Despite recognition by previous studies that climate change events relevant to orogen denudation are prevalent throughout the late Cenozoic, few studies have critically evaluated how different climate change events may, or may not, have affected the orogen climatology, weathering, and erosion. Furthermore, recent controversy exists concerning the spatial and temporal scales over which geologic and geochemical observations can record climate-driven changes in weathering and erosion (e.g. Whipple, 2009; von Blanckenburg et al., 2015; Braun, 2016). For example, the previous studies highlight that although palaeoclimate impacts on denudation rates are evident in some regions and measurable with some approaches, they are not always present (or detectable) and the spatial and temporal scale of climate change influences our ability to record climate-sensitive denudation histories. This study contributes to our understanding of the interactions among climate, weathering, and erosion by bridging the gap between the palaeoclimatology and surface process communities by documenting the magnitude and distribution of climate change over tectonically active orogens.

Motivated by the need to better understand climate impacts on Earth surface processes, especially the denudation of orogens, we model palaeoclimate for four time slices in the late Cenozoic, use descriptive statistics to identify the extent of different regional climates, quantify changes in temperature and precipitation, and discuss the potential impacts on fluvial and/or hillslope erosion. In this study, we employ the ECHAM5 global atmospheric general circulation model (GCM) and document climate and climate change for time slices ranging between the Pliocene (PLIO, ~ 3 Ma) to pre-industrial (PI) times for the St Elias Mountains of southeastern Alaska, the US Pacific Northwest (Olympic and Cascade ranges), western South America (Andes), and South Asia (including parts of central and East Asia). Our approach is twofold and includes

1. an empirical characterisation of palaeoclimates in these regions based on the covariance and spatial clustering of monthly precipitation and temperature, the monthly change in precipitation and temperature magnitude, and wind speeds where appropriate.

2. identification of changes in annual mean precipitation and temperature in selected regions for four time periods: (PLIO, Last Glacial Maximum (LGM), the mid-Holocene (MH), and PI) and subsequent validation of the simulated precipitation changes for the MH and LGM.

Our focus is on documenting climate and climate change in different locations with the intent of informing past and on-going palaeodenudation studies of these regions. The results presented here also provide a means for future work to formulate testable hypotheses and investigations into whether or not regions of large palaeoclimate change produced a measurable signal in denudation rates or other Earth surface processes. More specifically, different aspects of the simulated palaeoclimate may be used as boundary conditions for vegetation and landscape evolution models, such as LPJ-GUESS and Landlab, to bridge the gap between climate change and quantitative estimates for Earth surface system responses. In this study, we intentionally refrain from applying predicted palaeoclimate changes to predict denudation rate changes. Such a prediction is beyond the scope of this study because a convincing (and meaningful) calculation of climate-driven transients in fluvial erosion (e.g. via the kinematic wave equation), variations in frost cracking intensity, or changes in hillslope sediment production and transport at the large regional scales considered here is not tractable within a single paper and instead is the focus of our ongoing work. Merited discussion of climatically induced changes in glacial erosion, as is important in the Cenozoic, is also beyond the scope of this study. Instead, our emphasis lies on providing and describing a consistently set-up GCM simulation framework for future investigations of Earth surface processes and identifying regions in which late Cenozoic climate changes potentially have a significant impact on fluvial and hillslope erosion.

2 Methods: climate modelling and cluster analyses for climate characterisation

2.1 ECHAM5 simulations

The global atmospheric GCM ECHAM5 (Roeckner et al., 2003) has been developed at the Max Planck Institute for Meteorology and is based on the spectral weather forecast model of the ECMWF (Simmons et al., 1989). In the context of palaeoclimate applications, the model has been used mostly at lower resolution (T31, ca. $3.75° \times 3.75°$; T63, ca. $1.9° \times 1.9°$ in the case of Feng et al., 2016, and T106 in the case of Li et al., 2017 and Feng and Poulsen, 2016). The

studies performed are not limited to the last millennium (e.g. Jungclaus et al., 2010) but also include research in the field of both warmer and colder climates, at orbital (e.g. Gong et al., 2013; Lohmann et al., 2013; Pfeiffer and Lohmann, 2016; X. Zhang et al., 2013, 2014; Wei and Lohmann, 2012) and tectonic timescales (e.g. Knorr et al., 2011; Stepanek and Lohmann, 2012), and under anthropogenic influence (Gierz et al., 2015).

Here, the ECHAM5 simulations were conducted at a T159 spatial resolution (horizontal grid size ca. $80\,km \times 80\,km$ at the Equator) with 31 vertical levels (between the surface and 10 hPa). This high model resolution is admittedly not required for all of the climatological questions investigated in this study, and it should be noted that the skill of GCMs in predicting orographic precipitation remains limited at this scale (e.g. Meehl et al., 2007). However, simulations were conducted at this resolution so that future work can apply the results in combination with different dynamical and statistical downscaling methods to quantify changes at large catchment to orogen scales. The output frequency is relatively high (1 day) to enhance the usefulness of our simulations as input for landscape evolution and other models that may benefit from daily input. The simulations were conducted for five different time periods: present-day (PD), PI, MH, LGM, and PLIO.

A PD simulation (not shown here) was used to establish confidence in the model performance before conducting palaeosimulations and has been compared with the following observation-based datasets: European Centre for Medium-Range Weather Forecasts (ECMWF) reanalyses (ERA40, Uppala et al., 2005), National Centers for Environmental Prediction and National Center for Atmospheric Research (NCEP/NCAR) reanalyses (Kalnay et al., 1996; Kistler et al., 2001), NCEP Regional Reanalysis (NARR; Mesinger et al., 2006), the Climate Research Unit (CRU) TS3.21 dataset (Harris et al., 2013), High Asia Refined Analysis (HAR30; Maussion et al., 2014), and the University of Delaware dataset (UDEL v3.01; Legates and Wilmott, 1990). (See Mutz et al., 2016, for a detailed comparison with a lower-resolution model).

The PI climate simulation is an ECHAM5 experiment with PI (reference year 1850) boundary conditions. Sea surface temperatures (SSTs) and sea ice concentration (SIC) are derived from transient coupled ocean–atmosphere simulations (Lorenz and Lohmann, 2004; Dietrich et al., 2013). Following Dietrich et al. (2013), greenhouse gas (GHG) concentrations (CO_2 : 280 ppm) are taken from ice-core-based reconstructions of CO_2 (Etheridge et al., 1996), CH_4 (Etheridge et al., 1998) and N_2O (Sowers et al., 2003). Sea surface boundary conditions for the MH originate from a transient, low-resolution, coupled atmosphere–ocean simulation of the MH (6 ka) (Wei and Lohmann, 2012; Lohmann et al., 2013), where the GHG concentrations (CO_2 : 280 ppm) are taken from ice core reconstructions of GHGs by Etheridge et al. (1996, 1998) and Sowers et al. (2003). GHG concentrations for the LGM (CO_2 : 185 ppm) have been prescribed following Otto-Bliesner et al. (2006). Orbital parameters for the MH and LGM are set according to Dietrich et al. (2013) and Otto-Bliesner et al. (2006), respectively. LGM land–sea distribution and ice sheet extent and thickness are set based on the PMIP III (Palaeoclimate Modelling Intercomparison Project, phase 3) guidelines (elaborated on by Abe-Ouchi et al., 2015). Following Schäfer-Neth and Paul (2003), SST and SIC for the LGM are based on GLAMAP (Sarnthein et al., 2003) and CLIMAP (CLIMAP project members, 1981) reconstructions for the Atlantic Ocean and Pacific and Indian oceans, respectively. Global MH and LGM vegetation is based on maps of plant functional types by the BIOME 6000 Palaeovegetation Mapping Project (Prentice et al., 2000; Harrison et al., 2001; Bigelow et al., 2003; Pickett et al., 2004) and model predictions by Arnold et al. (2009). Boundary conditions for the PLIO simulation, including GHG concentrations (CO_2 : 405), orbital parameters and surface conditions (SST, SIC, sea land mask, topography, and ice cover) are taken from the PRISM (Pliocene Research, Interpretation and Synoptic Mapping) project (Haywood et al., 2010; Sohl et al., 2009; Dowsett et al., 2010), specifically PRISM3D. The PLIO vegetation boundary condition was created by converting the PRISM vegetation reconstruction to the JSBACH plant functional types as described by Stepanek and Lohmann (2012), but the built-in land surface scheme was used.

SST reconstructions can be used as an interface between oceans and atmosphere (e.g. Li et al., 2017) instead of conducting the computationally more expensive fully coupled atmosphere–ocean GCM experiments. While the use of SST climatologies comes at the cost of capturing decadal-scale variability, and the results are ultimately biased towards the SST reconstructions the model is forced with; the simulated climate more quickly reaches an equilibrium state and the means of atmospheric variables used in this study do no change significantly after the relatively short spin-up period. The palaeoclimate simulations (PI, MH, LGM, PLIO) using ECHAM5 are therefore carried out for 17 model years, of which the first 2 years are used for model spinup. The monthly long-term averages (multi-year means for individual months) for precipitation, temperature, and precipitation and temperature amplitude, i.e. the mean difference between the hottest and coldest months, have been calculated from the following 15 model years for the analysis presented below.

For further comparison between the simulations, the investigated regions were subdivided (Fig. 1). Western South America was subdivided into four regions: parts of tropical South America (80–60° W, 23.5–5° S); temperate South America (80–60° W, 50–23.5° S); tropical Andes (80–60° W, 23.5–5° S; high-pass filtered), i.e. most of the Peruvian Andes, Bolivian Andes, and northernmost Chilean Andes; and temperate Andes (80–60° W, 50–23.5° S, high-pass filtered). South Asia was subdivided into three regions: tropical South Asia (40–120° E, 0–23.5° N), temperate South Asia

(40–120° E, 23.5–60° N), and high-altitude South Asia (40–120° E, 0–60° N; high-pass filtered).

Our approach of using a single GCM (ECHAM5) for our analysis is motivated by, and differs from, previous studies where inter-model variability exists from the use of different GCMs due to different parameterisations in each model. The variability in previous inter-model GCM comparisons exists despite the use of the same forcings (e.g. see results highlighted in IPCC AR5). Similarities identified between these palaeoclimate simulations conducted with different GCMs using similar boundary conditions can establish confidence in the models when in agreement with proxy reconstructions. However, differences identified in inter-model GCM comparisons highlight biases by all or specific GCMs, or reveal sensitivities to one changed parameter, such as model resolution. Given these limitations of GCM modelling, we present in this study a comparison of a suite of ECHAM5 simulations to proxy-based reconstructions (where possible) and, to a lesser degree, comment on general agreement or disagreement of our ECHAM5 results with other modelling studies. A detailed inter-model comparison of our results with other GCMs is beyond the scope of this study and better suited for a different study in a journal with a different focus and audience. Rather, by using the same GCM and identical resolution for the time slice experiments, we reduce the number of parameters (or model parameterisations) varying between simulations and thereby remove potential sources of error or uncertainty that would otherwise have to be considered when comparing output from different models with different parameterisations of processes, model resolution, and in some cases model forcings (boundary conditions). Nevertheless, the reader is advised to use these model results with the GCM's shortcoming and uncertainties in boundary condition reconstructions in mind. For example, precipitation results may require dynamical or statistical downscaling to increase accuracy where higher-resolution precipitation fields are required. Furthermore, readers are advised to familiarise themselves with the palaeogeography reconstruction initiatives and associated uncertainties. For example, while Pliocene ice sheet volume can be estimated, big uncertainties pertaining to their locations remain (Haywood et al., 2010).

2.2 Cluster analysis to document temporal and spatial changes in climatology

The aim of the clustering approach is to group climate model surface grid boxes together based on similarities in climate. Cluster analyses are statistical tools that allow elements (i) to be grouped by similarities in the elements' attributes. In this study, those elements are spatial units, the elements' attributes are values from different climatic variables, and the measure of similarity is given by a statistical distance. The four basic variables used as climatic attributes of these spatial elements are near-surface (2 m) air temperature, seasonal 2 m air temperature amplitude, precipitation rate, and seasonal

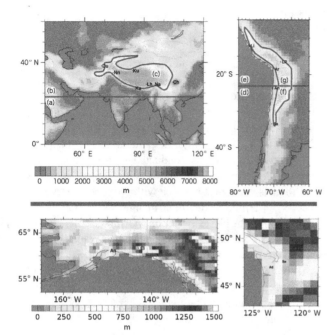

Figure 1. Topography for regions (a) tropical South Asia, (b) temperate South Asia, (c) high-altitude South Asia, (d) temperate South America, (e) tropical South America, (f) the temperate Andes, and (g) the tropical Andes, SE Alaska, and Cascadia.

precipitation rate amplitude. Since monsoonal winds are a dominant feature of the climate in the South Asia region, near-surface (10 m) speeds of u wind and v wind (zonal and meridional wind components, respectively) during the monsoon season (July) and outside the monsoon season (January) are included as additional variables in our analysis of that region. Similarly, u-wind and v-wind speeds during (January) and outside (July) the monsoon season in South America are added to the list of considered variables to take into account the South American Monsoon System (SASM) in the cluster analysis for this region. The long-term monthly means of those variables are used in a hierarchical clustering method, followed by a non-hierarchical k-means correction with randomised regroupment (Mutz et al., 2016; Wilks, 2011; Paeth, 2004; Bahrenberg et al., 1992).

The hierarchical part of the clustering procedure starts with as many clusters as there are elements (ni), then iteratively combines the most similar clusters to form a new cluster using centroids for the linkage procedure for clusters containing multiple elements. The procedure is continued until the desired number of clusters (k) is reached. One disadvantage of a pure hierarchical approach is that elements cannot be recategorised once they are assigned to a cluster, even though the addition of new elements to existing clusters changes the clusters' defining attributes and could warrant a recategorisation of elements. We address this problem by implementation of a (non-hierarchical) k-means clustering correction (e.g. Paeth, 2004). Elements are recategorised based

on the multivariate centroids determined by the hierarchical cluster analysis in order to minimise the sum of deviations from the cluster centroids. The Mahalanobis distance (e.g. Wilks, 2011) is used as a measure of similarity or distance between the cluster centroids since it is a statistical distance and thus not sensitive to different variable units. The Mahalanobis distance also accounts for possible multi-collinearity between variables.

The end results of the cluster analyses are subdivisions of the climate in the investigated regions into k subdomains or clusters based on multiple climate variables. The region-specific k has to be prescribed before the analyses. A large k may result in redundant additional clusters describing very similar climates, thereby defeating the purpose of the analysis to identify and describe the dominant, distinctly different climates in the region and their geographical coverage. Since it is not possible to know a priori the ideal number of clusters, k was varied between 3 and 10 for each region and the results presented below identify the optimal number of visibly distinctly different clusters from the analysis. Optimal k was determined by assessing the distinctiveness and similarities between the climate clusters in the systematic process of increasing k from 3 to 10. Once an increase in k no longer resulted in the addition of another cluster that was climatologically distinctly different from the others, and instead resulted in at least two similar clusters, k of the previous iteration was chosen as the optimal k for the region.

The cluster analysis ultimately results in a description of the geographical extent of a climate (cluster) characterised by a certain combination of mean values for each of the variables associated with the climate. For example, climate cluster 1 may be the most tropical climate in a region and thus be characterised by high precipitation values, high temperature values, and low seasonal temperature amplitude. Each of the results (consisting of the geographical extent of climates and mean vectors describing the climate) can be viewed as an optimal classification for the specific region and time. It serves primarily as a means for providing an overview of the climate in each of the regions at different times, reduces dimensionality of the raw simulation output, and identifies regions of climatic homogeneity that are difficult to notice by viewing simple maps of each climate variable. Its synoptic purpose is similar to that of the widely known Köppen–Geiger classification scheme (Peel et al., 2007), but we allow for optimal classification rather than prescribe classes, and our selection of variables is more restricted and made in accordance with the focus of this study.

3 Results

Results from our analysis are first presented for general changes in global temperature and precipitation for the different time slices (Figs. 2, 3), which is then followed by an analysis of changes in the climatology of selected orogens.

A more detailed description of temperature and precipitation changes in our selected orogens is presented in subsequent subsections (Fig. 4 and following). All differences in climatology are expressed relative to the PI control run. Changes relative to the PI rather than PD conditions are presented to avoid interpreting an anthropogenic bias in the results and focusing instead on pre-anthropogenic variations in climate. For brevity, near-surface (2 m) air temperature and total precipitation rate are referred to as temperature and precipitation.

3.1 Global differences in mean annual temperature

This section describes the differences between simulated MH, LGM, and PLIO annual mean temperature anomalies with respect to PI shown in Fig. 2b, and PI temperature absolute values shown in Fig. 2a. Most temperature differences between the PI and MH climate are within −1 to 1 °C. Exceptions to this are the Hudson Bay, Weddell Sea, and Ross Sea regions, which experience warming of 1–3, 1–5, and 1–9 °C, respectively. Continental warming is mostly restricted to low-altitude South America, Finland, western Russia, the Arabian peninsula (1–3 °C), and subtropical North Africa (1–5 °C). Simulation results show that LGM and PLIO annual mean temperature deviate from the PI means the most. The global PLIO warming and LGM cooling trends are mostly uniform in direction, but the magnitude varies regionally. The strongest LGM cooling is concentrated in regions where the greatest change in ice extent occurs (as indicated in Fig. 2), i.e. Canada, Greenland, the North Atlantic, northern Europe, and Antarctica. Central Alaska shows no temperature changes, whereas coastal southern Alaska experiences cooling of ≤ 9 °C. Cooling in the US Pacific Northwest is uniform and between 11 and 13 °C. Most of high-altitude South America experiences mild cooling of 1–3 °C, 3–5 °C in the central Andes, and ≤ 9 °C in the south. Along the Himalayan orogen, LGM temperature values are 5–7 °C below PI values. Much of central Asia and the Tibetan Plateau cools by 3–5 °C, and most of India, low-altitude China, and South East Asia cools by 1–3 °C.

In the PLIO climate, parts of Antarctica, Greenland, and the Greenland Sea experience the greatest temperature increase (≤ 19 °C). Most of southern Alaska warms by 1–5 and ≤ 9 °C near McCarthy, Alaska. The US Pacific Northwest warms by 1–5 °C. The strongest warming in South America is concentrated at the Pacific west coast and the Andes (1–9 °C), specifically between Lima and Chiclayo, and along the Chilean–Argentinian Andes south of Bolivia (≤ 9 °C). Parts of low-altitude South America to the immediate east of the Andes experience cooling of 1–5 °C. The Himalayan orogen warms by 3–9 °C, whereas Myanmar, Bangladesh, Nepal, northern India, and northeastern Pakistan cool by 1–9 °C.

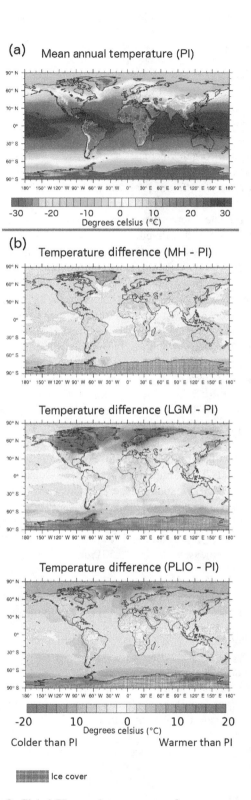

Figure 2. Global PI annual mean near-surface temperatures (a) and deviations of MH, LGM, and PLIO annual mean near-surface temperatures from PI values (b). Units are °C and insignificant (p < 99 %) differences (as determined by a t test) are greyed out.

3.2   Global differences in mean annual precipitation

Notable differences occur between simulated MH, LGM, and PLIO annual mean precipitation anomalies with respect to PI shown in Fig. 3b and the PI precipitation absolute values shown in Fig. 3a. Of these, MH precipitation deviates the least from PI values. The differences between MH and PI precipitation on land appear to be largest in northern tropical Africa (increase $\leq 1200\,\mathrm{mm\,a^{-1}}$), along the Himalayan orogen (increase $\leq 2000\,\mathrm{mm\,a^{-1}}$), and in central Indian states (decrease) $\leq 500\,\mathrm{mm}$. The biggest differences in western South America are precipitation increases in central Chile between Santiago and Puerto Montt. The LGM climate shows the largest deviation in annual precipitation from the PI climate, and precipitation on land mostly decreases. Exceptions are increases in precipitation rates in North American coastal regions, especially in coastal southern Alaska ($\leq 2300\,\mathrm{mm\,a^{-1}}$) and the US Pacific Northwest ($\leq 1700\,\mathrm{mm\,a^{-1}}$). Further exceptions are precipitation increases in low-altitude regions immediately east of the Peruvian Andes ($\leq 1800\,\mathrm{mm\,a^{-1}}$), central Bolivia ($\leq 1000\,\mathrm{mm\,a^{-1}}$), most of Chile ($\leq 1000\,\mathrm{mm\,a^{-1}}$), and northeastern India ($\leq 1900\,\mathrm{mm\,a^{-1}}$). Regions of notable precipitation decrease are northern Brazil ($\leq 1700\,\mathrm{mm\,a^{-1}}$), southernmost Chile and Argentina ($\leq 1900\,\mathrm{mm\,a^{-1}}$), coastal south Peru ($\leq 700\,\mathrm{mm\,a^{-1}}$), central India ($\leq 2300\,\mathrm{mm\,a^{-1}}$), and Nepal ($\leq 1600\,\mathrm{mm\,a^{-1}}$).

Most of the precipitation on land in the PLIO climate is higher than that in the PI climate. Precipitation is enhanced by ca. 100–200 $\mathrm{mm\,a^{-1}}$ in most of the Atacama Desert, by $\leq 1700\,\mathrm{mm\,a^{-1}}$ south of the Himalayan orogen, and by $\leq 1400\,\mathrm{mm\,a^{-1}}$ in tropical South America. Precipitation significantly decreases in central Peru ($\leq 2600\,\mathrm{mm}$), southernmost Chile ($\leq 2600\,\mathrm{mm}$), and from eastern Nepal to northernmost northeastern India ($\leq 250\,0\mathrm{mm}$).

3.3   Palaeoclimate characterisation from the cluster analysis and changes in regional climatology

In addition to the global changes described above, the PLIO to PI regional climatology changes substantially in the four investigated regions of South Asia (Sect. 3.3.1), the Andes (Sect. 3.3.2), southern Alaska (Sect. 3.3.3), and the Cascade Range (Sect. 3.3.4). Each climate cluster defines a separate distinct climate that is characterised by the mean values of the different climate variables used in the analysis. The clusters are calculated by taking the arithmetic means of all the values (climatic means) calculated for the grid boxes within each region. The regional climates are referred to by their cluster number $C_1, C_2, \ldots, C_k$, where k is the number of clusters specified for the region. The clusters for specific palaeoclimates are mentioned in the text as $C_{i[t]}$, where i corresponds to the cluster number ($i = 1, \ldots, k$) and t to the simulation time period ($t = $ PI, MH, LGM, PLIO). The descriptions first highlight the similarities and then the differences in

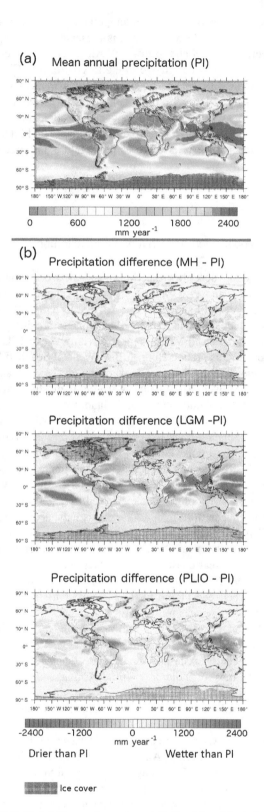

Figure 3. Global PI annual mean precipitation (**a**) and deviations of MH, LGM, and PLIO annual mean near-surface temperatures from PI values (**b**). Units are millimetres per year.

regional climate. The cluster means of seasonal near-surface temperature amplitude and seasonal precipitation amplitude are referred to as temperature and precipitation amplitude. The median, 25th percentile, 75th percentile, minimum, and maximum values for annual mean precipitation are referred to as P_{md}, P_{25}, P_{75}, P_{min}, and P_{max}, respectively. Likewise, the same statistics for temperature are referred to as T_{md}, T_{25}, T_{75}, T_{min}, and T_{max}. These are presented as box plots of climate variables in different time periods. When the character of a climate cluster is described as "high", "moderate", and "low", the climatic attribute's values are described relative to the value range of the specific region in time; thus high PLIO precipitation rates may be higher than high LGM precipitation rates. The character is presented in a raster plot to allow compact visual representation of it. The actual mean values for each variable in every time slice and region-specific cluster are included in tables in the Supplement.

3.3.1 Climate change and palaeoclimate characterisation in South, central, and East Asia

This section describes the regional climatology of the four investigated Cenozoic time slices and how precipitation and temperature changes from PLIO to PI times in tropical, temperate, and high-altitude regions. LGM and PLIO simulations show the largest simulated temperature and precipitation deviations (Fig. 4b) from PI temperature and precipitation (Fig. 4a) in the South Asia region. LGM temperatures are 1–7 °C below PI temperatures and the direction of deviation is uniform across the study region. PLIO temperature is mostly above PI temperatures by 1–7 °C. The cooling of 3–5 °C in the region immediately south of the Himalayan orogen represents one of the few exceptions. Deviations of MH precipitation from PI precipitation in the region are greatest along the eastern Himalayan orogeny, which experiences an increase in precipitation (≤ 2000 mm a^{-1}). The same region experiences a notable decrease in precipitation in the LGM simulation, which is consistent in direction with the prevailing precipitation trend on land during the LGM. PLIO precipitation on land is typically higher than PI precipitation.

Annual means of precipitation and temperature spatially averaged for the regional subdivisions and the different time slice simulations have been compared. The value range P_{25} to P_{75} of precipitation is higher for tropical South Asia than for temperate and high-altitude South Asia (Fig. 5a–c). The LGM values for P_{25}, P_{md}, and P_{75} are lower than for the other time slice simulations, most visibly for tropical South Asia (ca. 100 mm a^{-1}). The temperature range (both T_{75}–T_{25} and T_{max}–T_{min}) is smallest in hot (ca. 21 °C) tropical South Asia, wider in high-altitude (ca. -8 °C) South Asia, and widest in temperate (ca. 2 °C) South Asia (Fig. 5d–f). T_{md}, T_{25}, and T_{75} values for the LGM are ca. 1 °C, 1–2, and 2 °C below PI and MH temperatures in tropical, temperate, and high-altitude South Asia, respectively, whereas the same temperature statistics for the PLIO simulation are ca. 1 °C

above PI and MH values in all regional subdivisions (Fig. 5d–f). With respect to PI and MH values, precipitation and temperature are generally lower in the LGM and higher in the PLIO in tropical, temperate, and high-altitude South Asia.

In all time periods, the wettest climate cluster C_1 covers an area along the southeastern Himalayan orogen (Fig. 6a–d) and is defined by the highest precipitation amplitude (dark blue, Fig. 6e–h). $C_{5(PI)}$, $C_{3(MH)}$, $C_{4(LGM)}$, and $C_{5(PLIO)}$ are characterised by (dark blue, Fig. 6e–h) the highest temperatures and u-wind and v-wind speeds during the summer monsoon in their respective time periods, whereas $C_{4(PI)}$, $C_{5(MH)}$, and $C_{6(LGM)}$ are defined by low temperatures and the highest temperature amplitude and u-wind and v-wind speeds outside the monsoon season (in January) in their respective time periods (Fig. 6e–h). The latter three climate classes cover much of the more continental, northern landmass in their respective time periods and represent a cooler climate affected more by seasonal temperature fluctuations (Fig. 6a–d). The two wettest climate clusters C_1 and C_2 are more restricted to the eastern end of the Himalayan orogen in the LGM than during other times, indicating that the LGM precipitation distribution over the South Asia landmass is more concentrated in this region than in other time slice experiments.

3.3.2 Climate change and palaeoclimate characterisation in the Andes, western South America

This section describes the cluster-analysis-based regional climatology of the four investigated late Cenozoic time slices and illustrates how precipitation and temperature changes from PLIO to PI in tropical and temperate low- and high-altitude (i.e. Andes) regions in western South America (Figs. 7–9).

LGM and PLIO simulations show the largest simulated deviations (Fig. 7b) from PI temperature and precipitation (Fig. 7a) in western South America. The direction of LGM temperature deviations from PI temperatures is negative and uniform across the region. LGM temperatures are typically 1–3 °C below PI temperatures across the region and 1–7 °C below PI values in the Peruvian Andes, which also experience the strongest and most widespread increase in precipitation during the LGM (≤ 1800 mm a^{-1}). Other regions, such as much of the northern Andes and tropical South America, experience a decrease in precipitation in the same experiment. PLIO temperature is mostly elevated above PI temperatures by 1–5 °C. The Peruvian Andes experience a decrease in precipitation (≤ 2600 mm), while the northern Andes are wetter in the PLIO simulation compared to the PI control simulation.

PI, MH, LGM, and PLIO precipitation and temperature means for regional subdivisions have been compared. The P_{25} to P_{75} range is smallest for the relatively dry temperate Andes and largest for tropical South America and the tropical Andes (Fig. 8a–d). P_{max} is lowest in the PLIO in all

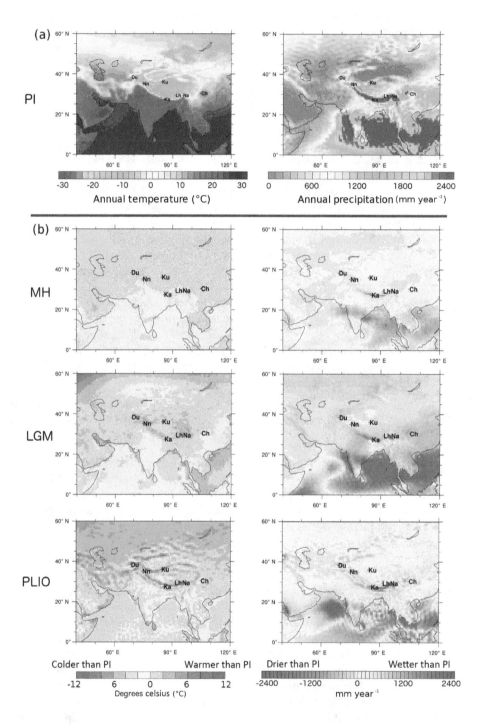

Figure 4. PI annual mean near-surface temperatures **(a)** and deviations of MH, LGM, and PLIO annual mean near-surface temperatures from PI values **(b)** for the South Asia region. Insignificant ($p < 99\%$) differences (as determined by a t test) are greyed out.

four regional subdivisions even though P_{md}, P_{25}, and P_{75} in the PLIO simulation are similar to the same statistics calculated for PI and MH time slices. P_{md}, P_{25}, and P_{75} for the LGM are ca. 50 mm a^{-1} lower in tropical South America and ca. 50 mm a^{-1} higher in the temperate Andes. Average PLIO temperatures are slightly warmer and LGM temperatures are

slightly colder than PI and MH temperatures in tropical and temperate South America (Fig. 8e and f). These differences are more pronounced in the Andes, however. T_{md}, T_{25}, and T_{75} are ca. 5 °C higher in the PLIO climate than in PI and MH climates in both the temperate and tropical Andes, whereas

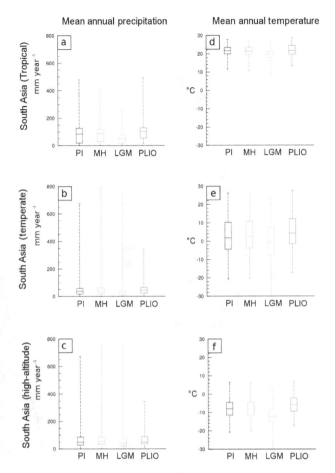

Figure 5. PI, MH, LGM, and PLIO annual mean precipitation in (**a**) tropical South Asia, (**b**) temperate South Asia, and (**c**) high-altitude South Asia; PI, MH, LGM, and PLIO annual mean temperatures in (**d**) tropical South Asia, (**e**) temperate South Asia, and (**f**) high-altitude South Asia. For each time slice, the minimum, lower 25th percentile, median, upper 75th percentile, and maximum are plotted.

the same temperatures for the LGM are ca. 2–4 °C below PI and MH values (Fig. 8g and h).

For the LGM, the model computes drier-than-PI conditions in tropical South America and the tropical Andes, enhanced precipitation in the temperate Andes, and a decrease in temperature that is most pronounced in the Andes. For the PLIO, the model predicts precipitation similar to PI, but with lower precipitation maxima. PLIO temperatures generally increase from PI temperatures, and this increase is most pronounced in the Andes.

The climate variability in the region is described by six different clusters (Fig. 9a–d), which have similar attributes in all time periods. The wettest climate C_1 is also defined by moderate to high precipitation amplitudes, low temperatures, and moderate to high u-wind speeds in summer and winter in all time periods (dark blue, Fig. 9e–h). $C_{2(PI)}$, $C_{2(MH)}$, $C_{3(LGM)}$, and $C_{2(PLIO)}$ are characterised by high temperatures and low seasonal temperature amplitude (dark blue, Fig. 9e–h), ge-

ographically cover the north of the investigated region, and represent a more tropical climate. $C_{5(PI)}$, $C_{5(MH)}$, $C_{6(LGM)}$, and $C_{6(PLIO)}$ are defined by low precipitation and precipitation amplitude, high temperature amplitude, and high u-wind speeds in winter (Fig. 9e–h), cover the low-altitude south of the investigated region (Fig. 9a–d), and represent dry, extratropical climates with more pronounced seasonality. In the PLIO simulation, the lower-altitude east of the region has four distinct climates, whereas the analysis for the other time slice experiments only yield three distinct climates for the same region.

3.3.3 Climate change and palaeoclimate characterisation in the St Elias Mountains, southeastern Alaska

This section describes the changes in climate and the results from the cluster analysis for southern Alaska (Figs. 10–12). As is the case for the other study areas, LGM and PLIO simulations show the largest simulated deviations (Fig. 10b) from PI temperature and precipitation (Fig. 10a). The sign of LGM temperature deviations from PI temperatures is negative and uniform across the region. LGM temperatures are typically 1–9 °C below PI temperatures, with the east of the study area experiencing the largest cooling. PLIO temperatures are typically 1–5 °C above PI temperatures and the warming is uniform for the region. In comparison to the PI simulation, LGM precipitation is lower on land but higher (≤ 2300 mm) in much of the coastal regions of southern Alaska. Annual PLIO precipitation is mostly higher (≤ 800 mm) than for PI.

P_{md}, P_{25}, P_{75}, P_{min}, and P_{max} for southern Alaskan mean annual precipitation do not differ much between PI, MH, and PLIO climates, while P_{md}, P_{25}, P_{75}, and P_{min} decrease by ca. 20–40 mm a^{-1} and P_{max} increases during the LGM (Fig. 11a). The Alaskan PLIO climate is distinguished from the PI and MH climates by its higher (ca. 2 °C) regional temperature means, T_{25}, T_{75}, and T_{md} (Fig. 11b). Mean annual temperatures, T_{25}, T_{75}, T_{min}, and T_{max}, are lower in the LGM than in any other considered time period (Fig. 11b), and about 3–5 °C lower than during the PI and MH.

Distinct climates are present in the PLIO to PI simulations for southeastern Alaska. Climate cluster C_1 is always geographically restricted to coastal southeastern Alaska (Fig. 12a–d) and characterised by the highest precipitation, precipitation amplitude, and temperature and by relatively low temperature amplitude (dark blue, Fig. 12e–h). Climate C_2 is characterised by moderate to low precipitation, precipitation amplitude, and temperature and by low temperature amplitude. C_2 is either restricted to coastal southeastern Alaska (in MH and LGM climates) or coastal southern Alaska (in PI and PLIO climates). Climate C_3 is described by low precipitation, precipitation amplitude, and temperature and moderate temperature amplitude in all simulations. It covers coastal western Alaska and separates climate C_1 and C_2 from the northern C_4 climate. Climate C_4 is distin-

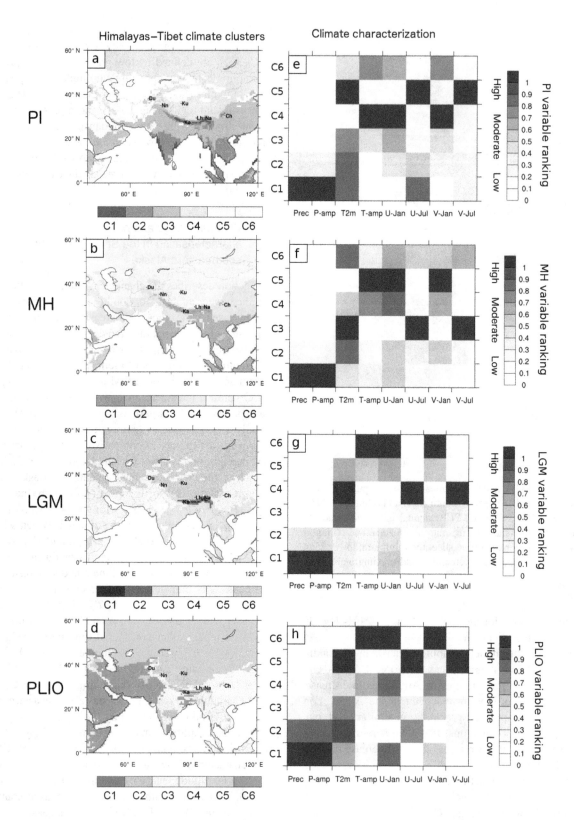

Figure 6. Geographical coverage and characterisation of climate classes C_1–C_6 based on cluster analysis of eight variables (near-surface temperature, seasonal near-surface temperature amplitude, total precipitation, seasonal precipitation amplitude, u wind in January and July, v wind in January and July) in the South Asia region. The geographical coverage of the climates C_1–C_6 is shown on the left for the PI (**a**), MH (**b**), LGM (**c**), and PLIO (**d**); the complementary, time-slice-specific characterisation of C_1–C_6 for the PI (**e**), MH (**f**), LGM (**g**), and PLIO (**h**) is shown on the right.

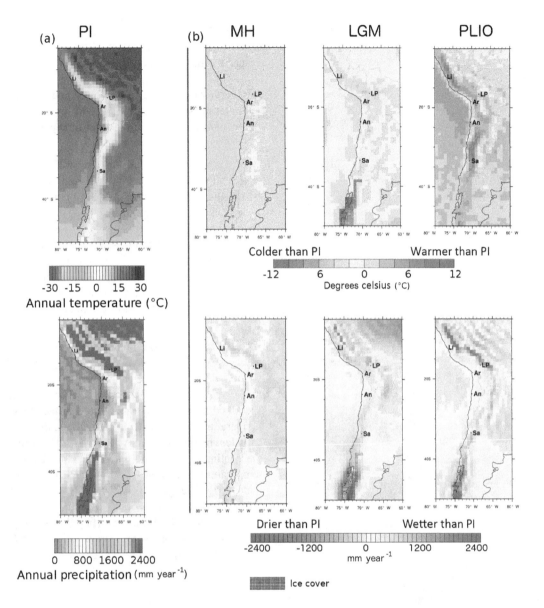

Figure 7. PI annual mean near-surface temperatures **(a)** and deviations of MH, LGM, and PLIO annual mean near-surface temperatures from PI values **(b)** for western South America. Insignificant ($p < 99\,\%$) differences (as determined by a t test) are greyed out.

guished by the highest mean temperature amplitude, by low temperature and precipitation amplitude, and by the lowest precipitation.

The geographical ranges of PI climates C_1–C_4 and PLIO climates C_1–C_4 are similar. $C_{1(PI/PLIO)}$ and $C_{2(PI/PLIO)}$ spread over a larger area than $C_{1(MH/LGM)}$ and $C_{2(MH/LGM)}$. $C_{2(PI/PLIO)}$ are not restricted to coastal southeastern Alaska, but also cover the coastal southwest of Alaska. The main difference in characterisation between PI and PLIO climates C_1–C_4 lies in the greater difference (towards lower values) in precipitation, precipitation amplitude, and temperature from $C_{1(PLIO)}$ to $C_{2(PLIO)}$ compared to the relatively moderate decrease in those means from $C_{1(PI)}$ to $C_{2(PI)}$.

3.3.4 Climate change and palaeoclimate characterisation in the Cascade Range, US Pacific Northwest

This section describes the character of regional climatology in the US Pacific Northwest and its change over time (Figs. 13–15). The region experiences cooling of typically 9–11 °C on land during the LGM and warming of 1–5 °C during the PLIO (Fig. 13b) when compared to PI temperatures (Fig. 13a). LGM precipitation increases over water, decreases on land by $\leq 800\,\mathrm{mm\,a^{-1}}$ in the north and in the vicinity of Seattle, and increases on land by $\leq 1400\,\mathrm{mm\,a^{-1}}$ on Vancouver Island and around Portland and the Olympic Mountains. Conversely, PLIO precipitation does not deviate

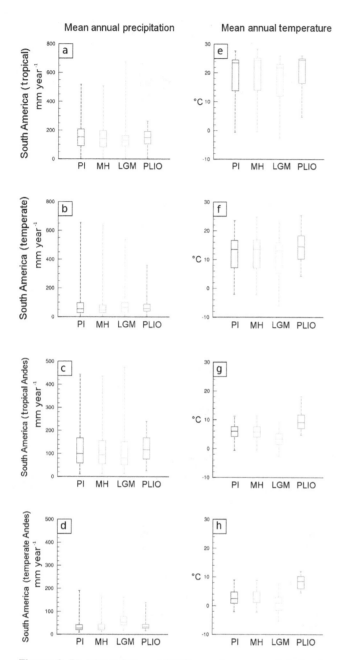

Mean annual precipitation   Mean annual temperature

Figure 8. PI, MH, LGM, and PLIO annual mean precipitation in
(**a**) tropical South America, (**b**) temperate South America, (**c**) the
tropical Andes, and (**d**) the temperate Andes; PI, MH, LGM, and
PLIO annual mean temperatures in (**e**) tropical South America,
(**f**) temperate South America, (**g**) the tropical Andes, and (**h**) the
temperate Andes. For each time slice, the minimum, lower 25th percentile, median, upper 75th percentile, and maximum are plotted.

much from PI values over water and varies in the direction of
deviation on land. MH temperature and precipitation deviation from PI values is negligible.

P_{md}, P_{25}, P_{75}, P_{min}, and P_{max} for the Cascade Range do
not notably differ between the four time periods (Fig. 14a).

The LGM range of precipitation values is slightly larger than
that of the PI and MH with slightly increased P_{md}, while the
respective range is smaller for simulation of the PLIO. The
T_{md}, T_{25}, T_{75}, and T_{max} values for the PLIO climate are ca.
2 °C higher than those values for PI and MH (Fig. 14b). All
temperature statistics for the LGM are notably (ca. 13 °C)
below their analogues in the other time periods (Fig. 14b).

PI, LGM, and PLIO clusters are similar in both their geographical patterns (Fig. 15a, c, d) and their characterisation
by mean values (Fig. 15e, g, h). C_1 is the wettest cluster and
shows the highest amplitude in precipitation. The common
characteristics of the C_2 cluster are moderate to high precipitation and precipitation amplitude. C_4 is characterised by
the lowest precipitation and precipitation amplitudes and the
highest temperature amplitudes. Regions assigned to clusters
C_1 and C_2 are in proximity to the coast, whereas C_4 is geographically restricted to more continental settings.

In the PI and LGM climates, the wettest cluster C_1 is also
characterised by high temperatures (Fig. 10e, g). However,
virtually no grid boxes were assigned to $C_{1(LGM)}$. $C_{1(MH)}$ differs from other climate states' C_1 clusters in that it is also
described by moderate to high near-surface temperature and
temperature amplitude (Fig. 10f), and in that it is geographically less restricted and covers much of Vancouver Island and
the continental coastline north of it (Fig. 10b). Near-surface
temperatures are highest for C_2 in PI, LGM, and PLIO climates (Fig. 10e, g, h) and low for $C_{2(MH)}$ (Fig. 10f). $C_{2(MH)}$
is also geographically more restricted than C_2 clusters in
PI, LGM, and PLIO climates (Fig. 10a–d). $C_{2(PI)}$, $C_{2(MH)}$,
and $C_{2(LGM)}$ have a low temperature amplitude (Fig. 10e–g),
whereas $C_{2(PLIO)}$ is characterised by a moderate temperature
amplitude (Fig. 10h).

4 Discussion

In the following, we synthesise our results and compare to
previous studies that investigate the effects of temperature
and precipitation change on erosion. Since our results do not
warrant merited discussion of subglacial processes without
additional work that is beyond the scope of this study, we instead advise caution in interpreting the presented precipitation and temperature results in an erosional context in which
the regions are covered with ice. For convenience, ice cover
is indicated in Figs. 2, 3, 7, 10 and 13, and a summary of ice
cover used as boundary conditions for the different time slice
experiments is included in the Supplement. Where possible,
we relate the magnitude of climate change predicted in each
geographical study area with terrestrial proxy data.

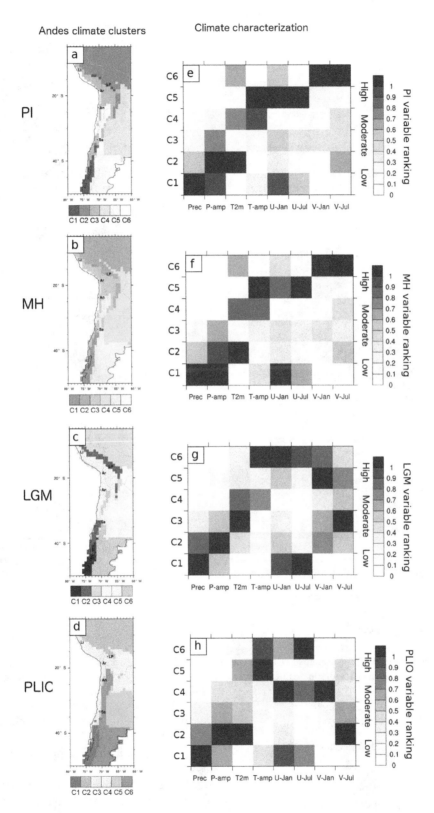

Figure 9. Geographical coverage and characterisation of climate classes C_1–C_6 based on cluster analysis of eight variables (near-surface temperature, seasonal near-surface temperature amplitude, precipitation, seasonal precipitation amplitude, u wind in January and July, v wind in January and July) in western South America. The geographical coverage of the climates C_1–C_6 is shown on the left for PI (**a**), MH (**b**), LGM (**c**), and PLIO (**d**); the complementary, time-slice-specific characterisation of C_1–C_6 for PI (**e**), MH (**f**), LGM (**g**), and PLIO (**h**) is shown on the right.

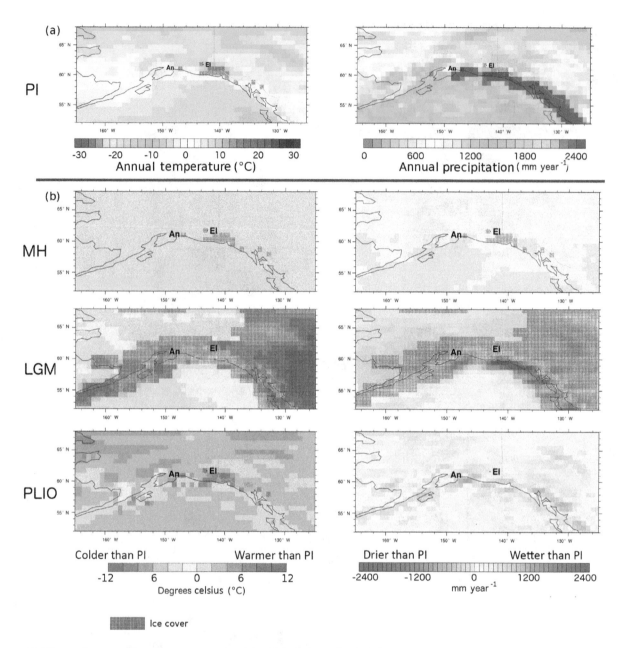

Figure 10. PI annual mean near-surface temperatures (**a**) and deviations of MH, LGM, and PLIO annual mean near-surface temperatures from PI values (**b**) for the southern Alaska region. Insignificant ($p < 99\%$) differences (as determined by a t test) are greyed out.

4.1 Synthesis of temperature changes

4.1.1 Temperature changes and implications for weathering and erosion

Changes in temperature can affect physical weathering due to temperature-induced changes in periglacial processes and promote frost cracking, frost creep (e.g. Matsuoka, 2001; Schaller et al., 2002; Matsuoka and Murton, 2008; Delunel et al., 2010; Andersen et al., 2015; Marshall et al., 2015), and biotic weathering and erosion (e.g. Moulton and Berner, 1998; Banfield et al., 1999; Dietrich and Perron, 2006).

Quantifying and understanding past changes in temperature is thus vital for our understanding of denudation histories. In the following, we highlight regions in the world where future observational studies might be able to document significant warming or cooling that would influence temperature-related changes in physical and chemical weathering over the last ~ 3 Myr.

Simulated MH temperatures show little deviation (typically $< 1\,^\circ$C) from PI temperatures in the investigated regions (Fig. 2b), suggesting little difference in MH temperature-related weathering. The LGM experiences widespread cool-

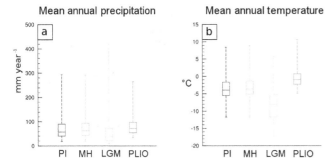

Figure 11. PI, MH, LGM, and PLIO annual mean precipitation **(a)** and mean annual temperatures **(b)** in southern Alaska. For each time slice, the minimum, lower 25th percentile, median, upper 75th percentile, and maximum are plotted.

ing, which is accentuated at the poles, increasing the Equator-to-pole pressure gradient and consequently strengthening global atmospheric circulation. Despite this global trend, cooling in coastal southern Alaska is higher ($\leq 9\,^\circ\text{C}$) than in central Alaska ($0 \pm 1\,^\circ\text{C}$). The larger temperature difference in southern Alaska geographically coincides with ice cover (Fig. 10b) and should thus be interpreted in the context of a different erosional regime. Cooling in most of the lower-latitude regions in South America and central to South East Asia is relatively mild. The greatest temperature differences in South America are observed for western Patagonia, which was mostly covered by glaciers. The Tibetan Plateau experiences more cooling ($3\text{–}5\,^\circ\text{C}$) than adjacent low-altitude regions ($1\text{–}3\,^\circ\text{C}$) during the LGM.

The PLIO simulation is generally warmer, and temperature differences accentuate warming at the poles. Warming in simulation PLIO is greatest in parts of Canada, Greenland, and Antarctica (up to $19\,^\circ\text{C}$), which geographically coincides with the presence of ice in the PI reference simulation and thus may be attributed to differences in ice cover. It should therefore also be regarded as areas in which process domain shifted from glacial to non-glacial. The warming in simulation PLIO in southern Alaska and the US Pacific Northwest is mostly uniform and in the range of $1\text{–}5\,^\circ\text{C}$. As before, changes in ice cover reveal that the greatest warming may be associated with the absence of glaciers relative to the PI simulation. Warming in South America is concentrated at the Pacific west coast and the Andes between Lima and Chiclayo and along the Chilean–Argentinian Andes south of Bolivia ($\leq 9\,^\circ\text{C}$).

Overall, annual mean temperatures in the MH simulation show little deviation from PI values. The more significant temperature deviations of the colder LGM and of the warmer PLIO simulations are accentuated at the poles, leading to higher and lower Equator-to-pole temperature gradients, respectively. The largest temperature-related changes (relative to PI conditions) in weathering and subsequent erosion, in many cases through a shift in the process domain from glacial

to non-glacial or vice versa, are therefore to be expected in the LGM and PLIO climates.

4.1.2 Temperature comparison to other studies

LGM cooling is accentuated at the poles, thus increasing the Equator-to-pole pressure gradient and consequently strengthens global atmospheric circulation, and is in general agreement with studies such as Otto-Bliesner et al. (2006) and Braconnot et al. (2007). The PLIO simulation shows little to no warming in the tropics and accentuated warming at the poles, as do findings of Salzmann et al. (2011), Robinson (2009), and Ballantyne (2010), respectively. This would reduce the Equator-to-pole sea and land surface temperature gradient, as also reported by Dowsett et al. (2010), and also weaken global atmospheric circulation. Agreement with proxy-based reconstructions, as is the case of the relatively little warming in lower latitudes, is not surprising given that SST reconstructions (derived from previous coarse resolution coupled ocean–atmosphere models) are prescribed in this uncoupled atmosphere simulation. It should be noted that coupled ocean–atmosphere simulations do predict more low-latitude warming (e.g. Stepanek and Lohmann, 2012; R. Zhang et al., 2013). The PLIO warming in parts of Canada and Greenland (up to $19\,^\circ\text{C}$) is consistent with values based on multi-proxy studies (Ballantyne et al., 2010). Due to a scarcity of palaeobotanical proxies in Antarctica, reconstruction-based temperature and ice sheet extent estimates for a PLIO climate have high uncertainties (Salzmann et al., 2011), making model validation difficult. Furthermore, controversy about relatively little warming in the south polar regions compared to the north polar regions remains (e.g. Hillenbrand and Fütterer, 2002; Wilson et al., 2002). Mid-latitude PLIO warming is mostly in the $1\text{–}3\,^\circ\text{C}$ range with notable exceptions of cooling in the northern tropics of Africa and on the Indian subcontinent, especially south of the Himalayan orogen.

4.2 Synthesis of precipitation changes

4.2.1 Precipitation and implications for weathering and erosion

Changes in precipitation affects erosion through river incision, sediment transport, and erosion due to extreme precipitation events and storms (e.g. Whipple and Tucker, 1999; Hobley et al., 2010). Furthermore, vegetation type and cover also co-evolve with variations in precipitation and with changes in geomorphology (e.g. Marston, 2010; Roering et al., 2010). These vegetation changes in turn modify hillslope erosion by increasing root mass and canopy cover and decreasing water-induced erosion via surface run-off (e.g. Gyssels et al., 2005). Therefore, understanding and quantifying changes in precipitation in different palaeoclimates is necessary for a more complete reconstruction of orogen denudation histories. A synthesis of predicted precipitation changes is provided below and highlights regions where changes in

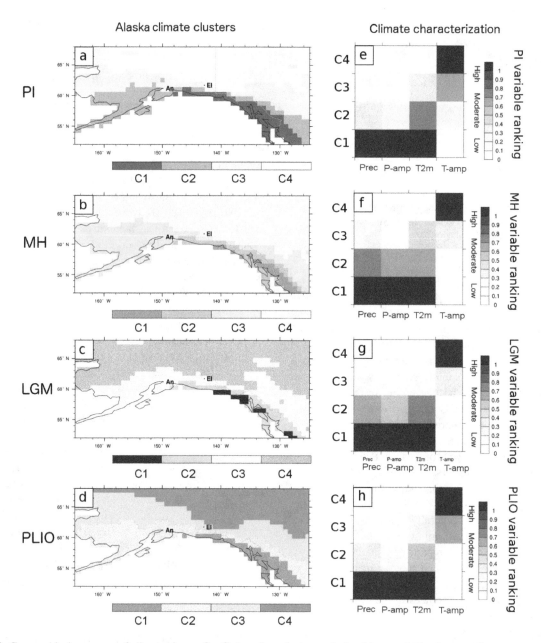

Figure 12. Geographical coverage of climate classes C_1–C_4 based on cluster analysis of four variables (near-surface temperature, seasonal near-surface temperature amplitude, total precipitation, seasonal total precipitation amplitude) in southern Alaska. The geographical coverage of the climates C_1–C_4 is shown on the left for PI (**a**), MH (**b**), LGM (**c**), and PLIO (**d**); the complementary, time-slice-specific characterisation of C_1–C_6 for PI (**e**), MH (**f**), LGM (**g**), and PLIO (**h**) is shown on the right.

river discharge and hillslope processes might be impacted by climate change over the last ∼ 3 Myr.

Most of North Africa is notably wetter during the MH, which is characteristic of the African Humid Period (Sarnthein, 1978). This pluvial regional expression of the Holocene Climatic Optimum is attributed to sudden changes in the strength of the African monsoon caused by orbital-induced changes in summer insolation (e.g. deMenocal et al., 2000). Southern Africa is characterised by a wetter climate to the east and drier climate to the west of the approximate lo-

cation of the Congo Air Boundary (CAB), the migration of which has previously been cited as a cause for precipitation changes in East Africa (e.g. Juninger et al., 2014). In contrast, simulated MH precipitation rates show little deviation from the PI in most of the investigated regions, suggesting little difference in MH precipitation-related erosion. The Himalayan orogen is an exception and shows a precipitation increase of up to 2000 mm a^{-1}. The climate's enhanced erosion potential, which could result from such a climatic change, should be taken into consideration when palaeoerosion rates

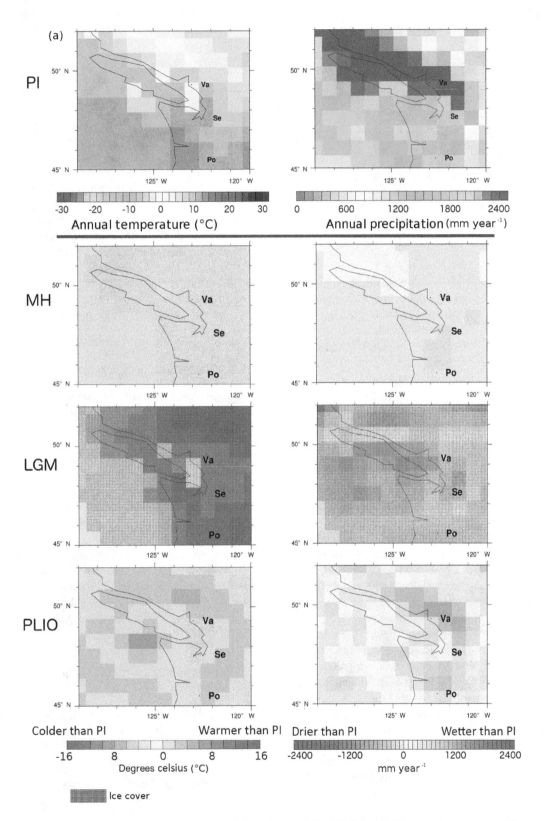

Figure 13. PI annual mean near-surface temperatures **(a)** and deviations of MH, LGM, and PLIO annual mean near-surface temperatures from PI values **(b)** for the US Pacific Northwest. Insignificant ($p < 99\%$) differences (as determined by a t test) are greyed out.

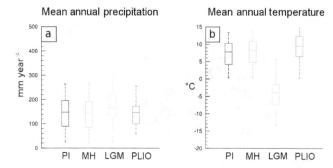

Figure 14. PI, MH, LGM, and PLIO annual mean precipitation (**a**) and annual mean temperatures (**b**) in the Cascades, US Pacific Northwest. For each time slice, the minimum, lower 25th percentile, median, upper 75th percentile, and maximum are plotted.

estimated from the geological record in this area are interpreted (e.g. Bookhagen et al., 2005). Specifically, higher precipitation rates (along with differences in other rainfall-event parameters) could increase the probability of mass movement events on hillslopes, especially where hillslopes are close to the angle of failure (e.g. Montgomery, 2001), and modify fluxes to increase shear stresses exerted on river beds and increase stream capacity to enhance erosion on river beds (e.g. by abrasion).

Most precipitation on land is decreased during the LGM due to large-scale cooling and decreased evaporation over the tropics, resulting in an overall decrease in inland moisture transport (e.g. Braconnot et al., 2007). North America, south of the continental ice sheets, is an exception and experiences increases in precipitation. For example, the investigated US Pacific Northwest and the southeastern coast of Alaska experience strongly enhanced precipitation of ≤ 1700 and $\leq 2300\,\mathrm{mm\,a^{-1}}$, respectively. These changes geographically coincide with differences in ice extent. An increase in precipitation in these regions may have had direct consequences on the glaciers' mass balance and equilibrium line altitudes, where the glaciers' effectiveness in erosion is highest (e.g. Egholm et al., 2009; Yanites and Ehlers, 2012). The differences in the direction of precipitation changes, and accompanying changes in ice cover would likely result in more regionally differentiated variations in precipitation-specific erosional processes in the St Elias Mountains rather than causing systematic offsets for the LGM. Although precipitation is significantly reduced along much of the Himalayan orogen ($\leq 1600\,\mathrm{mm\,a^{-1}}$), northeastern India experiences strongly enhanced precipitation ($\leq 1900\,\mathrm{mm\,a^{-1}}$). This could have large implications for studies of uplift and erosion at orogen syntaxes, where highly localised and extreme denudation has been documented (e.g. Koons et al., 2013; Bendick and Ehlers, 2014).

Overall, the PLIO climate is wetter than the PI climate, in particular in the (northern) mid-latitudes and is possibly related to a northward shift of the northern Hadley cell

boundary that is ultimately the result of a reduced Equator-to-pole temperature gradient (e.g. Haywood et al., 2000, 2013; Dowsett et al., 2010). Most of the PLIO precipitation over land increases, especially at the Himalayan orogen by $\leq 1400\,\mathrm{mm\,a^{-1}}$, and decreases from eastern Nepal to Namcha Barwa ($\leq 2500\,\mathrm{mm\,a^{-1}}$). Most of the Atacama Desert experiences an increase in precipitation by $100\text{--}200\,\mathrm{mm\,a^{-1}}$, which may have to be considered in erosion and uplift history reconstructions for the Andes. A significant increase ($\sim 2000\,\mathrm{mm\,a^{-1}}$) in precipitation from simulation PLIO to modern conditions is simulated for the eastern margin of the Andean Plateau in Peru and for northern Bolivia. This is consistent with recent findings of a pulse of canyon incision in these locations in the last $\sim 3\,\mathrm{Myr}$ (Lease and Ehlers, 2013).

Overall, the simulated MH precipitation varies least from PI precipitation. The LGM is generally drier than the PI simulation, even though pockets of a wetter-than-PI climate do exist, such as much of coastal North America. Extratropical increased precipitation of the PLIO simulation and decreased precipitation of the LGM climate may be the result of decreased and increased Equator-to-pole temperature gradients, respectively.

4.2.2 Precipitation comparison to other studies

The large-scale LGM precipitation decrease on land, related to cooling and decreased evaporation over the tropics, and greatly reduced precipitation along much of the Himalayan orogeny, is consistent with previous studies by, for example, Braconnot et al. (2007). The large-scale PLIO precipitation increase due to a reduced Equator-to-pole temperature gradient has previously been pointed out by Haywood et al. (2000, 2013) and Dowsett et al. (2010), for example. A reduction of this gradient by ca. 5 °C is indeed present in the PLIO simulation of this study (Fig. 2b). This precipitation increase over land agrees well with simulations performed at a lower spatial model resolution (see Stepanek and Lohmann, 2012). Section 4.4 includes a more in-depth discussion of how simulated MH and LGM precipitation differences compare with proxy-based reconstructions in South Asia and South America.

4.3 Trends in late Cenozoic changes in regional climatology

This section describes the major changes in regional climatology and highlights their possible implications on erosion rates.

4.3.1 Himalayas–Tibet, South Asia

In South Asia, cluster-analysis-based categorisation and description of climates (Fig. 6) remains similar throughout time. However, the two wettest climates (C_1 and C_2) are geographically more restricted to the eastern Himalayan oro-

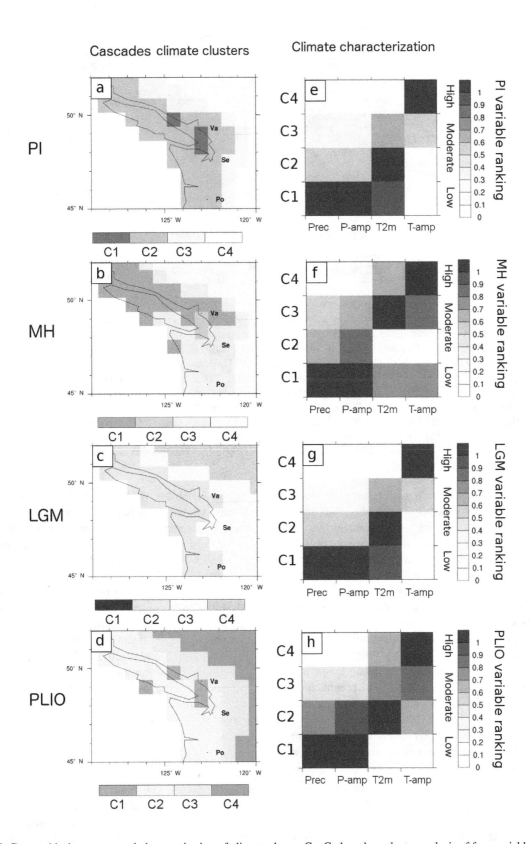

Figure 15. Geographical coverage and characterisation of climate classes C_1–C_4 based on cluster analysis of four variables (near-surface temperature, seasonal near-surface temperature amplitude, total precipitation, seasonal total precipitation amplitude) in the Cascades, US Pacific Northwest. The geographical coverage of the climates C_1–C_4 is shown on the left for PI **(a)**, MH **(b)**, LGM **(c)**, and PLIO **(d)**; the complementary, time-slice-specific characterisation of C_1–C_6 for PI **(e)**, MH **(f)**, LGM **(g)**, and PLIO **(h)** is shown on the right.

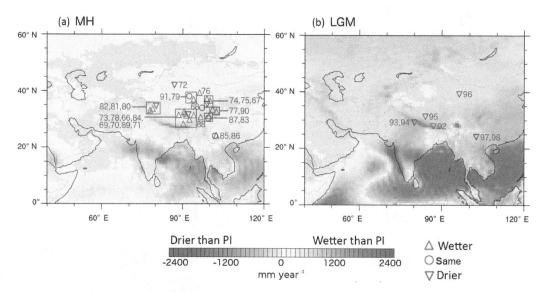

Figure 16. Simulated annual mean precipitation deviations of MH **(a)** and LGM **(b)** from PI values in South Asia, and temporally corresponding proxy-based reconstructions, indicating wetter (upward facing blue triangles), drier (downward facing red triangles), or similar (grey circles) conditions in comparison with modern climate. MH proxy-based precipitation differences are taken from Mügler et al. (2010) (66), Wischnewski et al. (2011) (67), Mischke et al. (2008), Wischnewski et al. (2011), Herzschuh et al. (2009) (68), Yanhong et al. (2006) (69), Morrill et al. (2006) (70), Wang et al. (2002) (71), Wuennemann et al. (2006) (72), Zhang et al. (2011), Morinaga et al. (1993), Kashiwaya et al. (1995) (73), Shen et al. (2005) (74), Liu et al. (2014) (75), Herzschuh et al. (2006a) (76), Zhang and Mischke (2009) (77), Nishimura et al. (2014) (78), Yu and Lai (2014) (79), Gasse et al. (1991) (80), Van Campo et al. (1996) (81), Demske et al. (2009) (82), Kramer et al. (2010) (83), Herzschuh et al. (2006b) (84), Hodell et al. (1999) (85), Hodell et al. (1999) (86), Shen et al. (2006) (87), Tang et al. (2000) (88), Tang et al. (2000) (89), Zhou et al. (2002) (90), Liu et al. (1998) (91), Asashi (2010) (92), Kotila et al. (2009) (93), Kotila et al. (2000) (94), Wang et al. (2002) (95), Hu et al. (2014) (96), Hodell et al. (1999) (97), and Hodell et al. (1999) (98).

gen in the LGM simulation. Even though precipitation over the South Asia region is generally lower, this shift indicates that rainfall on land is more concentrated in this region and that the westward drying gradient along the orogen is more accentuated than during other time periods investigated here. While there is limited confidence in the global atmospheric GCM's abilities to accurately represent mesoscale precipitation patterns (e.g. Cohen, 1990), the simulation warrants careful consideration of possible, geographically non-uniform offsets in precipitation in investigations of denudation and uplift histories.

MH precipitation and temperature in tropical, temperate, and high-altitude South Asia are similar to PI precipitation and temperature, whereas LGM precipitation and temperature are generally lower (by ca. $100 \, \mathrm{mm \, a^{-1}}$ and 1–2 °C, respectively), possibly reducing precipitation-driven erosion and enhancing frost-driven erosion in areas pushed into a near-zero temperature range during the LGM.

4.3.2   Andes, South America

Clusters in South America (Fig. 9), which are somewhat reminiscent of the Köppen and Geiger classification (Kraus, 2001), remain mostly the same over the last 3 Myr. In the PLIO simulation, the lower-altitude east of the region is characterised by four distinct climates, which suggests enhanced

latitudinal variability in the PLIO climate compared to PI with respect to temperature and precipitation.

The largest temperature deviations from PI values are derived for the PLIO simulation in the (tropical and temperate) Andes, where temperatures exceed PI values by 5 °C. Conversely, LGM temperatures in the Andes are ca. 2–4 °C below PI values in the same region (Fig. 7g and h). In the LGM simulation, tropical South America experiences ca. $50 \, \mathrm{mm \, a^{-1}}$ less precipitation; the temperate Andes receive ca. $50 \, \mathrm{mm \, a^{-1}}$ more precipitation than in PI and MH simulations. These latitude-specific differences in precipitation changes ought to be considered in attempts to reconstruct precipitation-specific palaeoerosion rates in the Andes on top of longitudinal climate gradients highlighted by Montgomery et al. (2001), for example.

4.4   St Elias Mountains, southern Alaska

Southern Alaska is subdivided into two wetter and warmer clusters in the south and two drier, colder clusters in the north. The latter are characterised by increased seasonal temperature variability due to being located at higher latitudes (Fig. 12). The different Equator-to-pole temperature gradients for LGM and PLIO may affect the intensity of the Pacific–North American teleconnection (PNA; Barnston and Livzey, 1987), which has significant influence on temper-

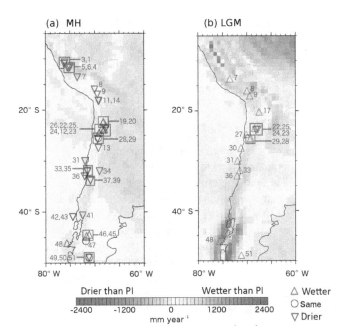

Figure 17. Simulated annual mean precipitation deviations of MH **(a)** and LGM **(b)** from PI values in South America and temporally corresponding proxy-based reconstructions, indicating wetter (upward-facing blue triangles), drier (downward-facing red triangles), or similar (grey circles) conditions in comparison with modern climate. MH proxy-based precipitation differences are taken from Bird et al. (2011) (1), Hansen et al. (1994) (2), Hansen et al. (1994) (3), Hansen et al. (1994) (4), Hansen et al. (1994) (5), Hansen et al. (1994) (6), Hillyer et al. (2009) (7), D'Agostino et al. (2002) (8), Baker et al. (2001) (9), Schwalb et al. (1999) (10), Schwalb et al. (1999) (11), Schwalb et al. (1999) (12), Schwalb et al. (1999) (13), Moreno et al. (2009) (14), Pueyo et al. (2011) (15), Mujica et al. (2015) (16), Fritz et al. (2004) (17), Gayo et al. (2012) (18), Latorre et al. (2006) (19), Latorre et al. (2003) (20), Quade et al. (2008) (21), Bobst et al. (2001) (22), Grosjean et al. (2001) (23), Betancourt et al. (2000) (24), Latorre et al. (2002) (25), Rech et al. (2003) (26), Diaz et al. (2012) (27), Maldonado et al. (2005) (28), Diaz et al. (2012) (29), Lamy et al. (2000) (30), Kaiser et al. (2008) (31), Maldonado et al. (2010) (32), Villagrán et al. (1990) (33), Méndez et al. (2015) (34), Maldonado and Villagrán (2006) (35), Lamy et al. (1999) (36), Jenny et al. (2002b) (37), Jenny et al. (2002b) (38), Villa-Martínez et al. (2003) (39), Bertrand et al. (2008) (40), De Basti et al. (2008) (41), Lamy et al. (2009) (42), Lamy et al. (2002) (43), Szeicz et al. (2003) (44), de Porras et al. (2012) (45), de Porras et al. (2014) (46), Markgraf et al. (2007) (47), Siani et al. (2010) (48), Gilli et al. (2001) (49), Markgraf et al. (2003) (50), and Stine and Stine (1990) (51).

atures and precipitation, especially in southeastern Alaska, and may in turn result in changes in regional precipitation and temperature patterns and thus on glacier mass balance. Changes in the Pacific Decadal Oscillation, which is related to the PNA pattern, has previously been connected to differences in late Holocene precipitation (Barron and Anderson, 2011). While this climate cluster pattern appears to be

a robust feature for the considered climate states, and hence over the recent geologic history, the LGM sets itself apart from PI and MH climates by generally lower precipitation (20–40 mm) and lower temperatures (3–5 °C; Figs. 10, 11), which may favour frost-driven weathering during glacial climate states (e.g. Andersen et al., 2015; Marshall et al., 2015) in unglaciated areas, whereas glacial processes would have dominated most of this region as it was covered by ice. Simulation PLIO is distinguished by temperatures that exceed PI and MH conditions by ca. 2 °C and by larger temperature and precipitation value ranges, possibly modifying temperature- and precipitation-dependent erosional processes in the region of southern Alaska.

4.5 Cascade Range, US Pacific Northwest

In all time slices, the geographic climate patterns, based on the cluster analysis (Fig. 15), represent an increase in the degree of continentality from the wetter coastal climates to the further inland climates with greater seasonal temperature amplitude and lower precipitation and precipitation amplitude (Fig. 15e–h). The most notable difference between the time slices is the strong cooling during the LGM, when temperatures are ca. 13 °C (Figs. 13, 14) below those of other time periods. Given that the entire investigated region was covered by ice (Fig. 13), we can assume a shift to glacially dominated processes.

4.6 Comparison of simulated and observed precipitation differences

The predicted precipitation differences reported in this study were compared with observed (proxy record) palaeoprecipitation change. Proxy-based precipitation reconstructions for the MH and LGM are presented for South Asia and South America for the purpose of assessing ECHAM5 model performance and for identifying inconsistencies among neighbouring proxy data. Due to the repeated glaciations, detailed terrestrial proxy records for the time slices investigated here are not available, to the best of our knowledge, for the Alaskan and Pacific NW USA studies. Although marine records and records of glacier extent are available in these regions, the results from them do not explicitly provide estimates of wetter–drier or colder–warmer conditions that can be spatially compared to the simulation estimates. For these two areas with no available records, the ECHAM5-predicted results therefore provide predictions from which future studies can formulate testable hypotheses to evaluate.

The palaeoclimate changes in terrestrial proxy records compiled here are reported as "wetter than today", "drier than today", or "the same as today" for each of the study locations and plotted on top of the simulation-based difference maps as upward-facing blue triangles, downward-facing red triangles, and grey circles, respectively (Figs. 16, 17). The numbers listed next to those indicators are the ID numbers assigned to the studies compiled for this comparison and are associated with a citation provided in the figure captions.

In South Asia, 14 out of 26 results from local studies agree with the model-predicted precipitation changes for the MH. The model seems able to reproduce the predominantly wetter conditions on much of the Tibetan Plateau, but predicts slightly drier conditions north of Chengdu, which is not reflected in local reconstructions. The modest mismatch between ECHAM5-predicted and proxy-based MH climate change in South Asia was also documented by Li et al. (2017), whose simulations were conducted at a coarser (T106) resolution. Despite these model–proxy differences, we note that there are significant discrepancies among the proxy data themselves in neighbouring locations in the MH, highlighting caution in relying solely upon these data for regional palaeoclimate reconstructions. These differences could result from either poor age constraints on the reported values or systematic errors in the transfer functions used to convert proxy measurements to palaeoclimate conditions. The widespread drier conditions on the Tibetan Plateau and immediately north of Laos are confirmed by seven out of seven of the palaeoprecipitation reconstructions. Of the reconstructed precipitation changes, 23 out of 39 agree with model predictions for South America during the MH. The model-predicted wetter conditions in the central Atacama Desert, as well as the drier conditions northwest of Santiago are confirmed by most of the reconstructions. The wetter conditions in southernmost Peru and the border to Bolivia and Chile cannot be confirmed by local studies. Of the precipitation reconstructions, 11 out of 17 for the LGM are in agreement with model predictions. These include wetter conditions in most of Chile. The most notable disagreement can be seen in northeastern Chile at the border to Argentina and Bolivia, where model-predicted wetter conditions are not confirmed by reported reconstructions from local sites.

Model performance is, in general, higher for the LGM than for the MH and overall satisfactory given that it cannot be expected to resolve sub-grid-scale differences in reported palaeoprecipitation reconstructions. However, as mentioned above, it should be noted that some location (MH of South Asia, and MH of norther Chile) discrepancies exist among neighbouring proxy samples and highlight the need for caution in how these data are interpreted. Other potential sources of error resulting in disagreement of simulated and proxy-based precipitation estimates are the model's shortcomings in simulating orographic precipitation at higher resolutions, and uncertainties in palaeoclimate reconstructions at the local sites. In summary, although some differences are evident in both the model–proxy data comparison and among neighbouring proxy data themselves, the comparison above highlights an overall good agreement between the model and data for the South Asia and South American study areas. Thus, although future advances in GCM model parameterisations and new or improved palaeoclimate proxy techniques are likely, the palaeoclimate changes documented here are found to be in general robust and provide a useful framework for future studies investigating how these predicted changes in palaeoclimate impact denudation.

5 Conclusions

We present a statistical cluster-analysis-based description of the geographic coverage of possible distinct regional expressions of climates from four different time slices (Figs. 6, 9, 12, 15). These are determined with respect to a selection of variables that characterise the climate of the region and may be relevant to weathering and erosional processes. While the geographic distribution of climate remains similar throughout time (as indicated by results of four different climate states representative for the climate of the last 3 Myr), results for the PLIO simulation suggests more climatic variability east of the Andes (with respect to near-surface temperature, seasonal temperature amplitude, precipitation, seasonal precipitation amplitude and seasonal u-wind and v-wind speeds). Furthermore, the wetter climates in the South Asia region retreat eastward along the Himalayan orogen for the LGM simulation; this is due to decreased precipitation along the western part of the orogen and enhanced precipitation on the eastern end, possibly signifying more localised high erosion rates.

Most global trends of the high-resolution LGM and PLIO simulations conducted here are in general agreement with previous studies (Otto-Bliesner et al., 2006; Braconnot et al., 2007; Wei and Lohmann, 2012; Lohmann et al., 2013; R. Zhang et al., 2013, 2014; Stepanek and Lohmann, 2012). The MH does not deviate notably from the PI, the LGM is relatively dry and cool, while the PLIO is comparably wet and warm. While the simulated regional changes in temperature and precipitation usually agree with the sign (or direction) of the simulated global changes, there are region-specific differences in the magnitude and direction. For example, the LGM precipitation of the tropical Andes does not deviate significantly from PI precipitation, whereas LGM precipitation in the temperate Andes is enhanced.

Comparisons to local, proxy-based reconstructions of MH and LGM precipitation in South Asia and South America reveal satisfactory performance of the model in simulating the reported differences. The model performs better for the LGM than the MH. We note however that compilations of proxy data such as we present here also identify inconsistences among neighbouring proxy data themselves, warranting caution in the extent to which both proxy data and palaeoclimate models are interpreted for MH climate change in South Asia and western South America.

The changes in regional climatology presented here are manifested, in part, by small to large magnitude changes in fluvial and hillslope relevant parameters such as precipitation and temperature. For the regions investigated here we find that precipitation differences among the PI, MH, LGM, and PLIO are in many areas around ± 200–$600\,\mathrm{mm\,yr^{-1}}$, and locally can reach maximums of ± 1000–$2000\,\mathrm{mm\,yr^{-1}}$ (Figs. 4, 7, 10, 13). In areas where significant precipitation increases are accompanied by changes in ice extent, such as parts of southern Alaska during the LGM, we would expect a shift in the erosional regime to glacier-dominated processes. Temperature differences between these same time periods are around 1–4 °C in many places, but reach maximum values of 8–10 °C. Many of these maxima in the temperature differences geographically coincide with changes in ice sheet extent and must therefore be interpreted as part of a different erosional process domain. However, we also observe large temperature differences (~ 5 °C) in unglaciated areas that would be affected by hillslope, frost cracking, and fluvial processes. The magnitude of these differences is not trivial, and will likely impact fluvial and hillslope erosion and sediment transport, as well as biotic and abiotic weathering. The regions of large-magnitude changes in precipitation and temperature documented here (Figs. 4, 7, 10, 13) offer the highest potential for future observational studies interested in quantifying the impact of climate change on denudation and weathering rates.

Competing interests. The authors declare that they have no conflict of interest.

Acknowledgements. European Research Council (ERC) Consolidator Grant number 615703 provided support for Sebastian G. Mutz. Additional support is acknowledged from the German science foundation (DFG) priority research program 1803 (EarthShape: Earth Surface Shaping by Biota; grants EH329/14-1 and EH329/17-1). We thank Byron Adams and Jessica Starke for constructive discussions. We also thank the reviewers (including Andrew Wickert) for their constructive feedback on this paper, which helped to significantly improve it. The DKRZ is thanked for computer time used for some of the simulations presented here. Christian Stepanek, Martin Werner, and Gerrit Lohmann acknowledge funding by the Helmholtz Climate Initiative Reklim and the Alfred Wegener Institute's research programme Marine, Coastal and Polar Systems.

References

Abe-Ouchi, A., Saito, F., Kageyama, M., Braconnot, P., Harrison, S. P., Lambeck, K., Otto-Bliesner, B. L., Peltier, W. R., Tarasov, L., Peterschmitt, J.-Y., and Takahashi, K.: Ice-sheet configuration in the CMIP5/PMIP3 Last Glacial Maximum experiments, Geosci. Model Dev., 8, 3621–3637, https://doi.org/10.5194/gmd-8-3621-2015, 2015.

Andersen, J. L., Egholm, D. L., Knudsen, M. F., Jansen, J. D., and Nielsen, S. B.: The periglacial engine of mountain erosion – Part 1: Rates of frost cracking and frost creep, Earth Surf. Dynam., 3, 447–462, https://doi.org/10.5194/esurf-3-447-2015, 2015.

Arnold, L., Breon, F. M., and Brewer, S: The earth as an extra-solar planet: the vegetation spectral signature today and during the last Quaternary climatic extrema, Int. J. Astrobiol., 8, 81–94, https://doi.org/10.1017/S1473550409004406, 2009.

Asahi, K.: Equilibrium-line altitudes of the present and Last Glacial Maximum in the eastern Nepal Himalayas and their implications for SW monsoon climate, Quatern. Int., 212, 26–34, https://doi.org/10.1016/j.quaint.2008.08.004, 2010.

Bahrenberg, G., Giese, E., and Nipper, J.: Multivariate Statistik, Statistische Methoden in der Geographie 2, Stuttgart, 1992.

Baker, P. A., Seltzer, G. O., Fritz, S. C., Dunbar, R. B., Grove, M. J., Tapia, P. M., Cross L., Rowe H. D., and Baroda, J. P.: The History of South America Tropical Precipitation for the Past 25,000 Years, Science, 291, 640–643, https://doi.org/10.1126/science.291.5504.640, 2001.

Ballantyne, A. P., Greenwood, D. R., Sinninghe Damste, J. S., Csank, A. Z., Eberle, J. J., and Rybczynski, N.: Significantly warmer Arctic surface temperatures during the Pliocene indicated by multiple independent proxies, Geology, 38, 603–606, 2010.

Banfield, J. F., Barker, W. W., Welch, S. A., and Taunton, A.: Biological impact on mineral dissolution: application of the lichen model to understanding mineral weathering in the rhizosphere, P. Natl. Acad. Sci. USA, 96, 3404–3411, 1999.

Barnston, A. G. and Livezey, R. E.: Classification, Seasonality and Persistence of Low-Frequency Atmospheric Circulation Patterns, Mon. Weather Rev., 115, 1083–1126, https://doi.org/10.1175/1520-0493(1987)115<1083:CSAPOL>2.0.CO;2, 1987.

Barron, J. A. and Anderson, L.: Enhanced Late Holocene ENSO/PDO expression along the margins of the eastern North Pacific, Quatern. Int., 235, 3–12, 2011.

Bendick, R. and Ehlers, T. A.: Extreme localized exhumation at syntaxes initiated by subduction geometry, Geophys. Res. Lett., 41, 2014GL061026, https://doi.org/10.1002/2014GL061026, 2014.

Bertrand, S., Charlet, F., Charlier, B., Renson, V., and Fagel, N.: Climate variability of southern Chile since the Last Glacial Maximum: a continuous sedimentlogical records from Lago Puyehue (40° S), J. Paleolimnol., 39, 179–195, https://doi.org/10.1007/s10933-007-9117-y, 2008.

Betancourt, J. L., Latorre C., Rech J. A., Quade J., and Rylander K. A.: A 22,000 Year Record of Monsoonal Precipitation from Northern Chile's Atacama Desert, Science, 289, 1542–1546, https://doi.org/10.1126/science.289.5484.1542, 2000.

Bigelow, N. H., Brubaker, L. B., Edwards, M. E., Harrison, S. P., Prentice, I. C., Anderson, P. M., Andreev, A. A., Bartlein, P. J., Christensen, T. R., Cramer, W., Kaplan, J. O., Lozhkin, A. V., Matveyeva, N. V., Murray, D. V., McGuire, A. D., Razzhivin, V. Y., Ritchie, J. C., Smith, B., Walker, D. A., Gajewski, K., Wolf, V., Holmqvist, B. H., Igarashi, Y., Kremenetskii, K., Paus, A., Pisaric, M. F. J., and Vokova, V. S.: Climate change and Arctic ecosystems I. Vegetation changes north of 55° N between the last glacial maximum, mid-Holocene and present, J. Geophys.

Res.-Atmos., 108, 1–25, https://doi.org/10.1029/2002JD002558, 2003.

Bird, B. W., Abbott M. B., Rodbell D. T., and Vuille M.: Holocene tropical South American hydroclimate revealed from a decadally resolved lake sediment $\delta18O$ record, Earth Planet. Sc. Lett., 310, 192–202, https://doi.org/10.1016/j.epsl.2011.08.040, 2011.

Bobst, A. L., Lowenstein, T. K., Jordan, T. E., Godfrey, L. V., Ku, T.-L., and Luo, S.: A 106 ka paleoclimate record from drill core of the Salar de Atacama, northern Chile, Palaeogeogr. Palaeocl., 173, 21–42, https://doi.org/10.1016/S0031-0182(01)00308-X, 2001.

Bohner, J.: General climatic controls and topoclimatic variations in Central and High Asia, Boreas, 35, 279–295, 2006.

Bookhagen, B., Thiede, R. C., and Strecker, M. R.: Late Quaternary intensified monsoon phases control landscape evolution in the northwest Himalaya, Geology, 33, 149–152, https://doi.org/10.1130/G20982.1, 2005.

Boos, W. R. and Kuang, Z.: Dominant control of the South Asian monsoon by orographic insulation versus plateau heating, Nature, 463, 218–222, 2010.

Braconnot, P., Otto-Bliesner, B., Harrison, S., Joussaume, S., Peterchmitt, J.-Y., Abe-Ouchi, A., Crucifix, M., Driesschaert, E., Fichefet, Th., Hewitt, C. D., Kageyama, M., Kitoh, A., Laîné, A., Loutre, M.-F., Marti, O., Merkel, U., Ramstein, G., Valdes, P., Weber, S. L., Yu, Y., and Zhao, Y.: Results of PMIP2 coupled simulations of the Mid-Holocene and Last Glacial Maximum – Part 1: experiments and large-scale features, Clim. Past, 3, 261–277, https://doi.org/10.5194/cp-3-261-2007, 2007.

Braun, J.: A simple model for regolith formation by chemical weathering: Regolith Formation, J. Geophys. Res.-Earth, 121, 2140–2171, https://doi.org/10.1002/2016JF003914, 2016.

Caves, J.: Late Miocene Uplift of the Tian Shan and Altai and Reorganization of Central Asia Climate, GSA Today, https://doi.org/10.1130/gsatg305a.1, 2017.

CLIMAP Project Members: Seasonal Reconstruction of the Earth's Surface at the Last Glacial Maximum, Map and Chart Series, Vol. 36, Geological Society of America, 18 pp., 1981.

Cohen, S. J.: Bringing the global warming issue close to home: The challenge of regional impact studies, B. Am. Meteorol. Soc., 71, 520–526, 1990.

D'Agostino, K., Seltzer, G., Baker, P., Fritz, S., and Dunbar, R.: Late-Quaternary lowstands of Lake Titicaca: evidence from high-resolution seismic data, Palaeogeogr. Palaeocl., 179, 97–111, https://doi.org/10.1016/S0031-0182(01)00411-4, 2002.

De Batist, M., Fagel N., Loutre M. F., and Chapron E.: A 17,900-year multi-proxy lacustrine record of Lago Puyehue (Chilean Lake District): introduction, J. Paleolimnol., 39, 151–161, https://doi.org/10.1007/s10933-007-9113-2, 2007.

Delunel, R., van der Beek, P. A., Carcaillet, J., Bourlès, D. L., and Valla, P. G.: Frost-cracking control on catchment denudation rates: Insights from in situ produced 10Be concentrations in stream sediments (Ecrins–Pelvoux massif, French Western Alps), Earth Planet. Sc. Lett., 293, 72–83, https://doi.org/10.1016/j.epsl.2010.02.020, 2010.

DeMenocal, P., Ortiz, J., Guilderson, T., Adkins, J., Samthein, M., Baker, L., and Yarusinsky, M.: Abrupt onset and termination of the African Humid Period:: rapid climate responses to gradual insolation forcing, Quaternary Sci. Rev., 19, 347–361, https://doi.org/10.1016/S0277-3791(99)00081-5, 2000.

Demske, D., Tarasov, P. E., Wünnemann, B., and Riedel, F.: Late glacial and Holocene vegetation, Indian monsoon and westerly circulation in the Trans-Himalaya recorded in the lacustrine pollen sequence from Tso Kar, Ladakh, NW India, Palaeogeogr. Palaeocl., 279, 172–185, https://doi.org/10.1016/j.palaeo.2009.05.008, 2009.

de Porras, M. E., Maldonado, A., Abarzua, A. M., Cardenas, M. L., Francois, J. P., Martel-Cea, A., Stern, C. R., Mendez, C., and Reyes, O.: Postglacial vegetation, fire and climate dynamics at Central Chilean Patagonia (Lake Shaman, 44° S), Quaternary Sci. Rev., 50, 71–85, https://doi.org/10.1016/j.quascirev.2012.06.015 2012.

de Porras, M. E., Maldonado, A., Quintana, F. A., Martel-Cea, A., Reyes, O., and Méndez, C.: Environmental and climatic changes in central Chilean Patagonia since the Late Glacial (Mallín El Embudo, 44° S), Clim. Past, 10, 1063–1078, https://doi.org/10.5194/cp-10-1063-2014, 2014.

Dettman, D. L., Fang, X. M., Garzione, C. N., and Li, J. J.: Uplift-driven climate change at 12 Ma: a long delta O-18 record from the NE margin of the Tibetan plateau, Earth Planet. Sc. Lett., 214, 267–277, 2003.

Diaz, F. P., Latorre, C., Maldonado, A., Quade, J., and Betancourt, J. L.: Rodent middens reveal episodic, long-distance plant colonizations across the hyperarid Atacama Desert over the last 34,000 years, J. Biogeogr., 39, 510–525, https://doi.org/10.1111/j.1365-2699.2011.02617.x, 2012.

Dietrich, S., Werner, M., Spangehl, T., and Lohmann, G.: Influence of orbital forcing and solar activity on water isotopes in precipitation during the mid- and late Holocene, Clim. Past, 9, 13–26, https://doi.org/10.5194/cp-9-13-2013, 2013.

Dietrich, W. E. and Perron, J. T.: The search for a topographic signature of life, Nature, 439, 411–418, https://doi.org/10.1038/nature04452, 2006.

Dowsett, H. J., Robinson, M., Haywood, A., Salzmann, U., Hill, D., Sohl, L., Chandler, M., Williams, M., Foley, K., and Stoll, D.: The PRISM3D paleoenvironmental reconstruction, Stratigraphy, 7, 123–139, 2010.

Egholm, D. L., Nielsen, S. B., Pedersen, V. K., and Lesemann, J.,: Glacial efects limiting moutain height, Nature, 460, 884–887, 2009.

Ehlers, T. A. and Poulsen, C. J.: Influence of Andean uplift on climate and paleoaltimetry estimates, Earth Planet. Sc. Lett., 281, 238–248, 2009.

Etheridge, D. M., Steele, L., Langenfelds, R., Francey, R., Barnola, J., and Morgan, V.,: Natural and anthropogenic changes in atmospheric CO_2 over the last 1000 years from air in Antarctic ice and firn, J. Geophys. Res., 101, 4115–4128, 1996.

Etheridge, D. M., Steele, L., Francey, R., and Langenfelds, R.: Atmospheric methane between 1000 a.d. and present: evidence of anthropogenic emissions and climatic variability, J. Geophys. Res., 103, 15979–15993, 1998.

Feng, R. and Poulsen, C. J.: Refinement of Eocene lapse rates, fossil-leaf altimetry, and North American Cordilleran surface elevation estimates, Earth Planet. Sc. Lett., 436, 130–141, https://doi.org/10.1016/j.epsl.2015.12.022, 2016.

Feng, R., Poulsen, C. J., Werner, M., Chamberlain, C. P., Mix, H. T., and Mulch, A.: Early Cenozoic evolution of topography, climate, and stable isotopes in precipitation in the North American Cordillera, Am. J. Sci., 313, 613–648, 2013.

Feng, R., Poulsen, C. J., and Werner, M.: Tropical circulation intensification and tectonic extension recorded by Neogene terrestrial d18O records of the western United States, Geology, 44, 971–974, https://doi.org/10.1130/G38212.1, 2016.

Fritz, S. C., Baker, P. A., Lowenstein, T. K., Seltzer, G. O., Rigsby, C. A., Dwyer, G. S., Tapia, P. M., Arnold, K. K., Ku, T. L., and Luo, S: Hydrologic variation during the last 170,000 years in the southern hemisphere tropics of South America, Quaternary Res., 61, 95–104, https://doi.org/10.1016/j.yqres.2003.08.007, 2004.

Gasse, F., Arnold, M., Fontes, J. C., Fort, M., Gibert, E., Huc, A., Li, B., Li, Y., Liu, Q., Melleres, F., Van Campo, E., Wang, F., and Zhang, Q.: A 13,000-year climate record from western Tibet, Nature, 353, 742–745, https://doi.org/10.1016/j.quaint.2006.02.001, 1991.

Gayo, E. M., Latorre, C., Santoro, C. M., Maldonado, A., and De Pol-Holz, R.: Hydroclimate variability in the low-elevation Atacama Desert over the last 2500 yr, Clim. Past, 8, 287–306, https://doi.org/10.5194/cp-8-287-2012, 2012.

Gierz, P., Lohmann, G., and Wei, W.: Response of Atlantic Overturning to future warming in a coupled atmosphere-ocean-ice sheet model, Geophys. Res. Lett., 42, 6811–6818, https://doi.org/10.1002/2015GL065276, 2015.

Gilli, A., Ariztegui, D., Bradbury, J. P., Kelts, K. R., Markgraf, V., and McKenzie, J. A.: Tracking abrupt climate change in the Southern Hemisphere: a seismic stratigraphic study of Lago Cardiel, Argentina (49° S), Terra Nova, 13, 443–448, https://doi.org/10.1046/j.1365-3121.2001.00377.x, 2001.

Glotzbach, C., van der Beek, P., Carcaillet, J., and Delunel, R.: Deciphering the driving forces of erosion rates on millennial to million-year timescales in glacially impacted landscapes: An example from the Western Alps, J. Geophys. Res.-Earth, 118, 1491–1515, 2013.

Gong, X., Knorr, G., Lohmann, G., and Zhang, X.: Dependence of abrupt Atlantic meridional ocean circulation changes on climate background states, Geophys. Res. Lett., 40, 3698–3704, https://doi.org/10.1002/grl.50701, 2013.

Grosjean, M., Van Leeuwen, J., Van der Knaap, W., Geyh, M., Ammann, B., Tanner, W., Messerli, B., Núñez, L., Valero-Garcés, B., and Veit, H.: A 22,000 14C year BP sediment and pollen record of climate change from Laguna Miscanti (23° S), northern Chile, Global Planet. Change, 28, 35–51, https://doi.org/10.1016/S0921-8181(00)00063-1, 2001.

Gyssels, G., Poesen, J., Bochet, E., and Li, Y.: Impact of plant roots on the resistance of soils to erosion by water: a review, Prog. Phys. Geog., 29, 189–217, https://doi.org/10.1191/0309133305pp443ra, 2005.

Hansen, B. C. S., Seltzer, G. O., and Wright Jr., H. E.: Late Quaternary vegetational change in the central Peruvian Andes, Palaeogeogr. Palaeocl., 109, 263–285, https://doi.org/10.1016/0031-0182(94)90179-1, 1994.

Harris, I., Jones, P. D., Osborn, T. J., and Lister, D. H.: Updated high-resolution grids of monthly climatic observations – the CRU TS3.10 Dataset, Int. J. Climatol., 34, 623–642, https://doi.org/10.1002/joc.3711, 2013.

Harrison, S. P., Yu, G., Takahara, H., and Prentice, I. C.: Palaeovegetation – Diversity of temperate plants in east Asia, Nature, 413, 129–130, 2001.

Harrison, S. P., Bartlein, P. J., Brewer, S., Prentice, I. C., Boyd, M., Hessler, I., Holmgren, K., Izumi, K., and Willis, K.: Climate model benchmarking with glacial and mid-Holocene climates, Clim. Dynam., 43, 671–688, doi 10.1007/s00382-013-1922-6, 2013.

Haywood, A. M., Valdes, P. J., and Sellwood, B. W.: Global scale palaeoclimate reconstruction of the middle Pliocene climate using the UKMO GCM: initial results, Global Planet. Change, 25, 239–256, 2000.

Haywood, A. M., Dowsett, H. J., Otto-Bliesner, B., Chandler, M. A., Dolan, A. M., Hill, D. J., Lunt, D. J., Robinson, M. M., Rosenbloom, N., Salzmann, U., and Sohl, L. E.: Pliocene Model Intercomparison Project (PlioMIP): experimental design and boundary conditions (Experiment 1), Geosci. Model Dev., 3, 227–242, https://doi.org/10.5194/gmd-3-227-2010, 2010.

Haywood, A. M., Hill, D. J., Dolan, A. M., Otto-Bliesner, B. L., Bragg, F., Chan, W.-L., Chandler, M. A., Contoux, C., Dowsett, H. J., Jost, A., Kamae, Y., Lohmann, G., Lunt, D. J., Abe-Ouchi, A., Pickering, S. J., Ramstein, G., Rosenbloom, N. A., Salzmann, U., Sohl, L., Stepanek, C., Ueda, H., Yan, Q., and Zhang, Z.: Large-scale features of Pliocene climate: results from the Pliocene Model Intercomparison Project, Clim. Past, 9, 191–209, https://doi.org/10.5194/cp-9-191-2013, 2013.

Herman, F., Seward, D., Valla, P. G., Carter, A., Kohn, B., Willett, S. D., and Ehlers, T. A.: Worldwide acceleration of mountain erosion under a cooling climate, Nature, 504, 423–426, https://doi.org/10.1038/nature12877, 2013.

Herzschuh, U., Kuerschner, H., and Mischke, S.: Temperature variability and vertical vegetation belt shifts during the last ~50,000 yr in the Qilian Mountains (NE margin of the Tibetan Plateau, China), Quaternary Res., 66, 133–146, https://doi.org/10.1016/j.yqres.2006.03.001, 2006a.

Herzschuh, U., Winter, K., Wuennemann, B., and Li, S.: A general cooling trend on the central Tibetan Plateau throughout the Holocene recorded by the lake Zigetang pollen spectra, Quatern. Int., 154, 113–121, https://doi.org/10.1016/j.quaint.2006.02.005, 2006b.

Herzschuh, U., Kramer A., Mischke S., and Zhang C.: Quantitative climate and vegetation trends since the late glacial on the northeastern Tibetan Plateau deduced from Koucha lake pollen spectra, Quaternary Res., 71, 162–171, https://doi.org/10.1016/j.yqres.2008.09.003, 2009.

Hillenbrand, C.-D. and Fütterer, D. K.: Neogene to Quaternary deposition of opal on the continental rise west of the Antarctic Peninsula, ODP Leg 178, Sites 1095, 1096, and 1101, in: Proceedings of the Ocean Drilling Programme, Scientific Results, 178, edited by: Barker, P. F., Camerlenghi, A., Acton, G. D., and Ramsay, A. T. S., Texas A and M University, College Station, Texas, 1–40 (CD-ROM), 2002.

Hillyer, R., Valencia, B. G., Bush, M. B., Silman, M. R., and Steinitz-Kannan, M.: A 24,700-yr paleolimnological history from the Peruvian Andes, Quaternary Res., 71, 71–82, https://doi.org/10.1016/j.yqres.2008.06.006, 2009.

Hobley, D. E., Sinclair, H. D., and Cowie, P. A.: Processes, rates, and time scales of fluvial response in an ancient postglacial landscape of the northwest Indian Himalaya, Geol. Soc. Am. Bull., 122, 1569–1584, 2010.

Hodell, D. A., Brenner, M., Kanfoush, S. L., Curtis, J. H., Stoner, J. S., Song, X., Wu, Y., and Whitmore, T. J.: Paleoclimate of southwestern China for the past 50,000 yr inferred from lake sediment records, Quaternary Res., 52, 369–380, https://doi.org/10.1006/qres.1999.2072, 1999.

Hu, G., Yi, C.-L., Zhang, J.-F., Liu, J.-L., Jiang, T., and Qin, X.: Optically stimulated luminescence dating of a moraine and a terrace in Laohugou valley, western Qilian Shan, northeastern Tibet, Quatern. Int., 321, 37–49, https://doi.org/10.1016/j.quaint.2013.12.019, 2014.

Insel, N., Poulsen, C. J., and Ehlers, T. A.: Influence of the Andes Mountains on South American moisture transport, convection, and precipitation, Clim. Dynam., 35, 1477–1492, 2010.

Jeffery, M. L., Ehlers, T. A., Yanites, B. J., and Poulsen, C. J.: Quantifying the role of paleoclimate and Andean Plateau uplift on river incision: paleoclimate role in river incision, J. Geophys. Res.-Earth, 118, 852–871, https://doi.org/10.1002/jgrf.20055, 2013.

Jenny, B., Valero-Garces, B. L., Urrutia, R., Kelts, K., Veit, H., and Geyh, M.: Moisture changes and fluctuations of the Westerlies in Mediterranean Central Chile during the last 2000 years: The Laguna Aculeo record (33° 50' S), Quatern. Int., 87, 3–18, https://doi.org/10.1016/S1040-6182(01)00058-1, 2002a.

Jenny, B., Valero-Garcés, B. L., Villa-Martínez, R., Urrutia, R., Geyh, M. A., and Veit, H.: Early to Mid-Holocene Aridity in Central Chile and the Southern Westerlies: The Laguna Aculeo Record (34° S), Quaternary Res., 58, 160–170, https://doi.org/10.1006/qres.2002.2370, 2002b.

Jungclaus, J. H., Lorenz, S. J., Timmreck, C., Reick, C. H., Brovkin, V., Six, K., Segschneider, J., Giorgetta, M. A., Crowley, T. J., Pongratz, J., Krivova, N. A., Vieira, L. E., Solanki, S. K., Klocke, D., Botzet, M., Esch, M., Gayler, V., Haak, H., Raddatz, T. J., Roeckner, E., Schnur, R., Widmann, H., Claussen, M., Stevens, B., and Marotzke, J.: Climate and carbon-cycle variability over the last millennium, Clim. Past, 6, 723–737, https://doi.org/10.5194/cp-6-723-2010, 2010.

Junginger, A., Roller, S., Olaka, L. A., and Trauth, M. H.: The effects of solar irradiation changes on the migration of the Congo Air Boundary and water levels of paleo-Lake Suguta, Northern Kenya Rift, during the African Humid Period (15–5 ka BP), Palaeogeogr. Palaeocl., 396, 1–16, https://doi.org/10.1016/j.palaeo.2013.12.007, 2014.

Kaiser, J., Schefuss, E., Lamy, F., Mohtadi, M., and Hebbeln, D.: Glacial to Holocene changes in sea surface temperature and coastal vegetation in north central Chile: high *versus* low latitude forcing, Quaternary Sci. Rev., 27, 2064–2075, https://doi.org/10.1016/j.quascirev.2008.08.025, 2008.

Kalnay, E., Kanamitsu, M., Kistler, R., Collins, W., Deaven, D., Gandin, L., Iredell, M., Saha, S., White, G., Woollen, J., Zhu, Y., Chelliah, M., Ebisuzaki, W., Higgins, W., Janowiak, J., Mo, K. C., Ropelewski, C., Wang, J., Leetmaa, A., Reynolds, R., Jenne, R., and Joseph, D.: The NCEP/NCAR 40-year reanalysis project, B. Am. Meteorol. Soc., 77, 437–471, 1996.

Kashiwaya, K., Masuzawa, T., Morinaga, H., Yaskawa, K., Yuan, B. Y., Liu, J. Q., and Gu, Z.: Changes in hydrological conditions in the central Qing-Zang (Tibetan) Plateau inferred from lake bottom sediments, Earth Planet. Sc. Lett., 135, 31–39, https://doi.org/10.1016/0012-821X(95)00136-Z, 1995.

Kent-Corson, M., Sherman, L., Mulch, A., and Chamberlain, C.: Cenozoic topographic and climatic response to changing tectonic boundary conditions in Western North America, Earth Planet. Sc. Lett., 252, 453–466, https://doi.org/10.1016/j.epsl.2006.09.049, 2006.

Kirchner, J. W., Finkel, R. C., Riebe, C. S., Granger, D. E., Clayton, J. L., King, J. G., and Megahan, W. F.: Mountain erosion over 10 yr, 10 k.y., and 10 m.y. time scales, Geology, 29, 591–594, 2001.

Kistler, R., Collins, W., Saha, S., White, G., Woollen, J., Kalnay, E., Chelliah, M., Ebisuzaki, W., Kanamitsu, M., Kousky, V., Van den Dool, H., Jenne, R., and Fiorino, M.: The NCEP–NCAR 50–Year Reanalysis: Monthly Means CD–ROM and Documentation, B. Am. Meteorol. Soc., 82, 247–267, 2001.

Knorr, G., Butzin, M., Micheels, A., and Lohmann, G.: A Warm Miocene Climate at Low Atmospheric CO_2 levels, Geophys. Res. Lett., 38, L20701, https://doi.org/10.1029/2011GL048873, 2011.

Koons, P. O., Zeitler, P. K., and Hallet, B.: 5.14 Tectonic Aneurysms and Mountain Building, in: Treatise on Geomorphology, Elsevier, 318–349, 2013.

Kotlia, B. S., Sharma, C., Bhalla, M. S., Rajagopalan, G.,. Subrahmanyam, K., Bhattacharya, A., and Valdiya, K. S.: Paleoclimatic conditions in the late Pleistocene Wadda Lake, eastern Kumaun Himalaya (India), Palaeogeogr. Palaeocl., 162, 105–118, https://doi.org/10.1016/S0031-0182(00)00107-3, 2000.

Kramer, A., Herzschuh, U., Mischke, S., and Zhang, C.: Late glacial vegetation and climate oscillations on the southeastern Tibetan Plateau inferred from the lake Naleng pollen profile, Quaternary Res., 73, 324–335, https://doi.org/10.1016/j.yqres.2009.12.003, 2010.

Kraus, H.: Die Atmosphäre der Erde, Eine Einführung in die Meteorologie, Berlin, 2001.

Kutzbach, J. E., Guetter, P. J., Ruddiman, W. F., and Prell, W. L.: Sensitivity of Climate to Late Cenozoic Uplift in Southern Asia and the American West – Numerical Experiments, J. Geophys. Res.-Atmos., 94, 18393–18407, 1989.

Kutzbach, J. E., Prell, W. L., and Ruddiman, W. F.: Sensitivity of Eurasian Climate to Surface Uplift of the Tibetan Plateau, J. Geol., 101, 177–190, 1993.

Lamy, F. and Kaiser. J.: Past Climate Variability in South America and Surrounding Regions, Chapter 6: Glacial to Holocene Paleoceanographic and Continental Paleoclimate Reconstructions Based on ODP Site 1233/GeoB 3313 Off southern Chile, 2009.

Lamy, F., Hebbeln, D., and Wefer, G.: High Resolution Marine Record of Climatic Change in Mid-latitude Chile during the Last 28,000 Years Based on Terrigenous Sediment Parameters, Quaternary Res., 51, 83–93, https://doi.org/10.1006/qres.1998.2010, 1999.

Lamy, F., Klump, J., Hebbeln, D., and Wefer, G.: Late Quaternary rapid climate change in northern Chile, Terra Nova, 12, 8–13, https://doi.org/10.1046/j.1365-3121.2000.00265.x, 2000.

Lamy, F., Rühlemann, C., Hebbeln, D., and Wefer, G.: High- and low-latitude climate control on the position of the southern Peru-

Cihle Current during the Holocene, Paleoceanography, 17, 16-1–16-10, https://doi.org/10.1029/2001PA000727, 2002.

Latorre, C., Betancourt, J., Rylander, K., and Quade, J.: Vegetation invasions into absolute desert: A 45 000 yr rodent midden record from the Calama-Salar de Atacama basins, northern Chile (lat 22°–24° S), Geol. Soc. Am. Bull., 114, 349–366, https://doi.org/10.1130/0016-7606(2002)114<0349:VIIADA>2.0.CO;2, 2002.

Latorre, C., Betancourt, J. L., Rylander, K. A., Quade, J., and Matthei, O.: A vegetation history from the arid prepuna of northern Chile (22–23° S) over the last 13500 years, Palaeogeogr. Palaeocl., 194, 223–246, https://doi.org/10.1016/S0031-0182(03)00279-7, 2003.

Latorre, C., Betancourt, J. L., and Arroyo, M. T. K.: Late Quaternary vegetation and climate history of a perennial river canyon in the Río Salado basin (22° S) of northern Chile, Quaternary Res., 65, 450–466, https://doi.org/10.1016/j.yqres.2006.02.002, 2006.

Lease, R. O. and Ehlers, T. A.: Incision into the Eastern Andean Plateau During Pliocene Cooling, Science, 341, 774–776, https://doi.org/10.1126/science.1239132, 2013.

Lechler, A. and Niemi, N.: Sedimentologic and isotopic constraints on the Paleogene paleogeography and paleotopography of the southern Sierra Nevada, California, Geology, 39, 379–382, https://doi.org/10.1130/g31535.1, 2011.

Lechler, A., Niemi, N., Hren, M., and Lohmann, K.: Paleoelevation estimates for the northern and central proto-Basin and Range from carbonate clumped isotope thermometry, Tectonics, 32, 295–316, https://doi.org/10.1002/tect.20016, 2013.

Legates, D. R. and Willmott, C. J.: Mean Seasonal and Spatial Variability in Gauge-Corrected, Global Precipitation, Int. J. Climatol., 10, 111–127, https://doi.org/10.1002/joc.3370100202, 1990.

Li, J., Ehlers, T. A., Werner, M., Mutz, S. G., Steger, C., and Paeth, H.: Late quarternary climate, precipitation $\delta^{18}O$, and Indian monsoon variations over the Tibetan Plateau, Earth Planet. Sc. Lett., 457, 412–422, 2017.

Licht, A., Quade, J., Kowler, A., de los Santos, M., Hudson, A., Schauer, A., Huntington, K., Copeland, P., and Lawton, T.: Impact of the North American monsoon on isotope paleoaltimeters: Implications for the paleoaltimetry of the American southwest, Am. J. Sci., 317, 1–33, https://doi.org/10.2475/01.2017.01, 2017.

Liu, K., Yao, Z., and Thompson, L. G.: A pollen record of Holocene climatic changes from the Dunde ice cap, Qinghai-Tibetan Plateau, Geology, 26, 135–138, 1998.

Liu, X., Colman, S. M., Brown, E. T., An, Z., Zhou, W., Jull, A. J. T., Huang, Cheng, Y., P., Liu, W., and Xu, H.: A climate threshold at the eastern edge of the Tibetan plateau, Geophys. Res. Lett., 41, 5598–5604, https://doi.org/10.1002/2014GL060833, 2014.

Lohmann, G., Pfeiffer, M., Laepple, T., Leduc, G., and Kim, J.-H.: A model-data comparison of the Holocene global sea surface temperature evolution, Clim. Past, 9, 180—1839, https://doi.org/10.5194/cp-9-1807-2013, 2013.

Lorenz, S. J. and Lohmann, G.: Acceleration technique for Milankovitch type forcing in a coupled atmosphere-ocean circulation model: method and application for the Holocene, Clim. Dynam., 23, 727–743, https://doi.org/10.1007/s00382-004-0469-y, 2004.

Maldonado, A. and Villagrán, C.: Climate variability over the last 9900 cal yr BP from a swamp forest pollen record along the semiardi coast of Chile, Quaternary Res., 66, 146–258, https://doi.org/10.1016/j.yqres.2006.04.003, 2006.

Maldonado, A. J., Betancourt, J. L., Latorre, C., and Villagrán, C.: Pollen analyses from a 50000-yr rodent midden series in the southern Atacama Desert (25° 30' S), J. Quaternary Sci., 20, 493–507, https://doi.org/10.1002/jqs.936, 2005.

Maldonado, A., Méndez, C., Ugalde, P., Jackson, D., Seguel, R., and Latorre, C.: Early Holocene climate change and human occupation along the semiarid coast of north-central Chile, J. Quaternary Sci., 25, 985–988, https://doi.org/10.1002/jqs.1385, 2010.

Markgraf, V., Bradbury, J. P., Schwalb, A., Burns, S., Stern, C., Ariztegui, D., Gilli, D., Anselmetti, F. S., Stine, S., and Maidana, N.: Holocene palaeoclimates of southern Patagonia: limnological and environmental history of Lago Cardiel, Argentina, Holocene, 13, 581–591, https://doi.org/10.1191/0959683603hl648rp, 2003.

Markgraf, V., Whitlock, C., and Haberle, S.: Vegetation and fire history during the last 18,000 cal yr B.P. in Southern Patagonia: Mallín Pollux, Coyhaique, Province Aisén (45° 41'30" S, 71° 50'30" W, 640 m elevation), Palaeogeogr. Palaeocl., 254, 492–507, https://doi.org/10.1016/j.palaeo.2007.07.008, 2007.

Maroon, E. A., Frierson, D. M. W., and Battisti, D. S.: The tropical precipitation response to Andes topography and ocean heat fluxes in an aquaplanet model, J. Climate, 28, 381–398, https://doi.org/10.1175/JCLI-D-14-00188.1, 2015.

Maroon, E. A., Frierson, D. M. W., Kang, S. M., and Scheff, J.: The precipitation response to an idealized subtropical continent, J. Climate, 29, 4543–4564, https://doi.org/10.1175/JCLI-D-15-0616.1, 2016.

Marshall, J. A., Roering, J. J., Bartlein, P. J., Gavin, D. G., Granger, D. E., Rempel, A. W., Praskievicz, S. J., and Hales, T. C.: Frost for the trees: Did climate increase erosion in unglaciated landscapes during the late Pleistocene?, Sci. Adv., 1, 1–10, 2015.

Marston, R. A.: Geomorphology and vegetation on hillslopes: Interactions, dependencies, and feedback loops, Geomorphology, 116, 206–217, https://doi.org/10.1016/j.geomorph.2009.09.028, 2010.

Matsuoka, N.: Solifluction rates, processes and landforms: A global review, Earth Sci. Rev., 55, 107–134, 2001.

Matsuoka, N. and Murton, J.: Frost weathering: Recent advances and future directions, Permafrost Periglac., 19, 195–210, 2008.

Maussion, F., Scherer, D., Mölg, T., Collier, E., Curio, J., and Finkelnburg, R.: Precipitation seasonality and variability over the Tibetan Plateau as resolved by the High Asia Reanalysis, J. Climate, 27, 1910–1927, https://doi.org/10.1175/JCLI-D-13-00282.1, 2014.

Méndez, C., Gil, A., Neme, G., Nuevo Delaunay, A., Cortegoso, V., Huidobro, C., Durán, and Maldano, A.: Mid Holocene radiocarbon ages in the Subtropical Andes (∼29°–35° S), climatic change an implicaton for human space organization, Quatern. Int., 356, 15–26, https://doi.org/10.1016/j.quaint.2014.06.059, 2015.

Meehl, G. A., Covey, C., Delworth, T., Latif, M., McAvaney, B., Mitchell, J. F. B., Stouffer, R. J., and Taylor, K. E.: The WCRP CMIP3 multi-model dataset: A new era in climate change research, B. Am. Meteorol. Soc., 88, 1383–1394, 2007.

Mesinger, F., DiMego, G., Kalnay, E., Mitchell, K., Shafran, P.C., Ebisuzaki, W., Jovic, D., Woollen, J., Rogers, E., Berbery, E.H., Ek, M.B., Fan, Y., Grumbine, R., Higgins, W., Li, H., Lin, Y., Manikin, G., Parrish, D., and Shi, W.: North American Regional Reanalysis, B. Am. Meteorol. Soc., 87, 343–360, https://doi.org/10.1175/BAMS-87-3-343, 2006.

Methner, K., Fiebig, J., Wacker, U., Umhoefer, P., Chamberlain, C., and Mulch, A.: Eocene-Oligocene proto-Cascades topography revealed by clumped (Δ47) and oxygen isotope (δ18O) geochemistry (Chumstick Basin, WA, USA), Tectonics, 35, 546–564, https://doi.org/10.1002/2015tc003984, 2016.

Mischke S., Kramer M., Herzschuh U., Shang H., Erzinger J., and Zhang C.: Reduced early holocene moisture availability in the Bayan Har Mountains, northeastern Tibetan Plateau, inferred from a multi-proxy lake record, Palaeogeogr. Palaeocl., 267, 59–76, https://doi.org/10.1016/j.palaeo.2008.06.002, 2008.

Molnar, P. and England, P.: Late Cenozoic uplift of mountain ranges and global climate change: chicken or egg?, Nature, 346, 29–34, 1990.

Molnar, P., Boos, W. R., and Battisti, D. S.: Orographic Controls on Climate and Paleoclimate of Asia: Thermal and Mechanical Roles for the Tibetan Plateau, Annu. Rev. Earth Planet. Sc., 38, 77–102, 2010.

Montgomery, D. R.: Slope distributions, threshold hillslopes, and steady-state topography, Am. J. Sci., 301, 432–454, 2001.

Montgomery, D. R., Balco, G., and Willett, S.D.: Climate, tectonics, and the morphology of the Andes, Geology, 29, 579–582, 2001.

Moon, S., Chamberlain, C. P., Blisniuk, K., Levine, N., Rood, D. H., and Hilley, G. E.: Climatic control of denudation in the deglaciated landscape of the Washington Cascades, Nat. Geosci., 4, 469–473, 2011.

Moreno, A., Santoro, C. M., and Latorre, C.: Climate change and human occupation in the northernmost Chilean Altiplano over the last ca. 11500 cal. a BP, Quaternary Sci., 24, 373–382, https://doi.org/10.1002/jqs.1240, 2009.

Morinaga, H., Itota, C., Isezaki, N., Goto, H., Yaskawa, K., Kusakabe, M., Liu, J., Gu, Z., Yuan, B., and Cong, S.: Oxygen-18 and carbon-13 records for the last 14,000 years from lacustrine carbonates of Siling-Co (lake) in the Qinghai-Tibetan Plateau, Geophys. Res. Lett., 20, 2909–2912, https://doi.org/10.1029/93GL02982, 1993.

Morrill, C., Overpeck, J. T., Cole, J. E., Liu, K., Shen, C., and Tang, L.: Holocene variations in theAsian monsoon inferred from the geochemistry of lake sediments in central Tibet, Quaternary Res., 65, 232–243, https://doi.org/10.1016/j.yqres.2005.02.014, 2006.

Moulton, K. L. and Berner, R. A.: Quantification of the effect of plants on weathering: studies in Iceland, Geology, 26, 895–898, 1998.

Muegler, I., Gleixner, G., Guenther, F., Maeusbacher, R., Daut, G., Schuett, B., Berking, J., Schwalb, A., Schwark, L,, Xu, B., Yao, T., Zhu, L., and Yi, C.: A multi-proxy approach to reconstruct hydrological changes and Holocene climate development of Nam Co, Central Tibet, J. Paleolimnol., 43, 625–648, https://doi.org/10.1007/s10933-009-9357-0, 2010.

Mujica, M. I., Latorre, C., Maldonado, A., González-Silvestre, L., Pinto, R., De Pol-Holz, R., and Santoro, C. M.: Late Quaternary climate change, relict populations and present-day refugia in the northern Atacama Desert: a case study from Quebrada La Higuera (18° S), J. Biogeogr., 42, 76–88, https://doi.org/10.1111/jbi.12383, 2015.

Mulch, A., Sarna-Wojcicki, A., Perkins, M., and Chamberlain, C.: A Miocene to Pleistocene climate and elevation record of the Sierra Nevada (California), P. Natl. Acad. Sci. USA, 105, 6819–6824, https://doi.org/10.1073/pnas.0708811105, 2008.

Mulch, A., Chamberlain, C., Cosca, M., Teyssier, C., Methner, K., Hren, M., and Graham, S.: Rapid change in high-elevation precipitation patterns of western North America during the Middle Eocene Climatic Optimum (MECO), Am. J. Sci., 315, 317–336, https://doi.org/10.2475/04.2015.02, 2015.

Mutz, S. G., Ehlers, T. A., Li, J., Steger, C., Peath, H., Werner, M., and Poulsen, C. J.: Precipitation $\delta^{18}O$ over the South Asia Orogen from ECHAM5-wiso Simulation: Statistical Analysis of Temperature, Topography and Precipitation, J. Geophys. Res.-Atmos., 121, 9278–9300, https://doi.org/10.1002/2016JD024856, 2016.

Nishimura, M., Matsunaka, T., Morita, Y., Watanabe, T., Nakamura, T., Zhu, L., Nara, W. F., Imai, A., Izutsu, Y., and Hasuike, N.: Paleoclimatic changes on the southern Tibetan Plateau over the past 19,000 years recorded in lake Pumoyum Co, and their implications for the southwest monsoon evolution, Palaeogeogr. Palaeocl., 396, 75–92, https://doi.org/10.1016/j.palaeo.2013.12.015, 2014.

Otto-Bliesner, B. L., Brady, C. B., Clauzet, G., Tomas, R., Levis, S., and Kothavala, Z.: Last Glacial Maximum and Holocene Climate in CCSM3, J. Climate, 19, 2526–2544, 2006.

Paeth, H.: Key Factors in African Climate Change Evaluated by a Regional Climate Model, Erdkunde, 58, 290–315, 2004.

Peel, M. C., Finlayson, B. L., and McMahon, T. A.: Updated world map of the Köppen-Geiger climate classification, Hydrol. Earth Syst. Sci., 11, 1633–1644, https://doi.org/10.5194/hess-11-1633-2007, 2007.

Pfeiffer, M. and Lohmann, G.: Greenland Ice Sheet influence on Last Interglacial climate: global sensitivity studies performed with an atmosphere-ocean general circulation model, Clim. Past, 12, 1313–1338, https://doi.org/10.5194/cp-12-1313-2016, 2016.

Pickett, E. J., Harrison, S. P., Flenley, J., Grindrod, J., Haberle, S., Hassell, C., Kenyon, C., MacPhail, M., Martin, H., Martin, A. H., McKenzie, M., Newsome, J. C., Penny, D., Powell, J., Raine, J. I., Southern, W., Stevenson, J., Sutra, J.-P., Thomas, I., van der Kaars, S., and Ward, J.: Pollen-based reconstructions of biome distributions for Australia, South-East Asia and the Pacific (SEAPAC region) at 0, 6000 and 18,000 14C years B.P., J. Biogeogr., 31, 1381–1444, https://doi.org/10.1111/j.1365-2699.2004.01001.x, 2004.

Pingel, H., Mulch, A., Alonso, R., Cottle, J., Hynek, S., Poletti, J., Rohrmann, A., Schmitt, A., Stockli, D., and Strecker, M.: Surface uplift and convective rainfall along the southern Central Andes (Angastaco Basin, NW Argentina), Earth Planet. Sc. Lett., 440, 33–42, https://doi.org/10.1016/j.epsl.2016.02.009, 2016.

Prentice, I. C., Jolly, D., and BIOME 6000 Participants: Mid-Holocene and glacial-maximum vegetation geography of the northern continents and Africa, J. Biogeogr., 27, 507–519, 2000.

Pueyo, J. J., Sáez, A., Giralt, S., Valero-Garcés, B. L., Moreno, A., Bao, R., Schwalb, A., Herrera, C., Klosowska, B., and Taberner, C.: Carbonate and organic matter sedimentation and isotopic signatures in Lake Chungará, Chilean Altiplano, during the last 12.3 kyr, Palaeogeogr. Palaeocl., 307, 339–355, https://doi.org/10.1016/j.palaeo.2011.05.036, 2011.

Quade, J., Rech, J. A., Betancourt, J. L., Latorre, C., Quade, B., Rylander, K. A., and Fisher, T.: Paleowetlands and regional climate change in the central Atacama

Desert, northern Chile, Quaternary Res., 69, 343–360, https://doi.org/10.1016/j.yqres.2008.01.003, 2008.

Raymo, M. E. and Ruddiman, W. F.: Tectonic forcing of late cenozoic climate, Nature, 359, 117–122, 1992.

Rech, J. A., Pigati, J. S., Quade, J., and Betancourt, J. L.: Re-evaluation of mid-Holocene deposits at Quebrada Puripica, northern Chile, Palaeogeogr. Palaeocl., 194, 207–222, https://doi.org/10.1016/S0031-0182(03)00278-5, 2003.

Robinson, M. M.: New quantitative evidence of extreme warmth in the Pliocene Arctic, Stratigraphy, 6, 265–275, 2009.

Roeckner, E., Bäuml, G., Bonaventura, L., Brokoph, R., Esch, M., Giorgetta, M., Hagemann, S., Kirchner, I., Kornblueh, L., Manzini, E., Rhodin, A., Schlese, U., Schulzweida, U., and Tompkins, A.: The atmospheric general circulation model ECHAM5. Part I: Model description. Rep. 349Rep., Max Planck Institute for Meteorology, Hamburg, 127 pp., 2003.

Roering, J. J., Marshall, J., Booth, A. M., Mort, M., and Jin, Q.: Evidence for biotic controls on topography and soil production, Earth Planet. Sc. Lett., 298, 183–190, https://doi.org/10.1016/j.epsl.2010.07.040, 2010.

Salzmann, U., Williams, M., Haywood, A. M., Johnson, A. L. A., Kender, S., and Zalasiewicz, J.: Climate and environment of a Pliocene warm world, Palaeogeogr. Palaecl., 309, 1–8, https://doi.org/10.1016/j.palaeo.2011.05.044, 2011.

Sarnthein, M.: Sand deserts during glacial maximum and climatic optimum, Nature, 272, 43–46, https://doi.org/10.1038/272043a0, 1978.

Sarnthein, M., Gersonde, R., Niebler, S., Pflaumann, U., Spielhagen, R., Thiede, J., Wefer, G., and Weinelt, M.: Overview of Glacial Atlantic Ocean Mapping (GLAMAP 2000), Paleoceanography, 18, 1030, https://doi.org/10.1029/2002PA000769, 2003.

Schäfer-Neth, C. and Paul, A.: The Atlantic Ocean at the last glacial maximum: objective mapping of the GLAMAP sea-surface conditions, in: The South Atlantic in the late quaternary: reconstruction of material budgets and current systems, edited by: Wefer, G., Mulitza, S., and Ratmeyer, V., Springer, Berlin, 531–548, 2003.

Schaller, M. and Ehlers T. A.: Limits to quantifying climate driven changes in denudation rates with cosmogenic radionuclides, Earth Planet. Sc. Lett., 248, 153–167, https://doi.org/10.1016/j.epsl.2006.05.027, 2006.

Schaller, M., von Blanckenburg, F., Veldkamp, A., Tebbens, L. A., Hovius, N., and Kubik, P. W.: A 30 000 yr record of erosion rates from cosmogenic 10 Be in Middle European river terraces, Earth Planet. Sc. Lett., 204, 307–320, 2002.

Schwalb, A., Burns, S., and Kelts, K.: Holocene environments from stable isotope stratigraphy of ostracods and authigenic carbonate in Chilean Altiplano Lakes, Palaeogeogr. Palaeocl., 148, 153–168, https://doi.org/10.1016/S0031-0182(98)00181-3, 1999.

Shen J., Liu X. Q., Wang S. M., and Matsumoto R.: Palaeoclimatic changes in the Qinghai, Lake area during the last 18,000 years, Quatern. Int., 136, 131–140, https://doi.org/10.1016/j.quaint.2004.11.014, 2005.

Shen C., Liu K., Tang L., and Overpeck J. T.: Quantitative relationships between modern pollen rain and climate in the Tibetan Plateau, Rev. Palaeobot. Palyno., 140, 61–77, https://doi.org/10.1016/j.revpalbo.2006.03.001, 2006.

Siani, G., Colin, C., Michel, E., Carel, M., Richter, T., Kissel, C., and Dewilde, F.: Late Glacial to Holocene terrigenous sediment record in the Northern Patagonian margin: Paleoclimate implications, Palaeogeogr. Palaeocl., 297, 26–36, https://doi.org/10.1016/j.palaeo.2010.07.011, 2010.

Simmons, A. J., Burridge, D. M., Jarraud, M., Girard, C., and Wergen, W.: The ECMWF Medium-Range prediction models development of the numerical formulations and the impact of increased resolution, Meteorol. Atmos. Phys., 40, 28–60, 1989.

Sohl, L. E., Chandler, M. A., Schmunk, R. B., Mankoff, K., Jonas, J. A., Foley, K. M., and Dowsett, H. J.: PRISM3/GISS topographic reconstruction, U.S. Geological Survey Data Series, 419, 6 pp., 2009.

Sowers T., Alley, R. B., and Jubenville, J.: Ice core records of atmospheric N_2O covering the last 106,000 years, Science, 301, 945–948, 2003.

Stepanek, C. and Lohmann, G.: Modelling mid-Pliocene climate with COSMOS, Geosci. Model Dev., 5, 1221–1243, https://doi.org/10.5194/gmd-5-1221-2012, 2012.

Stine, S. and Stine M.: A record from Lake Cardiel of climate change in southern South America, Nature, 345, 705–708, https://doi.org/10.1038/345705a0,1990.

Szeicz, J. M., Haberle, S. G., and Bennett, K. D.: Dynamics of North Patagonian rainforests from fine-resolution pollen, charcoal and tree-ring analysis, Chonos Archipelago, Southern Chile, Aust. Ecol., 28, 413–422, https://doi.org/10.1046/j.1442-9993.2003.01299.x, 2003.

Takahashi, K. and Battisti, D.: Processes controlling the mean tropical pacific precipitation pattern. Part I: The Andes and the eastern Pacific ITCZ, J. Climate, 20, 3434–3451, 2007a.

Takahashi, K. and Battisti, D.: Processes controlling the mean tropical pacific precipitation pattern. Part II: The SPCZ and the southeast pacific dry zone, J. Climate, 20, 5696–5706, 2007b.

Tang, L., Shen, S., Liu, K., and Overpeck, J. T.: Changes in south Asian monsoon: new high-resolution paleoclimatic records from Tibet, China, Chinese Sci. Bull., 45, 87–91, https://doi.org/10.1007/BF02884911, 2000.

Thiede, R. C. and Ehlers, T. A.: Large spatial and temporal variations in Himalayan denudation, Earth Planet. Sc. Lett., 374, 256–257, https://doi.org/10.1016/j.epsl.2013.03.004, 2013.

Thomas, A.: The climate of the Gongga Shan range, Sichuan Province, PR China, Arct. Alp. Res., 29, 226–232, 1997.

Uppala, S. M., Kållberg, P. W., Simmons, A. J., Andrae, U., da Costa Bechtold, V., Fiorino, M., Gibson, J. K., Haseler, J., Hernandez, A., Kelly, G. A., Li, X., Onogi, K., Saarinen, S., Sokka, N., Allan, R. P., Andersson, E., Arpe, K., Balmaseda, M. A., Beljaars, A. C. M., van de Berg, L., Bidlot, J., Bormann, N., Caires, S., Chevallier, F., Dethof, A., Dragosavac, M., Fisher, M. Fuentes, M., Hagemann, S., Hólm, E., Hoskins, B. J., Isaksen, L., Janssen, P. A. E. M., Jenne, R., McNally, A. P., Mahfouf, J.-F., Morcrette, J.-J., Rayner, N. A., Saunders, R. W., Simon, P., Sterl, A., Trenberth, K. E., Untch, A., Vasiljevic, D., Viterbo, P., and Woollen, J.: The ERA-40 re-analysis, Q. J. Roy. Meteor. Soc., 131, 2961–3012, 2005.

Valla, P. G., Shuster, D. L., and van der Beek, P. A.: Significance increase in relief of European Alps during mid-Pleistocene glaciations, Nat. Geosci., 4, 688–692, https://doi.org/10.1038/ngeo1242, 2011.

Van Campo, E., Cour, P., and Huang, S.: Holocene environmental changes in Bangong Co basin (Western Tibet). Part 2: The pollen record, Palaeogeogr. Palaeocl., 120, 49–63, https://doi.org/10.1016/0031-0182(95)00033-X, 1996.

Villagrán, C. and Varela, J.: Palynological Evidence for Increased Aridity on the Cenral Chilean Coast during the Holocene, Quaternary Res., 34, 198–207, https://doi.org/10.1016/0033-5894(90)90031-F, 1990.

Villa-Martínez, R. Villagrán, C., and Jenny, B.: The last 7500 cal yr B.P. sf westerly rainfall in Central Chile inferred from a high-resolution pollen record from Laguna Aculeo (34° S), Quaternary Res., 60, 284–293, https://doi.org/10.1016/j.yqres.2003.07.007, 2003.

von Blanckenburg, F., Bouchez, J., Ibarra, D.E., and Maher, K.: Stable runoff and weathering fluxes into the oceans over Quaternary climate cycles, Nat. Geosci., 8, 538–542, https://doi.org/10.1038/ngeo2452, 2015.

Wei, W. and Lohmann, G.: Simulated Atlantic Multidecadal Oscillation during the Holocene, J. Climate, 25, 6989–7002, https://doi.org/10.1175/JCLI-D-11-00667.1, 2012.

Wang, R. L., Scarpitta, S. C., Zhang, S. C., and Zheng, M. P.: Later Pleistocene/Holocene climate conditions of Qinghai-Xizhang Plateau (Tibet) based on carbon and oxygen stable isotopes of Zabuye Lake sediments, Earth Planet. Sc. Lett., 203, 461–477, https://doi.org/10.1016/S0012-821X(02)00829-4, 2002.

Whipple, K. X.: The influence of climate on the tectonic evolution of mountain belts, Nat. Geosci., 2, 97–104, https://doi.org/10.1038/ngeo413, 2009.

Whipple, K. X. and Tucker, G. E.: Dynamics of the stream-power river incision model: Implications for height limits of mountain ranges, landscape response timescales, and research needs, J. Geophys. Res.-Sol. Ea., 104, 17661–17674, 1999.

Whipple, K. X., Kirby, E., and Broecklehurst, S. H.: Geomorphic limits to climate-induced increases in topographic relief, Nature, 401, 39–43, https://doi.org/10.1038/43375, 1999.

Wilks, D. S.: Statistical methods in the atmospheric sciences, 3rd Edn., Academic Press, Oxford, 2011.

Willett, S. D., Schlunegger, F., and Picotti, V.: Messinian climate change and erosional destruction of the central European Alps, Geology, 34, 613–616, 2006.

Wilson, G. S., Barron, J. A., Ashworth, A. C., Askin, R. A., Carter, J. A., Curren, M. G., Dalhuisen, D. H., Friedmann, E. I., Fyodorov-Davidov, D. G., Gilichinsky, D. A., Harper, M. A., Harwood, D. M., Hiemstra, J. F., Janecek, T. R., Licht, K. J., Ostroumov, V. E., Powell, R. D., Rivkina, E. M., Rose, S. A., Stroeven, A. P., Stroeven, P., van der Meer, J. J. M., and Wizevich, M. C.: The Mount Feather Diamicton of the Sirius Group: an accumulation of indicators of Neogene Antarctic glacial and climatic history, Palaeogeogr. Palaeocl., 182, 117–131, 2002.

Wischnewski, J., Mischke, S., Wang, Y., and Herzschuh, U.: Reconstructing climate variability on the northeastern Tibetan Plateau since the last lateglacial – a multi-proxy, dual-site approach comparing terrestrial and aquatic signals, Quaternary Sci. Rev., 30, 82–97, https://doi.org/10.1016/j.quascirev.2010.10.001, 2011.

Wünnemann, B., Mischke, S., and Chen, F.: A Holocene sedimentary record from Bosten lake, China, Palaeogeogr. Palaeocl., 234, 223–238, https://doi.org/10.1016/j.palaeo.2005.10.016, 2006.

Yanhong, W., Lücke, A., Zhangdong, J., Sumin, W., Schleser, G. H., Battarbee, R. W., and Weilan, X.: Holocene climate development on the central Tibetan Plateau: A sedimentary record from Cuoe Lake, Palaeogeogr. Palaeocl., 234, 328–340, https://doi.org/10.1016/j.palaeo.2005.09.017, 2006.

Yanites, B. J. and Ehlers, T. A.: Global climate and tectonic controls on the denudation of glaciated mountains, Earth Planet. Sc. Lett., 325, 63–75, 2012.

Yu, L. and Lai, Z.: Holocene climate change inferred from stratigraphy and OSL chronology of Aeolian sediments in the Qaidam Basin, northeastern Qinghai-Tibetan Plateau, Quaternary Res., 81, 488–499, https://doi.org/10.1016/j.yqres.2013.09.006, 2014.

Zhang, C. and Mischke, S.: A lateglacial and Holocene lake record from the Nianbaoyeze Mountains and inferences of lake, glacier and climate evolution on the eastern Tibetan Plateau, Quaternary Sci. Rev., 28, 1970–1983, https://doi.org/10.1016/j.quascirev.2009.03.007, 2009.

Zhang, J., Chen, F., Holmes, J. A., Li, H., Guao, X., Wang, J., Li, S., Lu, Y., Zhao, Y., and Qiang, M.: Holocene monsoon climate documented by oxygen and carbon isotopes from lake sediments and peat bogs in China: a review and synthesis, Quaternary Sci. Rev., 30, 1973–1987, https://doi.org/10.1016/j.quascirev.2011.04.023, 2011.

Zhang, R., Yan, Q., Zhang, Z. S., Jiang, D., Otto-Bliesner, B. L., Haywood, A. M., Hill, D. J., Dolan, A. M., Stepanek, C., Lohmann, G., Contoux, C., Bragg, F., Chan, W.-L., Chandler, M. A., Jost, A., Kamae, Y., Abe-Ouchi, A., Ramstein, G., Rosenbloom, N. A., Sohl, L., and Ueda, H.: Mid-Pliocene East Asian monsoon climate simulated in the PlioMIP, Clim. Past, 9, 2085–2099, h ttps://doi.org/10.5194/cp-9-2085-2013, 2013.

Zhang, X., Lohmann, G., Knorr, G., and Xu, X.: Different ocean states and transient characteristics in Last Glacial Maximum simulations and implications for deglaciation, Clim. Past, 9, 2319–2333, https://doi.org/10.5194/cp-9-2319-2013, 2013.

Zhang, X., Lohmann, G., Knorr, G., and Purcell C.: Abrupt glacial climate shifts controlled by ice sheet changes, Nature, 512, 290–294, https://doi.org/10.1038/nature13592, 2014.

Zhisheng, A., Kutzbach, J. E., Prell, W. L., and Porter, S. C.: Evolution of Asian monsoons and phased uplift of the South Asian plateau since Late Miocene times, Nature, 411, 62–66, 2001.

Zhou, W. J., Lu, X. F., Wu, Z. K., Deng, L., Jull, A. J. T., Donahue, D. J., and Beck, W.: Peat record reflecting Holocene climate change in the Zoigê Plateau and AMS radiocarbon dating, Chinese Sci. Bull., 47, 66–70, https://doi.org/10.1360/02tb9013, 2002.

7

Impacts of a large flood along a mountain river basin: the importance of channel widening and estimating the large wood budget in the upper Emme River (Switzerland)

Virginia Ruiz-Villanueva[1], Alexandre Badoux[2], Dieter Rickenmann[2], Martin Böckli[2], Salome Schläfli[3], Nicolas Steeb[2], Markus Stoffel[1,4,5], and Christian Rickli[2]

[1]Institute for Environmental Sciences, University of Geneva,
Boulevard Carl-Vogt 66, 1205 Geneva, Switzerland
[2]Swiss Federal Research Institute WSL, Zürcherstrasse 111, 8903 Birmensdorf, Switzerland
[3]Institute of Geological Sciences, University of Bern, Baltzerstrasse 1+3, 3012, Bern, Switzerland
[4]Department of Earth Sciences, University of Geneva, 13 rue des Maraîchers, 1205 Geneva, Switzerland
[5]Department F.-A. Forel for Aquatic and Environmental Sciences, University of Geneva,
66 Boulevard Carl Vogt, 1205 Geneva, Switzerland

Correspondence: Virginia Ruiz-Villanueva (virginia.ruiz@unige.ch)

Abstract. On 24 July 2014, an exceptionally large flood (recurrence interval ca. 150 years) caused large-scale inundations, severe overbank sedimentation, and damage to infrastructure and buildings along the Emme River (central Switzerland). Widespread lateral bank erosion occurred along the river, thereby entraining sediment and large wood (LW) from alluvial forest stands. This work analyzes the catchment response to the flood in terms of channel widening and LW recruitment and deposition, but also identifies the factors controlling these processes. We found that hydraulic forces (e.g., stream power index) or geomorphic variables (e.g., channel width, gradient, valley confinement), if considered alone, are not sufficient to explain the flood response. Instead, the spatial variability of channel widening was first driven by precipitation and secondly by geomorphic variables (e.g., channel width, gradient, confinement, and forest length). LW recruitment was mainly caused by channel widening (lateral bank erosion) and thus indirectly driven by precipitation. In contrast, LW deposition was controlled by channel morphology (mainly channel gradient and width). However, we also observed that extending the analysis to the whole upper catchment of the Emme River by including all the tributaries and not only to the most affected zones resulted in a different set of significant explanatory or correlated variables. Our findings highlight the need to continue documenting and analyzing channel widening after floods at different locations and scales for a better process understanding. The identification of controlling factors can also contribute to the identification of critical reaches, which in turn is crucial for the forecasting and design of sound river basin management strategies.

1 Introduction

Floods in mountain river basins are characterized by complex, intense meteorological events and equally complex process coupling between the hillslopes and channels (i.e., debris flows, debris floods, and floods), resulting in a high spatial variability of morphological responses (Harvey, 1986; Miller, 1990; Lapointe et al., 1998; Magilligan et al., 1998; Heritage et al., 2004; Arnaud-Fassetta, 2013; Savi et al., 2013; Thompson and Croke, 2013; Rickenmann et al., 2016). During high-intensity events, mass-movement processes (e.g., landslides, debris flows) may affect channel morphology and sediment supply, influencing the total sediment load during a flood (Lin et al., 2008). In forested areas, mass movements and bank erosion not only deliver large amounts of inorganic sediment, but also introduce large quantities of wood into the channel corridor. As a load component in forested rivers, large wood (defined as wood pieces exceeding 10 cm in diameter and 1 m in length; LW) can be placed in a similar framework to that used for sediment, whereby LW recruitment, transport, and deposition are the main processes to be understood as part of the LW budgeting (Gurnell, 2007). Large wood is a key component of stream ecosystems (Ruiz-Villanueva et al., 2016, and references within); however, LW and sediment in channels can also favor the creation of temporary dams and subsequently produce secondary flood pulses, thereby enhancing erosion and/or leading to the destruction of infrastructure along the channel (Cenderelli and Kite, 1998; Wohl et al., 2010; Ruiz-Villanueva et al., 2014). Flood damage and flood losses are intrinsic to the occurrence of major floods (Merritts, 2011). However, urbanization, an increase in impervious surfaces (Hollis, 1975), and river channelization or embankment constructions (Wyżga, 1997) are frequently invoked to explain the high economic losses caused by major flood events (Hajdukiewicz et al., 2015). Under such conditions, even frequent floods (i.e., lower-magnitude events) can lead to unexpectedly high damage.

Over the last decades, several major flood events occurred in different parts of Switzerland (e.g., August 1978, August 1987, September 1993, May 1999, October 2000, August 2005, and August 2007; Hilker et al., 2009; Badoux et al., 2014), thereby causing significant financial damage and costs. The August 2005 flood, which was by far the costliest natural disaster in Switzerland since the start of systematic records in 1972 (Hilker et al., 2009), claimed six lives and caused total financial damage exceeding CHF 3 billion. The dominant processes observed during this event were flooding, bank erosion, overbank sedimentation, landslides, and debris flows (Rickenmann and Koschni, 2010; Rickenmann et al., 2016). Moreover, the transport and deposition of more than 69 000 m^3 of LW along alpine and pre-alpine rivers has been recorded (Steeb et al., 2017; Rickli et al., 2018). The consequences of events like the one in 2005 pose threats to important infrastructure such as roads and settlements, and therefore these processes need to be better understood and quantified to provide a process understanding and improved preparedness. However, predicting the impacts of major floods on the fluvial system is very challenging and requires a wide range of analyses (Rinaldi et al., 2016; Surian et al., 2016). Some of the most recent studies in the field have focused on the (i) reconstruction of the hydrological event (e.g., Gaume et al., 2004), an (ii) analysis of flood hydraulic variables (e.g., Howard and Dolan, 1981; Miller, 1990; Wohl et al., 1994; Benito, 1997; Heritage et al., 2004; Thompson and Croke, 2013), (iii) hillslope processes and channel connectivity (e.g., Bracken et al., 2015; Croke et al., 2013; Wohl, 2017), (iv) geomorphic and sedimentological analysis of flood deposits (e.g., Wells and Harvey, 1987; Macklin et al., 1992), (v) quantification of morphological changes (e.g., Arnaud-Fassetta et al., 2005; Krapesch et al., 2011;

Thompson and Croke, 2013; Comiti et al., 2016; Surian et al., 2016; Righini et al., 2017), (vi) sediment budgeting (e.g., Milan, 2012; Thompson and Croke, 2013), or, more recently, (vii) the study of LW dynamics and budgeting (e.g., Lucía et al., 2015; Steeb et al., 2017).

Post-event surveys are invaluable when it comes to improving insights on flood-related processes (Gaume and Borga, 2008; Marchi et al., 2009; Rinaldi et al., 2016) such as LW recruitment and factors controlling LW deposition, which are both crucial for the proper management of river basins and flood hazard mitigation (Comiti et al., 2016). Despite this fact, analyses of LW dynamics after flood events remain quite rare (Comiti et al., 2016). We added this important component (i.e., LW dynamics) to the hydrometeorological and geomorphological post-event survey after the July 2014 flood in the Emme River. We focused on morphological changes (in terms of channel widening), the coupling between hillslopes and headwaters to the main channel, the supply of large quantities of LW, and its deposition through the river corridor. We analyzed the whole upper catchment of the Emme River, including all tributaries and not only the ones that were most affected in July 2014. By doing so we aimed at unraveling diverging responses among the different tributaries and river segments in terms of channel widening and LW dynamics. In terms of morphology, similar river sub-reaches may have responded differently to the flood, and we hypothesized that these differences could be explained by morphological and hydrometeorological parameters. To test this hypothesis we selected different morphological and hydrometeorological variables, such as channel gradient, channel sinuosity, drainage area, confinement index, forested channel length, and stream power, to identify the factors controlling channel widening, LW recruitment, and LW deposition. The geomorphic response of the catchment and the initiation of processes such as LW recruitment due to mass movements or bank erosion might be driven by precipitation, among other variables (e.g., discharge, channel width, depth, and gradient). However, the rainfall patterns and subsequent disturbance regimes that influence the temporal variation in LW export in a given watershed network are not yet fully understood (Seo et al., 2012, 2015). Therefore, we include the event precipitation as an explanatory variable in our analysis. We hypothesize that differences in the spatial precipitation pattern would have led to differences in channel widening, thereby regulating LW dynamics.

2 Material and methods

2.1 The Emme River basin

The Emme River has its origin in the Swiss pre-Alps (1400 m a.s.l.) and runs through the Emmental in the cantons of Lucerne and Bern in central Switzerland. The total drainage area at its mouth with the Aare River (near the city of Solothurn) is 963 km^2, with a stream length of 80 km. This work focuses on the upper Emme River basin (Fig. 1), including the uppermost tributaries to the inlet of the Emme River into a gorge called Räbloch. At this point, the Emme River basin has a drainage area of 94 km^2 and the network is formed by 19 streams (18 tributaries and the main branch of the Emme; Table 1). These 19 streams were further divided into 64 sub-reaches (lengths ranging between 36 and 5238 m, mean value 837 m) as explained in Sect. 2.3.2 (see also Fig. S1 in the Supplement). The only existing stream gauge in the area is located at Eggiwil station (Fig. 1; 745 m a.s.l. with a drainage area of 124 km^2), which is several kilometers downstream of the Räbloch gorge.

The geology of the basin (Fig. 1c) is composed mainly of Helvetic marginal limestone, the Ultrahelvetian flysch (with marls and sandstones), and sub-alpine molasse composed of sandstone, molasse conglomerates, and marls (Lehmann, 2001). During the Pleistocene glaciation, a large part of the Emmental was covered by glaciers, and moraine remains are preserved in the areas of Eggiwil, Oberburg, and Burgdorf. The Emme basin is extensively occupied by agricultural lands (50 %, mostly downstream); 40 % of the surface remains forested today, and only about 10 % is urbanized (Fig. 1b). The climate is temperate with moderate warm summers (the mean temperature in July is 16 °C according to the Langnau data series from 1931–2015; Federal Office of Meteorology and Climatology MeteoSwiss) and cold winters (mean temperature −1 °C in January). Total annual precipitation averaged 1315 mm at the station of Langnau (1901–2015), with mean monthly peaks in June of 160 mm. The flow regime of the Emme River is characterized by a seasonally fluctuating flow due to snowmelt in spring and thunderstorms in summer.

A history of severe flooding led to intensive river management activities in the 19th and 20th centuries with the construction of dams and weirs. These measures also resulted in the isolation of tributaries and low sediment transport (Fig. S2). Additionally, poor riparian conditions and water extractions for irrigation strongly influenced Emme River hydrology (Burkhardt-Holm and Scheurer, 2007).

2.2 The 24 July 2014 flood event

July 2014 was a very wet month in Switzerland, with frequent and extensive rainfall in the first 3 weeks, interrupted by a few dry intervals. Data from MeteoSwiss showed that the western half of Switzerland registered 2 to 3 times the long-term precipitation average for the month of July 2014 (FOEN, 2015). These wet episodes led to saturated soils, especially in the western and northeastern parts of Switzerland (FOEN, 2015; ARGE LLE Schangnau-Eggiwil, 2015). Between 24 and 28 July 2014, several thunderstorms occurred over different Swiss regions. Until 27 July, the storms were related to a weak pressure system over western Europe (MeteoSwiss, 2017). Generally, such relatively uniform pressure distributions result in light and variable winds at

ground level, which allows for the formation of cumulonimbus clouds, typically over regions with rough topography such as the Swiss pre-Alps. On 24 July, an extremely violent stationary thunderstorm developed with a precipitation hotspot located over the upper Emmental. The storm cell caused intense rainfall in the headwater catchments of the upper Emme basin where it triggered very severe floods. According to hourly CombiPrecip data from MeteoSwiss (Sideris et al., 2014), the heavy precipitation yielded maximal hourly values of approximately 65 mm locally (with totals reaching 96 mm during the 7 h event; Fig. 2). Heavy rainfall was largely restricted to the upper Emme catchment with a local maximum just north of the Sädelgrabe catchment. The cantonal rain gauge Marbachegg (red dot II in Fig. 1) that recorded the highest event precipitation value of 76 mm is located roughly 2 km northwest of the confluence of the Sädelgrabe torrent with the Emme River. According to the study entitled ARGE LLE Schangnau-Eggiwil (2015), the rainfall event was associated with a recurrence interval between 100 and 200 years.

Due to the wet soil conditions caused by the antecedent rain, several of the small steep tributaries of the Emme River reacted very quickly to the 24 July 2014 rainstorm. The receiving Emme River produced an exceptionally large flood. The discharge station in Eggiwil (124 km^2 catchment area, 38 years of records) registered a peak discharge of 338 m^3 s^{-1}, which corresponds to a recurrence interval of $\sim$ 150 years (FOEN, 2017). Runoff in Eggiwil rose very quickly and reached a maximum within only a few hours. In the framework of the local post-event analysis, peak values of the Emme runoff upstream of the gauging station Eggiwil were estimated for this flood based on downstream measurements and by using local flood marks (ARGE LLE Schangnau-Eggiwil, 2015; Table 2).

Hydrographs were reconstructed for Schangnau and Räbloch (ARGE LLE Schangnau-Eggiwil, 2015). Peak discharge amounted to approximately 240 m^3 s^{-1} at Kemmeriboden (51 km^2 catchment area). Along our study reach, peak values probably increased until Schangnau where they reached about 330 m^3 s^{-1}. In a natural gorge between the villages of Schangnau and Eggiwil (a place called Räbloch; 94 km^2 catchment area), the Emme River was impounded due to clogging. A temporary lake formed and according to field surveys peak runoff was reduced to about 280 m^3 s^{-1} (ARGE LLE Schangnau-Eggiwil, 2015).

At the Eggiwil gauging station, a first slight increase in discharge was recorded just after 06:00 LT and runoff reached 50 m^3 s^{-1} (a discharge statistically reached during one day per year based on data from 1975–2016) at approximately 09:00 LT; 5.5 h later, at 14:30 LT, the runoff along the falling limb of the Emme hydrograph decreased below 50 m^3 s^{-1}. Peak discharge at Eggiwil was reached at approximately 10:30 LT, about half an hour after the peak occurred at Räbloch. Hence, the 24 July 2014 flood event in the Emme was short. Similarly short floods with a very steep rising

Table 1. Overview of the 19 study streams, the sub-reaches in each stream, morphological characteristics (mean: averaged or total values), and total maximum and mean precipitation registered during the event in 2014 at each sub-catchment (explained in the text).

Stream name	Sub-reaches	Drainage area (km²)	Total stream length (m)	Total forested channel length (m)	Mean stream gradient	Mean width before (m)	Mean width ratio	Mean sinuosity	Total max. precip. (mm)	Total mean precip. (mm)
Bärselbach	12, 14, 15, 17, 18, 21, 23, 26, 27	13.1	7017	6276	0.048	7	1.74	1.27	92	69
Buembachgrabe	42, 43	4.9	4553	3913	0.062	10	1.48	1.14	97	86
Bietschligrabe	50, 51	2.3	1000	841	0.051	5	1.00	2.54	65	46
Bütlerschwandgrabe	53	2.8	1580	1248	0.137	11	1.01	1.21	62	46
Chaltbach	45	1.5	563	493	0.090	4	1.07	1.37	70	48
Emme	3, 4, 7, 11, 28, 30, 33, 36, 37, 40, 41, 44, 46, 48, 49, 52, 54, 55, 58, 59, 61, 62, 63	93.7	20571	12361	0.023	15	1.40	1.36	97	87
Gärtelbach	38	0.7	1769	1769	0.185	5	3.11	1.07	94	91
Hombach	64	2.8	922	907	0.082	8	1.48	1.37	50	38
Leimbach	1, 2	9.2	4402	4167	0.054	6	1.35	1.27	49	36
Sädelgrabe	39	1.6	3434	3219	0.123	6	4.75	1.25	97	94
Schöniseibach	8, 9, 10	4.5	1476	1280	0.085	7	1.06	1.27	81	65
Schwarzbach	56, 57	5.4	768	768	0.080	7	1.18	1.28	59	41
Stream	16, 22, 24, 25, 31, 32	1.0	1915	1787	0.133	5	1.82	1.30	92	89
Stream (Gerbehüsi)	60	0.2	227	227	0.193	6	1.00	1.25	53	48
Stream (Kemmeriboden)	29	0.3	610	610	0.256	5	3.11	1.09	92	67
Stream (Kemmerli)	34, 35	0.5	647	564	0.142	5	2.42	1.15	93	90
Stream (Schneeberg)	19, 20	1.1	846	751	0.099	6	1.57	1.34	87	79
Stream (Unterlochseite)	47	1.2	607	79	0.073	3	2.08	1.31	76	71
Mürenbach	5, 6	2.4	650	650	0.092	5	0.99	1.47	67	52

Figure 1. Location of the basin in central Switzerland. The coordinate system used was CH1903 / LV03. **(a)** Hill shade of the upper Emme River basin (up to Eggiwil); red dots show the location of the rain gauges (I: Kemmeriboden; II: Marbachegg; III: Schallenberg), and blue lines show the 19 streams analyzed (18 tributaries and the main river Emme). **(b)** Land use. Agr.: agriculture; rock: bare soil; forest; bush: shrubs and bushes. **(c)** Geology: 1 Quaternary and Neogene molasses; 2 moraines; 3 Paleogene flysch; 4 Cretaceous and Jurassic sedimentary rocks. **(d)** Debris flood and LW deposits in the lower part of the Sädelgrabe torrent upstream of the small road bridge and its confluence with the Emme River (photograph: Virginia Ruiz-Villanueva). **(e)** Road and bridge washed away during the flood in Bumbach (photograph: Virginia Ruiz-Villanueva). **(f)** Räbeli bridge damaged during the flood (photograph: Virginia Ruiz-Villanueva). Numbers from 1 to 5 **(a)** show the cross sections described in Table 2. Arrows **(d, e, f)** show the flow direction.

Table 2. Peak discharges along the Emme during the 24 July 2014 flood, measured or estimated in the framework of the local event analysis for several sites along the Emme River (data source: ARGE LLE Schangnau-Eggiwil, 2015). Note that the drainage area given here does not precisely correspond to the data in Table 2 because estimates were carried out for which flood marks were available. The locations of these sites are shown in Fig. 1.

Number in Fig. 1	Point along the Emme	Drainage area (km^2)	Peak discharge (best estimate) (m^3 s^{-1})	Specific peak discharge (best estimate) (m^3 s^{-1} km^{-2})	Range of peak value (m^3 s^{-1})
1	Kemmeriboden	51	240	4.7	204–276
2	Bumbach	67	300	4.5	255–345
3	Schangnau	86	330	3.8	281–380
4	Räbloch	94	280*	3.0	238–336
5	Eggiwil (Heidbüel)	124	338	2.7	Stream gauge record

* The reduction in discharge at this section is due to the clogging of the Räbloch gorge and related backwater effects.

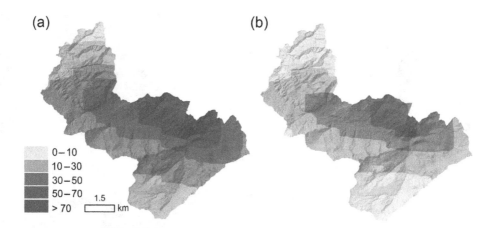

Figure 2. Map of the spatial distribution of precipitation (mm) on 24 July 2014 in the upper Emme catchment: **(a)** total event precipitation (mm) from 04:00 to 17:00 LT; **(b)** maximum hourly precipitation recorded at 07:00 LT.

hydrograph limb took place in June 1997 (245 m³ s⁻¹) and July 2012 (178 m³ s⁻¹), both caused by very intensive convective rainstorms as well. Further major floods that occurred in the 42 years of measurement were registered in 2005 and 2007 (both with peaks slightly above 175 m³ s⁻¹). However, these events were much longer due to the long-lasting nature of the triggering precipitation event (Bezzola and Hegg, 2007; Bezzola and Ruf, 2009).

The Emme River overflowed at various points in the upper catchment and caused large-scale inundations and severe overbank sedimentation (Fig. 1). Infrastructure, flood protection structures, and buildings were damaged and, in some cases, even destroyed. Moreover, widespread bank erosion occurred all along the Emme River, thereby entraining sediments and wood from alluvial forest stands. The steep torrents produced considerable debris floods and debris flows and transported large amounts of sediment and LW. The two most active torrents (Sädelgrabe and Gärtelbach) overtopped their channels and deposited ample amounts of material on their fans. Near the confluence of the Sädelgrabe and the Emme River, the road was obstructed by several meters of coarse material from the torrents. Furthermore, shallow landslides and hillslope debris flows occurred in steep locations of the upper Emme catchment. The lower part of the Gärtelbach (from an elevation around 1300 m a.s.l.) delivered around 2000 m³ of sediment to the Emme River, most of it recruited in the fluvial corridor, with 5000–7000 m³ of sediment deposited on the fan. The other main sediment source into the Emme River was the Sädelgrabe, where around 2000 m³ of material was deposited in the channel, and around 15 000 m³ of material was deposited on the cone (according to ARGE LLE Schangnau-Eggiwil, 2015). However, more detailed sediment budgeting was outside the scope of our work. Financial damage to private property and infrastructure (e.g., roads, bridges, hydraulic structures) in the worst-affected municipalities of Schangnau and Eggiwil was estimated at approximately CHF 20 million (Andres et al., 2015).

2.3 Methods

2.3.1 Field survey

A post-event survey was carried out right after the flood and during the following weeks. The Swiss Federal Office for the Environment (FOEN) initiated a project to study the recruitment, transport, and deposition of large wood in the upper catchment of the Emme River (Badoux et al., 2015; Böckli et al., 2016; Rickli et al., 2016), in which the main geomorphic effects of the flood were analyzed as well (Zurbrügg, 2015). This project was carried out in close collaboration with the local authorities (ARGE LLE Schangnau-Eggiwil, 2015).

The field survey after the flood focused on the quantification of deposited wood, the identification of recruitment sources, and the identification of changes in planform geometry (i.e., channel widening). The survey was carried out along 9.5 km of the Emme River (the section between 1250 m downstream of the confluence with the Bärselbach stream and the Räbloch gorge) and two of its main tributaries (Sädelgrabe and Gärtelbach), although other tributaries were visited as well. Regarding large wood, source areas (including landslides or debris floods and bank erosion) were identified in the field and mapped using aerial images (see next subsection), and wood deposits were measured in the field (details explained below). Moreover, we noted whether LW from hillslope processes reached the streams, as most of the mass movements were shallow landslides not directly connected to the channel network. However, mass movements were not very common and the main process recruiting LW was bank erosion.

Each piece of LW (length > 1 m and diameter > 10 cm; Wohl et al., 2010) deposited during the flood along the studied reaches was assigned to a class relative to its mid-length diameter and length (Marcus et al., 2002; Daniels, 2006; Lucía et al., 2015; Rickli et al., 2016); i.e., seven classes were distinguished from < 10 to > 40 cm in diameter and nine classes from < 2 to > 16 m for length. Log volume was calculated as solid cylinders (Thévenet et al., 1998). Wood accumulations (i.e., wood jams) were also measured. The wood volume of each jam was calculated geometrically through its area and height (measured in the field), considering a 50–80 % range in porosity (Thévenet et al., 1998). In the tributary catchments where large quantities of wood were deposited, mainly along the Sädelgrabe fan, the extension of wood deposits and the size of accumulations prevented the measurement of individual pieces. Areas with a similar density of wood were identified and plots were measured to estimate total wood volume in the area (see Fig. S3). Most of the recruited wood from the Gärtelbach was deposited along the Emme floodplain. Civil protection services removed some of the wood deposits immediately after the flood, storing the material at two sites close to the river, one near the confluence between the Sädelgrabe and the Emme and another near the Bumbach bridge (Fig. 1). These piles (five in total) were analyzed as well and wood samples were measured to estimate the stored wood volume and wood size distribution (Rickli et al., 2016).

2.3.2　GIS analysis

The field survey was complemented with GIS analyses (using ArcGIS 10.1; ESRI©) with the aim of extending the study to the upper catchment and including all tributaries. The entire upper catchment was analyzed by splitting the stream network into 64 sub-reaches according to the tributary junctions and the location of bridges, as bridges may act as obstacles to the downstream transfer of wood (see Fig. S1, Table 1). A total of 54.5 km of stream network length was analyzed.

For all sub-reaches, we calculated key morphological and hydrological parameters, such as maximum and minimum elevation, channel gradient, channel sinuosity (determined as the ratio between the actual sub-reach length and the straight distance), and drainage area, by using the available DEM (SwissALTI3D, 2 m spatial resolution) for the catchment, the GIS spatial analysis, and GIS hydrological geoprocessing. Other morphological parameters such as valley bottom width were extracted from the DEM using the fluvial corridor tool (Alber and Piégay, 2011; Roux et al., 2014). Moreover, the available aerial orthoimages (Swisstopo) were used to map the active channel before (image from March 2014, resolution 25 cm) and after (image from May 2015, resolution 25 cm) the flood. The post-flood units were also mapped in the field, with a focus on bank erosion and on measurements of the length and width of eroded banks (mostly along the Emme River). GIS measurements were compared and validated with field observations. The width of the active channel before and after the flood and the valley bottom (i.e., al-

luvial plain) width were calculated at several transects within each sub-reach. The centerline to the pre-flood and post-flood active channel polygon was obtained using the polygon-to-centerline tool (Dilts, 2015) and perpendicular transects were obtained with the transect tool (Ferreira, 2014); width was measured based on these transects. Transects were delineated at approximately regular intervals ranging between 20 and 50 m in length, with a total of 980 transects along the stream network.

We calculated the confinement index (C_i) as the ratio between the valley bottom width (W_{valley}) and the initial channel width (pre-flood; W_i),

$$C_i = W_{valley}/W_i, \tag{1}$$

and the width ratio (W_r) as the ratio between the width of the channel post-flood (W_f) and the pre-flood channel width (W_i), as proposed by Krapesch et al. (2011):

$$W_r = W_f/W_i. \tag{2}$$

Discharge was not measured except at the outlet of the basin (Eggiwil stream gauge station; Fig. 1), but estimations at other river sections were available (Table 2). Using these data and the drainage area (A) we used a potential equation to estimate peak discharges at all sub-reaches.

$$Q = 23 \cdot A^{0.6} \tag{3}$$

Because the estimates using Eq. (3) were relatively uncertain, stream power was not calculated using the estimated peak discharge of the flood; instead, we used the stream power index (SPI) proposed by Marchi and Dalla Fontana (2005) calculated as the product of the channel slope (S) and the square root of the drainage area (A).

$$SPI = S \cdot A^{0.5} \tag{4}$$

The spatial and temporal distribution of the precipitation was available from the CombiPrecip database recorded by MeteoSwiss, which is calculated using a geostatistical combination of rain gauge measurements and radar estimates with a regular grid of 1 km resolution (Sideris et al., 2014). For each sub-reach the drainage area was computed as explained above, and the hourly and cumulative total mean and total maximum values (i.e., the mean and maximum value of the total precipitation registered at each sub-catchment) were calculated.

The forest stand volumes ($m^3\,ha^{-1}$) before the event and eroded during the flood were assigned based on land use maps available for the study area and on information provided by the canton of Bern and the Swiss National Forest Inventory (NFI; Brassel and Lischke, 2001) to calculate recruited wood volume (in terms of eroded vegetation; see example in Fig. S4) and forested channel length. Forested channel length was determined by intersecting the forest cover with the river network. For this calculation, a wood buffer

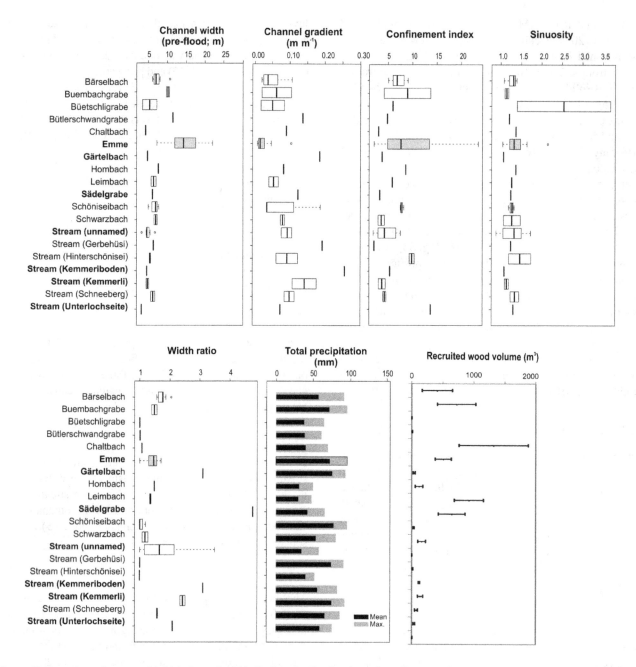

Figure 3. Box plots of averaged initial channel width (before the flood), channel gradient, confinement ratio, sinuosity, width ratio, precipitation, and recruited wood volumes for the 19 studied streams. Total maximum and mean total precipitation are calculated based on the 1 km precipitation grid cells in the respective catchments (maximum shows the highest value, mean shows the mean value recorded in each sub-catchment). Recruited wood volumes are given in ranges based on forest density ranges (as explained in the Methods section). In bold are the streams highlighted in Fig. 4.

strip of 10 m was added to the forest boundary to account for potential LW recruitment due to tree fall. The width of the strip was chosen to be half of the average tree height. The corresponding area comprises all possible locations of the centers of gravity of recruited wood logs (Mazzorana et al., 2011). The dataset used for this calculation is based on the digitized topological landscape model of Switzerland 1:

25 000 (source: Vector25 © 2007, Swisstopo, DV033594). Recruited wood volumes were normalized by initial channel area (i.e., $m^3\,ha^{-1}$) and channel length ($m^3\,km^{-1}$) to better compare sub-reaches and to compare our results with other studies in other regions. Detailed quantitative information about previously stored wood in the river channels was not available and we therefore had to assume that a value of

$100\,\mathrm{m^3\,ha^{-1}}$ was reliable for this catchment based on previous studies (Rickli und Bucher, 2006).

Deposited wood was directly measured in the field as explained above and by Rickli et al. (2016). Besides the field survey and the GIS analysis, all available media data (see the Supplement), including a video recorded from a helicopter on the day of the flood (http://www.heliweb.ch, last access: 21 April 2015), were also investigated (Zurbrügg, 2015). This analysis allowed for the mapping of the original depositional sites of the removed wood right after the flood and complemented the wood budget calculations. Deposited wood volumes were also normalized by initial channel area (i.e., $\mathrm{m^3\,ha^{-1}}$) and channel length ($\mathrm{m^3\,km^{-1}}$) for comparisons.

2.3.3   Statistical analysis

First, an exploratory analysis of the potential factors at the sub-reach scale was done by applying simple linear regression and correlation (nonparametric Spearman rank test). The explanatory variables analyzed were width ratio, wood recruited volume (total volume, $\mathrm{m^3}$; volume per area, $\mathrm{m^3\,ha^{-1}}$; volume per stream length, $\mathrm{m^3\,km^{-1}}$), and wood deposit volume (total volume, $\mathrm{m^3}$; volume per area, $\mathrm{m^3\,ha^{-1}}$; volume per stream length, $\mathrm{m^3\,km^{-1}}$). The controlling factors included were initial channel width, width ratio (for wood recruitment and deposition), channel gradient, sinuosity, confinement index, SPI, forested channel length, and total maximum and mean precipitation.

Sub-reaches were grouped according to their morphological characteristics, channel widening (using a value of width ratio $=> 1.2$ to characterize sub-reaches with important geomorphic changes in terms of channel widening), LW recruitment (sub-reaches with and without LW recruitment), and LW deposition (sub-reaches with and without LW deposition). Differences between groups of sub-reaches were tested using the nonparametric Mann–Whitney (i.e., Wilcoxon signed rank test for two groups) or Kruskal–Wallis (for more than two groups) tests. Significance of correlations and differences was set when p value < 0.1.

We hypothesize that one single variable may not explain the channel widening or LW dynamics, but that the combination of multiple variables would. Thus, we applied multivariate analysis to estimate the probability and factors controlling channel widening, LW recruitment, and LW deposition. We applied multiple linear regression and multivariate binary logistic regression by using a stepwise approach in both cases to identify the best model based on the Akaike information criterion (AIC) and the determination coefficient. The multivariate binary logistic regression estimates the probability of a binary response (e.g., high channel widening and low channel widening, presence or absence of LW recruitment) based on different predictor (or independent) variables (e.g., morphological variables). As the variables analyzed have very different units and different orders of magnitudes, the dataset

was standardized by mean-centering (the average value of each variable is calculated and then subtracted from the data, resulting in a transformed dataset such that the resulting variable has a zero mean; Becker et al., 1988) prior to computing (logistic and linear) multiple regressions. All analyses were done for all sub-reaches together, for sub-reaches along the Emme River only, and for sub-reaches along all tributaries. Variables were considered significant for p value < 0.1. Statistical analyses were carried out using the statistical software R (R Core Team, 2017) and the packages xlsx (Dragulescu, 2014), Rcmdr (Fox, 2005, 2017; Fox and Bouchet-Valat, 2017), corrgram (Wright, 2017), corrplot (Wei and Simko, 2017), and Hmisc (Harrell Jr., 2016).

3   Results

The morphology of the sub-reaches along the Emme River and tributaries is significantly different (see Fig. S5 in the Supplement); therefore, we analyzed their channel widening separately. Figure 3 shows the averaged values for different morphological variables, the calculated width ratio, and the precipitation for the 19 study streams including the Emme River reach (see Table 1). Looking at the different tributaries and the Emme River reach, we observe that the morphological response in terms of width ratio was very different and scatter in the data is very large (Fig. 3). The highest width ratio was observed in the Sädelgrabe, with nearly 5 times the initial channel width after the flood. The Gärtelbach and the tributary near Kemmeriboden also experienced significant channel widening. These streams were relatively narrow before the event, with initial channel widths smaller than $10\,\mathrm{m}$, very steep (with channel gradients higher than 0.1), and highly confined (with confinement indices smaller than or near 5).

3.1   Morphological flood response: channel widening

The exploratory analysis of the morphological characteristics (Fig. 4) showed that the relationships between width ratio and channel gradient, confinement index, initial channel width, SPI, sinuosity, and total maximum precipitation vary substantially. A large scatter exists in the data, and in some cases, relationships are very different for the Emme sub-reaches and for the tributary sub-reaches.

According to the Spearman rank test for all sub-reaches together (Fig. 5) and for the tributary sub-reaches (Fig. S7), a significant positive correlation was found between width ratio and the total maximum and mean precipitation. Forested channel length was also significantly correlated with channel widening along the Emme sub-reaches (Fig. S8).

When sub-reaches without significant widening (i.e., width ratio < 1.2; we consider a value larger than 20 % to be a reasonable threshold to distinguish significant widening) are removed from the correlation analysis, other significant correlations besides precipitation were observed (Table 3),

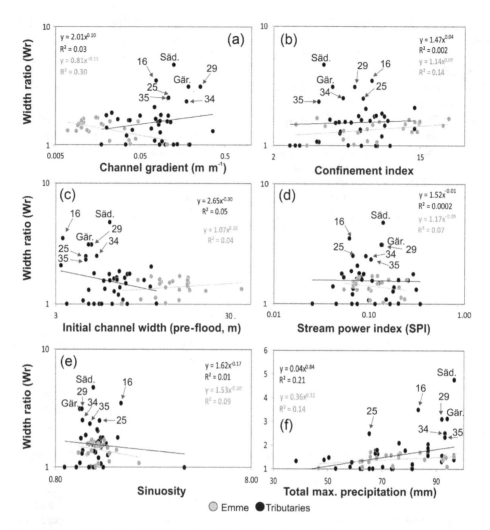

Figure 4. Relationships between width ratio and (**a**) channel gradient; (**b**) confinement; (**c**) channel width (pre-flood); (**d**) stream power index (in m m^{-1} km^{-2}); (**e**) sinuosity; (**f**) total maximum precipitation. Grey dots show sub-reaches along the Emme River, and black dots show sub-reaches along tributaries. Sub-reaches with the largest values are labeled. Säd.: Sädelgrabe; Gär.: Gärtelbach; numbers are sub-reaches as shown in Table 1. Grey and black lines show regression lines for the Emme River sub-reaches and tributary sub-reaches, respectively. Note that panel (**f**) has a linear x axis in contrast to the logarithmic x axis of panels (**a**)–(**e**).

such as channel gradient and initial channel width. Hence, the inclusion of sub-reaches that did not experience widening changed the results, a fact that is discussed further in Sect. 4.

We compared the sub-reaches showing widening (i.e., width ratio >=1.2) with the sub-reaches not showing widening (i.e., width ratio < 1.2), and results revealed significant differences between these two groups (see also Fig. S6) and between sub-reaches along the Emme and along tributaries (Fig. 6). We find that sub-reaches with a large width ratio were significantly less confined (high values of confinement index), less steep, and received much higher precipitation during the storm. By contrast, sub-reaches for which widening was important were also wider (channel width before the flood) and less forested; however, these differences were not significant. Interestingly, analysis of the sub-reaches along

the Emme and along the tributaries independently showed similar trends (Fig. 6).

The logistic regression points to an increase in the probability of widening occurrence with increasing precipitation and confinement index (Table S1). On the other hand, the probability of channel widening decreases with an increase in channel gradient, sinuosity, SPI, and forested channel length for all sub-reaches together. As with previous results, the sub-reaches along the Emme and along tributaries showed a contrasting behavior. Along the Emme, widening probability increased for wider, gentler, less sinuous, and less forested sub-reaches, whereas in the case of tributaries, the probability for the channels to widen was larger for narrower, steeper, sinuous forested sub-reaches.

The role of maximum precipitation is univocal in all cases, confirming our initial hypothesis about the role of the spa-

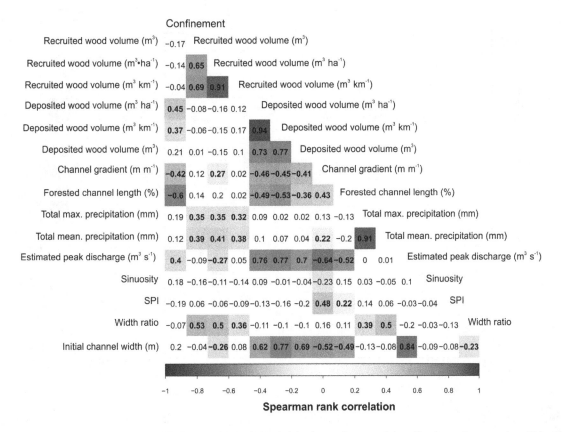

Figure 5. Spearman rank correlation matrix of all variables included in the analyses and for all sub-reaches together. Values show the Spearman rank results (significant correlations are in bold). Red colors show significant negative correlations, blue shows significant positive correlation, and white shows insignificant correlations.

Table 3. Spearman rank correlation matrix for the width ratio versus different variables and for all sub-reaches, only Emme sub-reaches, and only tributary sub-reaches showing widening (i.e., width ratio >=1.2). Bold indicates significant correlation.

Width ratio	Variables (all sub-reaches with width ratio => 1.2)	Variables (Emme sub-reaches with width ratio => 1.2)	Variables (tributary sub-reaches with width ratio => 1.2)
Confinement index	−0.12	0.22	−0.06
Channel gradient	**0.46**	0.33	0.26
Total max. precipitation (mm)	**0.35**	**0.40**	0.31
Total mean precipitation (mm)	**0.32**	**0.48**	**0.38**
Sinuosity	−0.06	0.08	−0.09
Forested channel length (%)	0.22	−0.18	−0.06
Initial channel width (m)	**−0.56**	**−0.49**	**−0.43**
SPI	−0.08	0.27	−0.23

tial distribution of precipitation. The logistic stepwise procedure revealed that the most significant variables explaining widening probability for all sub-reaches were total maximum precipitation, SPI, and estimated peak discharge (Table S1). Results obtained for sub-reaches along the Emme showed that forested channel length was also significant in explaining widening.

The multiple linear regression between width ratio values and the same explanatory variables for all sub-reaches identified precipitation, gradient, and SPI as significant variables. However, the obtained models explained only between 14 % and 19 % of the variability (Table S2). Separate multiple linear regression models for sub-reaches along the Emme and along tributaries further identify forested channel length, sinuosity, and initial channel width as significant variables;

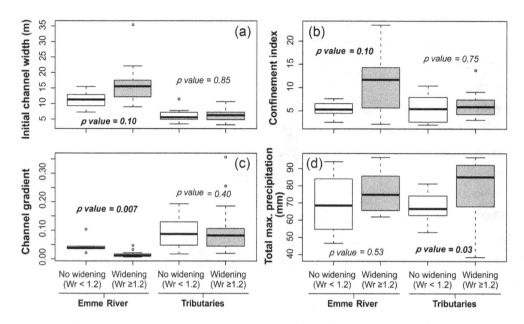

Figure 6. Box plots of morphological variables (initial channel width, confinement index, channel gradient) and total maximum precipitation for all sub-reaches showing widening (i.e., width ratio ≥ 1.2; grey boxes) and sub-reaches not showing widening (i.e., width ratio < 1.2). The bottom and top of the box indicate the first and third quartiles, respectively, the black line inside the box is the median, and circles are outliers. The Wilcoxon signed rank test result (p value) for the significance of differences is also shown; bold indicates significant differences.

overall, models explained between 20 % and 50 % of widening variability.

3.2 Large wood recruitment and deposition

3.2.1 Factors controlling large wood recruitment

The most important sources of LW were the tributaries Bärselbach, Buembachgrabe, Gärtelbach, Sädelgrabe, and Schöniseibach together with the main river Emme (see Fig. 3). To understand the factors controlling LW recruitment at the sub-reach scale better, we explored correlations between different variables and the total LW volume, as well as the normalized recruited wood volume per stream hectare (Fig. 7) and per channel length. In these analyses, we also included sub-reaches without LW recruitment.

Even though the results showed a large scatter, some relationships can be identified. For instance, we found a positive significant correlation between recruited wood volume (m³, m³ ha^{-1}, and m³ km^{-1}) and width ratio (Fig. 5). This confirms that bank erosion (i.e., channel widening) was the main recruitment process. Again, sub-reaches receiving larger amounts of precipitation recruited higher quantities of LW and we observe a statistically significant positive correlation between total maximum and mean precipitation and recruited wood volume (for all three recruited wood volume variables). This is explained by the control of precipitation driving discharge and thus driving the widening of channels and the wood recruitment process. Channel morphology may play a role in wood recruitment as well; we observe a signif-

icant negative correlation between recruited LW volume and initial channel width and a significant positive correlation with channel gradient (Fig. 5). However, these significant correlations were found only for wood volume per stream hectare and not for total wood volume or wood volume per stream length (Fig. 5), and thus conclusions should be taken with caution.

Independent analyses for sub-reaches along the Emme or along tributaries showed similar results (correlation matrices shown in Figs. S7 and S8). We also performed the same analysis with sub-reaches showing LW recruitment (i.e., removing those in which no LW was recruited) and found similar results in terms of significant correlations with the different variables (results not shown here). However, the comparative analysis of sub-reaches with and without LW recruitment (Fig. 8) revealed that LW recruitment was observed primarily in sub-reaches characterized by a significantly greater confinement index (i.e., unconfined sub-reaches) and significantly smaller slope. The results of all sub-reaches together, without grouping sub-reaches along the Emme and along tributaries, are shown in Fig. S9.

The logistic regression allowed for the calculation of the probability of LW recruitment occurrence; however, none of the analyzed variables were significant (Table S3), and the final stepwise logistic regression model selected just width ratio and confinement index as variables explaining LW recruitment probability. The multiple linear regression points to total maximum precipitation and width ratio as the most significant variables explaining total LW recruitment volume

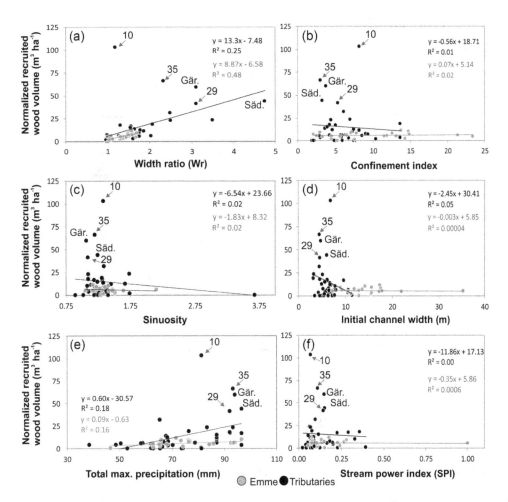

Figure 7. Relationships between recruited wood volume normalized by stream hectare (m³ ha⁻¹; mean value according to mean value of forest density) and (**a**) width ratio; (**b**) confinement index; (**c**) sinuosity; (**d**) initial channel width (m); (**e**) total maximum precipitation (mm); (**f**) SPI. Grey and black lines show regression lines for the Emme River and tributary sub-reaches, respectively. Säd.: Sädelgrabe; Gär.: Gärtelbach; numbers correspond to sub-reaches as shown in Table 1.

(total m³) variability, but forested channel length was also included in the final stepwise regression model for all sub-reaches. Between 10 % and 32 % of the variability was explained by these models (adjusted R^2) (Table S4).

3.2.2   Large wood deposition along the Emme River

LW deposits were analyzed along the Emme River and its tributary Sädelgrabe. However, because LW was mostly deposited on the Sädelgrabe fan and piled up nearby, only results obtained along the Emme sub-reaches can be provided here. The exploratory analysis of LW deposit distribution showed a positive relationship between deposited wood volume (normalized by initial stream area; m³ ha⁻¹) and width ratio, confinement index, initial channel width, and total precipitation; it showed a negative relationship with SPI (Fig. 9).

The Spearman test yielded a negative significant correlation of deposited LW with channel gradient and SPI and a positive correlation with estimated peak discharge (Fig. 5).

By contrast, the confinement index and initial channel width were only significantly correlated with deposited LW volume per hectare and per kilometer, respectively.

The comparison between Emme sub-reaches where LW was deposited or not showed statistically significant differences in terms of confinement index, channel gradient, and SPI (Fig. 10).

The probability of LW deposition estimated by logistic regression confirmed that LW deposition probability increases with increasing width ratio, confinement index, and initial channel width, whereas it decreases with increasing channel gradient and SPI. The multivariate stepwise logistic regression model identified both the confinement index and estimated peak discharge as significant variables explaining LW deposition, but also included the width ratio in the final model (Table S5).

The multiple linear regression of LW deposited volume (i.e., total m³, m³ m³ ha⁻¹, and m³ km⁻¹) showed

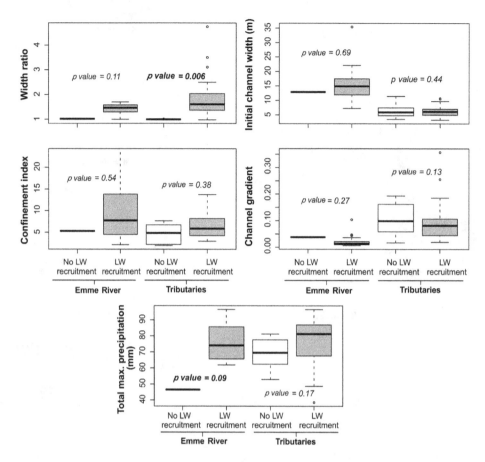

Figure 8. Box plots of morphological characteristics (width ratio, initial channel width, confinement index, channel gradient, and total max. precipitation of sub-reaches showing wood recruitment and not showing LW recruitment). The bottom and top of the box indicate the first and third quartiles, respectively, the black line inside the box is the median, and circles are outliers. The Wilcoxon signed rank test result (p value) for the significance of differences is also shown; bold indicates significant differences.

that the significant variables include channel gradient, estimated peak discharge, initial channel width, SPI, and confinement index (Table S6). The models explained between 51 % and 67 % of the variance. The largest variability (70 %) was explained for LW deposited volume per stream length ($m^3\,km^{-1}$).

3.3 Large wood budget and size distribution

LW budget was fully analyzed along (i) the lower part of the surveyed Emme River, in the section between Kemmeriboden (1.25 km downstream of the confluence with the Bärselbach stream) and the Räbloch, and (ii) the Sädelgrabe tributary. This tributary delivered large quantities of LW by mass movements, debris floods, and debris flows, which was mostly deposited along its fan and the Emme River.

Recruited LW volumes in the Sädelgrabe were due to landslides and bank erosion; the LW volume was estimated to be 331 m^3 (Table 4); together with the estimated volume of wood stored within the channel before the event (150 m^3), we obtained 481 m^3 of recruited and entrained wood. About

458 m^3 of wood was deposited at various locations (172 m^3 was deposited on the fan, 100 m^3 was piled up along the streambed and the municipal road, and 100 m^3 remained in the streambed of the Sädelgrabe after the event). Because the Sädelgrabe road bridge was completely blocked during the event (Fig. 1d), we estimated that only a small volume (about 40 m^3) was exported from the Sädelgrabe to the Emme River. Another source of LW was the Gärtelbach, which delivered large quantities of LW directly to the Emme River, a large portion of which (around 250 m^3) was deposited along the Emme floodplain in the vicinity of the bridge called Schwand downstream from the confluence.

Table 5 summarizes the partial wood budget computed along one segment of the Emme River.

As shown in Table 5 bank erosion along the surveyed Emme River segment recruited about 192 m^3 of wood, which together with the estimated previously stored wood (100 m^3) and the input from the Sädelgrabe was summed at 332 m^3. Roughly 250 m^3 was deposited in an area near Schwand and the rest along the Emme River (the sum of the deposited wood was approximately 360 m^3). In addition, about 300 m^3

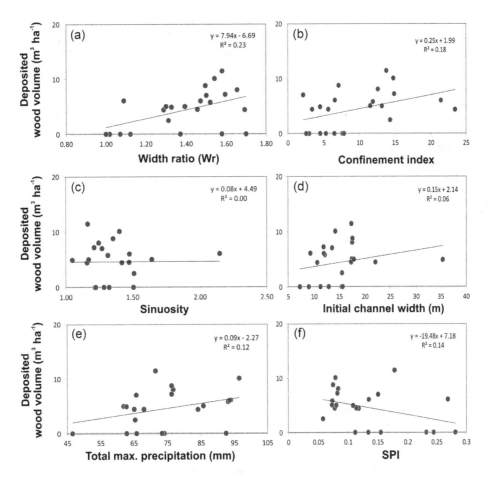

Figure 9. Relationships between deposited wood volume per initial stream hectare ($m^3\ ha^{-1}$) along the Emme River sub-reaches and **(a)** width ratio; **(b)** confinement index; **(c)** sinuosity; **(d)** channel width pre-flood; **(e)** total maximum precipitation (mm); and **(f)** SPI.

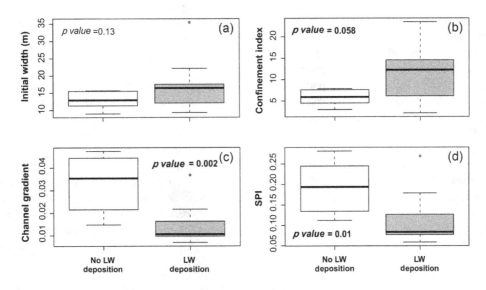

Figure 10. Box plots of morphological characteristics (initial channel width, confinement index, channel gradient, and SPI) of sub-reaches showing and not showing LW deposition along the Emme River. The bottom and top of the box indicate the first and third quartiles, respectively, the black line inside the box is the median, and circles are outliers. The Wilcoxon signed rank test result (*p* value) for the significance of differences is also shown; bold indicates significant differences.

Table 4. Wood budget along the Sädelgrabe. Uncertainties are included in the stated volumes.

Processes	Recruited (m^3)	Deposited (m^3)	Exported (m^3)
Landslides and/or bank erosion	331 ± 66		
Previously deposited in channel	150 ± 75		
Stored in the piles close to the confluence		100 ± 20	
Extracted before survey		32 ± 11	
Deposited on the fan (forests)		172 ± 34	
Deposited on the fan (pastures)		25 ± 9	
Subsequently deposited in channel (after event)		100 ± 50	
Stored in pile at fan apex		30 ± 15	
Exported to the Emme			40 ± 20
Total	481 ± 141	458 ± 139	40 ± 20

Table 5. Wood budget along the lower reach of the studied Emme River segment (reach between 1.25 km downstream of the Bärselbach stream and the Räbloch; Fig. 1). Uncertainties are included in the stated volumes.

Processes	Recruited (m^3)	Deposited (m^3)	Exported (m^3)
Bank erosion along the Emme studied reach	$192 \pm 38*$		
Previously deposited along the stream	100 ± 50		
Deposited along the river		360 ± 36	
Deposited but collected and piled in Bumbach		360 ± 36	
Stored jam in Räbloch gorge			480 ± 45
Input from Sädelgrabe	40 ± 20		
Input from Gärtelbach and other tributaries	unknown		
Total	332 ± 108	720 ± 72	480 ± 45

* This value is the estimated volume recruited by bank erosion along the surveyed Emme River reach only and does not include the LW recruitment upstream or in the tributaries, which is counted as input from Sädelgrabe ($40 \, m^3$) and other tributaries (unknown). See the text for details.

of deposited LW from flooded areas was collected and transported to a landfill and $60 \, m^3$ of LW was processed into firewood as part of cleanup work and post-event measures. Another important element of the balance is a large jam that formed about 1.6 km downstream of the investigated Emme section at Räbloch. According to eyewitness reports, a substantial amount of LW was transported and clogged at this narrow canyon, creating a dam 8 to 10 m in height with approximately $480 \, m^3$ of wood. Unfortunately, it is not known how much wood was transported from the upper Emme reach (e.g., from the Schöniseibach) or from the tributaries along the surveyed Emme reach (e.g., the Gärtelbach) where wood recruitment was important as well (Fig. 3); therefore, a mismatch exists between the estimations of recruited, deposited, and exported LW volumes (Table 5).

Pieces of LW were surveyed and measured both along the Emme sub-reaches between Kemmeriboden (1.25 km downstream of the confluence with the Bärselbach) and Räbloch (Fig. 1) and along the Sädelgrabe tributary. In total, 1995 (i.e., 1658 along the Emme and 297 on the Sädelgrabe fan and nearby piles) pieces were measured and the size distribution was further analyzed (Fig. 11). For both the Sädelgrabe and Emme River, piece frequency generally decreases with increasing piece length and diameter. Regarding the relative diameter distribution, almost no differences exist between the two sites, and in both cases the range class of 10–15 cm is the most frequent with approximately 50 % of the total. The mean and median values of piece length and diameter are very similar in the Emme River (mean D: 16.6 cm, mean L: 4.04 m, median D: 15 cm, median L: 2.32 m) and Sädelgrabe torrent (mean D: 17.4 cm, mean L: 3.06 m; median D: 15 cm, median L: 2.5 m). Regarding the relative length distribution, short wood pieces ($< 2 \, m$) were more frequently found along the Emme River (almost 60 %), whereas longer pieces ($> 2 \, m$) were more prevalent along the Sädelgrabe (around 60 %). However, the longest piece was found in the Emme (20.7 m), while the longest piece measured in the Sädelgrabe was substantially shorter with a value of 12.0 m.

4   Discussion

4.1   Channel widening during the 2014 flood

In this study, we presented an integration of different approaches and data sources (i.e., field survey, GIS remotely sensed data, and statistical analysis) at different spatial scales

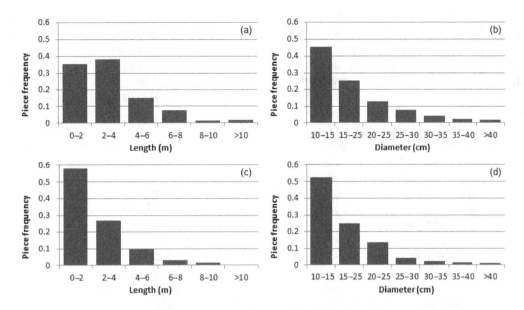

Figure 11. Size distribution (piece diameter and length) of deposited LW pieces in the Sädelgrabe (**a–b**) and in the Emme River (**c–d**). In all panels the bars relate to the relative frequency of pieces.

to better understand flood response in terms of channel widening and LW dynamics. We demonstrated the importance of performing an overall analysis of the entire catchment, although the flood event and responses to it were restricted to some areas of the catchment only. This approach allowed for the identification of hydrometeorological and geomorphic thresholds for channel widening, LW recruitment, and deposition. The inclusion of sub-reaches without important widening or without LW recruitment and deposition in the analysis showed that sub-reaches with similar characteristics may exhibit significantly different responses during the same event, and variables explaining these responses may not be identified properly if only one part of the dataset is analyzed. The threshold taken to distinguish sub-reaches with important widening, a width ratio => 1.2, was assumed to be reasonable considering that errors in the transect delineation, or in the delineation of the channel before and after the flood using the aerial images, could be up to 20 %.

Previous works have observed that hydraulic forces (e.g., stream power) are not sufficient to explain the geomorphic effects of floods (Nardi and Rinaldi, 2015), and other variables, such as initial channel width, confinement, and human interventions, should be included in assessments (Surian et al., 2016, and references therein). We confirmed with this study that the flood triggering precipitation is key in understanding the magnitude and spatial variability of catchment response (in terms of channel widening and LW dynamics) and that it should thus be included in future analyses. As hypothesized, differences in spatial precipitation patterns led to differences in the geomorphic response of the catchment, regulating channel widening, and thereby controlling LW dynamics. Although this observation may have been expected,

it has rarely been addressed in post-event surveys (Rinaldi et al., 2016) even in cases for which data were available at the proper spatial scale (e.g., Surian et al., 2016).

In general, we observed a large scatter in our dataset. However, precipitation was the univocal variable to explain channel widening in statistically significant terms, provided that all sub-reaches were included, whenever only sub-reaches along the Emme River or along tributaries were analyzed. A threshold value of around 80 mm of precipitation was observed in sub-reaches with the most important widening (Fig. 4). When sub-reaches without significant widening (illustrated here by sub-reaches with a width ratio < 1.2) were removed from analysis, channel morphology (in terms of initial channel width, confinement, and gradient) and hydraulic conditions (i.e., estimated peak discharge) were also significantly correlated with the width ratio. In fact, sub-reaches with a confinement index larger than 10 (i.e., unconfined channels) and wider than 10 m experienced less widening. This means that after intense precipitation events, channel morphology is a secondary driver for channel widening. Initial channel width was significantly negatively correlated with width ratio, as previously observed by Surian et al. (2016), Comiti et al. (2016), and Righini et al. (2017), who analyzed reaches that showed important widening in several streams in Italy. These authors also found the confinement index to be an important variable controlling channel widening. Regarding this variable, we observed contrasting behavior in the sub-reaches along the Emme River (where the width ratio was positively correlated with the confinement index) and along tributaries (where the width ratio was negatively correlated with the confinement index). This is because the largest widening was observed along tributary sub-reaches that are relatively more confined than the main river. In fact, only along the Emme River did we observe that sub-reaches showing channel widening were significantly less confined than sub-reaches not showing widening at all, as observed by Lucía et al. (2018). In contrast, differences were not significant along the tributaries. This apparently contradictory result might be explained by several factors. First, some large width ratios derived from aerial pictures (i.e., only on planimetric observations) may possibly include erosion of parts of the adjacent hillslopes, a process that also occurs during a flood. This was observed along the Sädelgrabe, where some highly confined transects showed widening ratios exceeding the confinement index (as also observed by Comiti et al., 2016). One may argue that these slope failures should not be considered as channel widening because the process here is more related to hillslope movements (e.g., falls, slips, slabs, slumps) than to channel processes. Second, some of the tributaries, especially the Sädelgrabe and Gärtelbach, received the highest amounts of precipitation, resulting in a more intensive response (e.g., large channel widening and debris flow triggering). Third, some uncertainties related to the use of the fluvial corridor tool to delineate the alluvial plain may affect the estimation of the confinement index. This tool uses the DEM and some predefined user parameters that may influence the final outcome (Roux et al., 2015). We carefully checked the results of this tool and adjusted the parameters to get a reliable valley bottom, which was verified using aerial imagery. However, small errors may remain, especially in steep and narrow sub-reaches. Besides channel confinement, lateral constraints, mainly artificial ripraps, artificial channelization, or natural bedrock, were present before the flood occurred, especially along the Emme River (see Fig. S2). These natural or artificial lateral constraints were not explicitly included in the analysis; however, they may have influenced results (Hajdukiewicz et al., 2015; Surian et al., 2016), therefore blurring factors controlling these processes and making their identification more difficult. In addition, major adjustments occurred during the last century, mostly channel narrowing and channel planform changes. These changes occurred at tributary confluences and along some of the Emme River sub-reaches, especially in unconfined sub-reaches where the stream changed from a braided to a single-thread pattern (Fig. S2). These anthropogenic changes may have an influence on current river response to floods and should thus be taken into account as well. Historical analyses were outside the scope of this study; however, they provided key information to assess whether and to what extent the response of a flood may involve channel segments that experienced significant changes in historical times (Rinaldi et al., 2016).

As shown in the results, sub-reaches along tributaries experiencing large channel widening were significantly steeper than those without widening, while along the Emme River channel, widening happened mostly along the gentler sub-reaches. This contrasting effect is explained by the same factors discussed above regarding widening and confinement. A negative correlation between width ratio and channel gradient, as found along the flatter Emme River reaches, was also observed by Lucía et al. (2015, 2018). The hydraulic conditions represented here by the estimated discharge and the SPI were not found to be significantly correlated with width ratio, although the multiple linear regression identified SPI as a significant variable explaining channel widening. Due to the large uncertainties related to the estimation of peak discharge at each transect and sub-reach, we preferred not to use total stream power or unit stream power for analysis, but selected SPI instead. The use of this index as a proxy for stream power is only based on the stream morphology and therefore also has some limitations (as shown by Lucía et al., 2018). The results should be treated cautiously. Additional hydrological modeling efforts could provide more robust estimates of discharge (Rinaldi et al., 2016) but are out of the scope of this study. Even when accurate discharge estimates are available, stream power has been shown to only partially explain channel changes, as other factors might be more relevant (Krapesch et al., 2011; Comiti et al., 2016; Surian et al., 2016; Righini et al., 2017; Lucía et al., 2018). Finally, another morphological variable included in our analysis was sinuosity. However, this variable was not significant and did not explain channel widening.

Besides channel morphology, the presence of vegetation also influenced channel widening. Forested channel length was negatively correlated with width ratio, and sub-reaches that experienced large widening were significantly less forested than those not experiencing channel widening. This illustrates the role of vegetation in protecting riverbanks from erosion (Abernethy and Rutherfurd, 1998). Other variables such as bank material (e.g., cohesive, non-cohesive, bedrock), type of vegetation, and vegetation density were not included in our analysis although they can be important factors affecting channel widening; they should therefore be considered in future analyses.

4.2 LW recruitment and deposition during the flood

LW recruitment was controlled primarily by bank erosion (i.e., channel widening), and thus factors controlling this process were identified as significant factors for LW recruitment. We observed a significant correlation between LW recruited volume and width ratio, precipitation, initial width, and channel gradient (i.e., the correlation with the last two variables was significant just for volume of wood recruited per initial channel hectare). The confinement index was also included in the final logistic regression model. During the Emme flood in 2014, we observed larger quantities of LW recruited in the smaller streams (i.e., the tributaries), which agrees with previous conceptual models (Seo and Nakamura, 2009). However, these models were defined for larger basins and longer times, not for single flood events. There are not many previous studies that analyzed LW recruitment after a single large flood. At the time of writing, the works of Lucía et al. (2015, 2018) and Steeb et al. (2017) were available, reporting results from northern Italy, the Swiss Alps, and southwestern Germany, respectively. The 2011 flood in the Magra River basin also recruited large amounts of LW, mostly by bank erosion. In their work, the authors did not find many significant correlations for total recruited wood volume, only a negative correlation with channel gradient (Lucía et al., 2015). In contrast, in the study of the 2016 flood in German streams, Lucía et al. (2018) observed significant correlations between recruited LW and drainage area and between stream power and confinement index. Our findings agree with these studies and previous observations in Switzerland (Steeb et al., 2017). They confirm the important role of bank erosion in recruiting wood material in mountain rivers, thereby highlighting the fact that hillslope processes were not the dominant LW supplier (contrary to what was proposed by Rigon et al., 2012).

This means that more attention should be paid to the understanding of bank erosion processes and the interactions with vegetation to predict or identify LW recruitment sources. Our findings also revealed that morphological variables alone may not explain or predict LW recruitment, and other factors should be considered as well, such as the triggering precipitation of the recruitment processes. As expected, the percentage of forested channel was also significant in the multiple linear regression model. However, other vegetation characteristics could play a role, such as the type and density of vegetation (Ruiz-Villanueva et al., 2014).

In our study, LW deposition was controlled mostly by channel morphology. We found significant correlations between LW deposited volume and initial channel width and between channel gradient and SPI. Sub-reaches where LW was deposited were significantly less confined (mainly in sub-reaches with confinement index higher than 7), wider (mainly in sub-reaches wider than 15 m, in agreement with the conceptual model proposed by Seo and Nakamura, 2009), and gentler than sub-reaches with no LW deposits. According to the multiple linear regression model, 67 % of the variance was explained by these variables. These results contrast with those found by Lucía et al. (2015), who did not find any statistically significant relationship with the controlling variables, although they observed that LW was more pronounced in the wider, gentle slope reaches typically located in the lower river sections. However, in their case, LW deposition was severely affected by the presence of several bridges and the formation of new in-channel islands due to bed aggradation. In two streams in Germany, Lucía et al. (2018) observed significant correlations between LW deposition and slope (negative correlation), drainage area, stream power, and confinement index. Our results partially agree with these observations.

Regarding the size of deposited logs, the median diameter observed in the field was 15 cm and the median length was 2.3 and 2.5 m in the Sädelgrabe and Emme River, respectively. These values were slightly smaller than values ob-

served after the flood in August 2005 in central Switzerland (Steeb et al., 2017; Rickli et al., 2018) and after the flood in the Magra River (although only log length was reported by Lucía et al., 2015), but in line with logs deposited along several streams in the Italian Alps (Rigon et al., 2012). We found smaller pieces along the Emme River compared to the Sädelgrabe, indicating that pieces in the Emme may have traveled longer distances and that pieces may have been broken during transport.

We could compute wood budgets just for one tributary (Sädelgrabe) and one segment of the Emme River. Similarly to what is commonly done for sediment transport, a wood budget for a river basin should be a quantitative statement of the rates of recruited (delivered), deposited, and transported wood volumes (Benda and Sias, 2003). The budget for the Emme River segment is not balanced, as we could not compute all elements (e.g., previously stored wood, deposited wood during the flood) of the budget in all sub-catchments upstream. Therefore, there is a mismatch between recruited and deposited LW volumes. This highlights the fact that computing wood budgets is very challenging and should be done at the catchment scale and not for a river segment only. However, it might be crucial for the proper management of river basins and when it comes to wood–flood hazard mitigation (Comiti et al., 2016).

The flood event analyzed here was a large flood, and although the recruited and transported LW resulted in significant damage (i.e., clogging bridges and damaging buildings), the exported volume was not extremely high. According to our estimations, most LW recruited in the Sädelgrabe (480 m^3) and along the lower reach in the Emme River (890 m^3) was not transported long distances downstream but deposited near its source. Part of the material was clogged in the Räbloch gorge (between the villages of Schangnau and Eggiwil), including the woody material from the bridge destroyed at Bumbach. Still, LW was transported further downstream and stored in several hydropower dams and reservoirs along the Aare (downstream of its confluence with the Emme River). According to the dam managers' estimations, a total of 1500 m^3 of wood was stored in five dams. However, it was not possible to compute precise budgets for the entire Emme catchment and its tributaries, and this value thus needs to be confirmed. Nevertheless, the exported LW volume in our study can be classified as very low when compared with volumes transported during the flood in August 2005 in Switzerland (Steeb et al., 2017) and with other events, as illustrated in the review by Ruiz-Villanueva et al. (2016).

Due to the complexity inherent to channel widening and LW dynamics, predictions on the location of major geomorphic changes and the magnitude of LW recruitment during large floods are very challenging (Buraas et al., 2014; Surian et al., 2016). Documenting events like the one reported here is fundamental for a better understanding of the processes involved and for the development of reliable and robust tools and approaches to facilitate the inclusion of such processes in flood hazard assessments (Comiti et al., 2016; Lucía et al., 2018). As such, a real need exists to complement current inundation mapping with a geomorphic approach (Rinaldi et al., 2015, 2016; Righini et al., 2017) and an integrative analysis of LW dynamics (Mazzorana et al., 2017).

5 Conclusions

Channel widening and LW dynamics are usually neglected in flood hazard mapping and river basin management. However, the present study clearly shows the importance of these processes during floods in mountain rivers. Still, the proper identification of factors controlling river basin response remains challenging. In that regard, our results also show that the identification of significant variables may be difficult, and depending on how the data are collected and analyzed (e.g., whether non-affected sub-reaches are included or not or which variables are considered), different outcomes are possible. However, we also showed that precipitation and variables such as forested channel length may play an important role in explaining channel widening, and they should thus be taken into consideration. Precipitation was the univocal statistically significant variable to explain channel widening, and only when sub-reaches without widening were removed from the analysis were channel morphology (i.e., initial channel width, confinement, and gradient) and hydraulic conditions (in terms of estimated peak discharge) also significantly correlated with width ratio. LW recruitment was controlled primarily by bank erosion and thus by the same variables controlling this process. This finding points to the need to better understand bank erosion processes and the interactions with vegetation to predict or identify LW recruitment sources. LW deposition was mostly controlled by channel morphology (i.e., initial channel width and gradient), and studies like this one are therefore crucial to identifying preferential reaches for wood deposition. This is an important component of the full wood budget, but not the only one. Further efforts in wood budgeting at the single event temporal scale are key to better understanding LW dynamics during floods in mountains rivers.

Author contributions. VRV, AB, DR, and CR contributed to the study conception and design. VRV, MB, SS, CR, and NS acquired the data. VRV, AB, DR, MB, SS, and CR analyzed and interpreted the data. VRV, AB, DR, and CR drafted the paper. NS and MS provided critical revisions.

Competing interests. The authors declare that they have no conflict of interest.

Acknowledgements. This study was funded by the Federal Office for the Environment (FOEN) in the framework of the "Large Wood during the July 2014 Emme Flood Project" (00.0157.PZ/N414-1285) and the "Research Project WoodFlow" (15.0018.PJ/O192-3154). Part of the work presented here was carried out within the scope of the local, solution-oriented event analysis of the 24 July 2014 Emme flood event (Lokale, lösungsorientierte Ereignisanaylse (LLE) Schangnau-Eggiwil) commissioned by the canton of Bern. We thank ARGE GEOTEST AG and Geo7 AG for the good collaboration. The Forest Division of the Canton of Bern (KAWA) provided the forest information for the canton of Bern. Massimiliano Zappa helped with the precipitation data handling, and Ramon Stalder and Karl Steiner assisted in the field.

References

Abernethy, B. and Rutherfurd, I. D.: Where along a river's length will vegetation most effectively stabilise stream banks?, Geomorphology, 23, 55–75, https://doi.org/10.1016/S0169-555X(97)00089-5, 1998.

Alber, A. and Piégay, H.: Spatial disaggregation and aggregation procedures for characterizing fluvial features at the network-scale: Application to the Rhone basin (France), Geomorphology, 125, 343–360, https://doi.org/10.1016/j.geomorph.2010.09.009, 2011.

Andres, N., Badoux, A., and Hegg, C.: Unwetterschäden in der Schweiz im Jahre 2014, Wasser Energie Luft, 107, 47–54, 2015.

ARGE LLE Schangnau-Eggiwil: Lokale, lösungsorientierte Ereignisanalyse (LLE) Schangnau-Eggiwil, Unwetter 24 Juli 2014, Report no. 1414 126.1, 2015.

Arnaud-Fassetta, G.: Dyke breaching and crevasse-splay sedimentary sequences of the Rhône Delta, France, caused by extreme river- flood of December 2003, Geogr. Fis. Din. Quat., 36, 7–26, 2013.

Arnaud-Fassetta, G., Cossart, E., and Fort, M.: Hydro-geomorphic hazards and impact of man-made structures during the catastrophic flood of June 2000 in the Upper Guil catchment (Queyras, Southern French Alps), Geomorphology, 66, 41–67, https://doi.org/10.1016/j.geomorph.2004.03.014, 2005.

Badoux, A., Andres, N., and Turowski, J. M.: Damage costs due to bedload transport processes in Switzerland, Nat. Hazards Earth Syst. Sci., 14, 279–294, https://doi.org/10.5194/nhess-14-279-2014, 2014.

Badoux, A., Böckli, M., Rickenmann, D., Rickli, C., Ruiz-Villanueva, V., Zurbrügg, S., and Stoffel, M.: Large wood transported during the exceptional flood event of 24 July 2014 in the Emme catchment (Switzerland), In: Proceedings of the 3rd International Conference Wood in World Rivers, Padova, Italy, July, 2015.

Becker, R. A., Chambers, J. M., and Wilks, A. R.: The New S Language. Wadsworth & Brooks/Cole, 1988.

Benda, L. E. and Sias, J. C.: A quantitative framework for evaluating the mass balance of in-stream organic debris, Forest Ecol. Manag., 172, 1–16, https://doi.org/10.1016/S0378-1127(01)00576-X, 2003.

Benito, G.: Energy expenditure and geomorphic work of the cataclysmic Missoula flooding in the Columbia river Gorge, USA, Earth Surf. Proc. Land., 22, 457–472, 1997.

Bezzola, G. R. and Hegg, C. (Eds.): Ereignisanalyse Hochwasser 2005, Teil 1 – Prozesse, Schäden und erste Einordnung, Umwelt-Wissen Nr. 0707, Bundesamt für Umwelt BAFU, Bern), Eidg. Forschungsanstalt WSL, Birmensdorf, Switzerland, 215 pp., 2007 (in German).

Bezzola, G. R. and Ruf, W.: Ereignisanalyse Hochwasser August 2007, Umwelt-Wissen Nr. 0927, Bundesamt für Umwelt BAFU, Bern, Switzerland, 209 pp., 2009 (in German).

Böckli, M., Badoux, A., Rickli, C., Forsting, D., Rickenmann, D., Ruiz-villanueva, V., Zurbrügg, S., and Stoffel, M.: Large wood-related hazards during the extreme flood of 24 July 2014 in the Emme catchment (Switzerland), in: Interpraevent 2016, 190–191, 2016.

Bracken, L. J., Turnbull, L., Wainwright, J., and Bogaart, P.: Sediment connectivity: a framework for understanding sediment transfer at multiple scales, Earth Surf. Proc. Land., 188, 177–188, https://doi.org/10.1002/esp.3635, 2015.

Brassel, P. and Lischke, H.: Swiss National Forest Inventory: Methods and models of the second assessment, Swiss Federal Research Institute WSL, Birmensdorf, Switzerland, 336 pp., 2001.

Buraas, E. M., Renshaw, C. E., Magilligan, F. J., and Dade, W. B.: Impact of reach geometry on stream channel sensitivity to extreme floods, Earth Surf. Proc. Land., 1789, 1778–1789, https://doi.org/10.1002/esp.3562, 2014.

Burkhardt-Holm, P. and Scheurer, K.: Application of the weight-of-evidence approach to assess the decline of brown trout (Salmo trutta) in Swiss rivers, Aquat. Sci., 69, 51–70, https://doi.org/10.1007/s00027-006-0841-6, 2007.

Cenderelli, D. A. and Kite, J. S.: Geomorphic effects of large debris flows on channel morphology at North Fork Mountain, eastern West Virginia, USA, Earth Surf. Proc. Land., 23, 1–19, 1998.

Comiti, F., Lucía, A., and Rickenmann, D.: Large wood recruitment and transport during large floods: a review, Geomorphology, 23–39, https://doi.org/10.1016/j.geomorph.2016.06.016, in press, 2016.

Croke, J., Fryirs, K., and Thompson, C.: Channel-floodplain connectivity during an extreme flood event: Implications for sediment erosion, deposition, and delivery, Earth Surf. Proc. Land., 38, 1444–1456, https://doi.org/10.1002/esp.3430, 2013

Daniels, M. D.: Distribution and dynamics of large woody debris and organic matter in a low-energy meandering stream, Geomorphology, 77, 286–298, https://doi.org/10.1016/j.geomorph.2006.01.011, 2006.

Dilts, T. E.: Polygon to Centerline Tool for ArcGIS, University of Nevada Reno, available at: http://www.arcgis.com/home/item.html?id=bc642731870740aabf48134f90aa6165 (last access: 29 January 2018), 2015.

Dragulescu, A. A.: xlsx: Read, write, format Excel 2007 and Excel 97/2000/XP/2003 files, R package version 0.5.7, available at: https://CRAN.R-project.org/package=xlsx (last access: 21 July 2018), 2014.

Ferreira, M.: Perpendicular Transects, GIS 4 Geomorphology, available at: http://gis4geomorphology.com/stream-transects-partial/ (last access: 29 January 2018), 2014.

FOEN: Hydrological Yearbook of Switzerland 2014, Federal Office for the Environment, Bern, Environmental Status no. UZ-1511-E, 36 pp., 2015.

FOEN: Hochwasserwahrscheinlichkeiten (Jahreshochwasser) Emme – Eggiwil, Heidbüel (EDV: 2409), Federal Office for the Environment, Bern, available at: https://www.hydrodaten.admin.ch/lhg/sdi/hq_studien/hq_statistics/2409hq.pdf (last access: 9 February 2018), 2017.

Fox, J.: The R Commander: A Basic Statistics Graphical User Interface to R, J. Stat. Softw., 14, 1–42, 2005.

Fox, J.: Using the R Commander: A Point-and-Click Interface or R, Boca Raton FL, Chapman and Hall/CRC Press, 2017.

Fox, J. and Bouchet-Valat, M.: Rcmdr: R Commander, R package version 2.3-2, 2017.

Gaume, E. and Borga, M.: Post-flood field investigations in upland catchments after major flash floods: proposal of a methodology and illustrations, J. Flood Risk Manag., 1, 175–189, https://doi.org/10.1111/j.1753-318X.2008.00023.x, 2008.

Gaume, E., Livet, M., Desbordes, M., and Villeneuve, J. P.: Hydrological analysis of the river Aude, France, flash flood on 12 and 13 November 1999, J. Hydrol., 286, 135–154, https://doi.org/10.1016/j.jhydrol.2003.09.015, 2004.

Gurnell, A. M.: Analogies between mineral sediment and vegetative particle dynamics in fluvial systems, Geomorphology, 89, 9–22, https://doi.org/10.1016/j.geomorph.2006.07.012, 2007.

Hajdukiewicz, H., Wyżga, B., Mikuś, P., Zawiejska, J., and Radecki-Pawlik, A.: Impact of a large flood on mountain river habitats, channel morphology, and valley infrastructure, Geomorphology, 272, 55–67, https://doi.org/10.1016/j.geomorph.2015.09.003, 2015.

Harrell Jr., F. E.: Hmisc: Harrell Miscellaneous, R package version 4.0-2, available at: https://CRAN.R-project.org/package=Hmisc (last access: 21 July 2018), 2016.

Harvey, A. M.: Geomorphic effects of a 100-year storm in the Howgill Fells, northwest England, Z. Geomorphol., 30, 71–91, 1986.

Heritage, G. L., Large, A. R. G., Moon, B. P., and Jewitt, G.: Channel hydraulics and geomorphic effects of an extreme flood event on the Sabie River, South Africa, Catena, 58, 151–181, 2004.

Hilker, N., Badoux, A., and Hegg, C.: The Swiss flood and landslide damage database 1972–2007, Nat. Hazards Earth Syst. Sci., 9, 913–925, https://doi.org/10.5194/nhess-9-913-2009, 2009.

Hollis, G. E.: The effect of urbanization on floods of different recurrence interval, Water Resour. Res., 11, 431–435, 1975.

Howard, A. and Dolan, R.: Geomorphology of the Colorado River in the Grand Canyon, J. Geol., 89, 269–298, 1981.

Krapesch, G., Hauer, C., and Habersack, H.: Scale orientated analysis of river width changes due to extreme flood hazards, Nat. Hazards Earth Syst. Sci., 11, 2137–2147, https://doi.org/10.5194/nhess-11-2137-2011, 2011.

Lapointe, M. F., Secretan, Y., Driscoll, S. N., Bergeron, N., and Leclerc, M.: Response of the River to the flood of July 1996 in the Saguenay Region of Quebec: large-scale avulsion in a glaciated valley, Water Resour. Res., 349, 2383–2392, 1998.

Lehmann, C.: Geschiebestudie hinteres Lauterbrunnental. Unveröffentlicht, Bericht, TBA-OIK-II, Thun, 2001.

Lin, G. W., Chen, H., Hovius, N., Horng, M. J., Dadson, S., Meunier, P., and Lines, M.: Effects of earthquake and cyclone sequencing on landsliding and fluvial sediment transfer in a mountain catchment, Earth Surf. Proc. Land., 33, 1354–1373, 2008.

Lucía, A., Comiti, F., Borga, M., Cavalli, M., and Marchi, L.: Dynamics of large wood during a flash flood in two mountain catchments, Nat. Hazards Earth Syst. Sci., 15, 1741–1755, https://doi.org/10.5194/nhess-15-1741-2015, 2015.

Lucía, A., Schwientek, M., Eberle, J., and Zarfl, C.: Planform changes and large wood dynamics in two torrents during a severe flash flood in Braunsbach, Germany 2016, Sci. Total Environ., 640–641, 315–326, https://doi.org/10.1016/j.scitotenv.2018.05.186, 2018.

Macklin, M. G., Rumsby, B. T., and Heap, T.: Flood alluviation and entrenchment: Holocene valley floor development and transformation in the British uplands, Geol. Soc. Am. Bull., 104, 631–643, 1992.

Magilligan, F. J., Phillips, J. D., James, L. A., and Gomez, B.: Geomorphic and Sedimentological Controls on the Effectiveness of an Extreme Flood, J. Geol., 106, 87–96, https://doi.org/10.1086/516009, 1998.

Marchi, L. and Dalla Fontana, G.: GIS morphometric indicators for the analysis of sediment dynamics in mountain basins, Environ. Geol., 48, 218–228, 2005.

Marchi, L., Borga, M., Preciso, E., Sangati, M., Gaume, E., Bain, V., Delrieu, G., Bonnifait, L., and Pogačnik, N.: Comprehensive post-event survey of a flash flood in Western Slovenia: Observation strategy and lessons learned, Hydrol. Process., 23, 3761–3770, https://doi.org/10.1002/hyp.7542, 2009.

Marcus, W. A., Marston, R. A., Colvard, C. R., and Gray, R. D.: Mapping the spatial and temporal distributions of woody debris in streams of the Greater Yellowstone Ecosystem, USA, Geomorphology, 44, 323–335, https://doi.org/10.1016/S0169-555X(01)00181-7, 2002.

Mazzorana, B., Hübl, J., Zischg, A., and Largiader, A.: Modelling woody material transport and deposition in alpine rivers, Nat. Hazards, 56, 425–449, https://doi.org/10.1007/s11069-009-9492-y, 2011.

Mazzorana, B., Ruiz-Villanueva, V., Marchi, L., Cavalli, M., Gems, B., Gschnitzer, T., and Valdebenito, G.: Assessing and mitigating large wood-related hazards in mountain streams: recent approaches, J. Flood Risk Manag., https://doi.org/10.1111/jfr3.12316, in press, 2017.

Merritts, D.: The effects of variable river flow on human communities, in: Inland Flood Hazards: Human, Riparian, and Aquatic Communities, edited by: Wohl, E. E., Cambridge Univ. Press, New York, 271–290, 2011.

MeteoSwiss: 24–28 July 2014 Weather situation, available at: http://www.meteoswiss.admin.ch/home/climate/past/climate-extremes/extreme-value-analyses/high-impact-precipitation-events/24-28-july-2014.html, last access: 20 April 2017.

Milan, D.: Geomorphic impact and system recovery following an extreme flood in an upland stream, Thinhope Burn, northern England, UK, Geomorphology, 138, 319–328, 2012.

Miller, A. J.: Flood hydrology and geomorphic effectiveness in the central Appalachians, Earth Surf. Proc. Land., 15, 119–134, 1990.

Nardi, L. and Rinaldi, M.: Spatio-temporal patterns of channel changes in response to a major flood event: The case of the Magra River (central-northern Italy), Earth Surf. Proc. Land., 40, 326–339, https://doi.org/10.1002/esp.3636, 2015.

R Core Team: R: A language and environment for statistical computing, R Foundation for Statistical Computing, Vienna, Austria, available at: https://www.R-project.org/ (last access: 21 July 2018), 2017.

Rickenmann, D. and Koschni, A.: Sediment loads due to fluvial transport and debris flows during the 2005 flood events in Switzerland, Hydrol. Process., 24, 993–1007, https://doi.org/10.1002/hyp.7536, 2010.

Rickenmann, D., Badoux, A., and Hunzinger, L.: Significance of sediment transport processes during piedmont floods: The 2005 flood events in Switzerland, Earth Surf. Proc. Land., 230, 224–230, https://doi.org/10.1002/esp.3835, 2016.

Rickli, C. and Bucher, H.: Einfluss ufernaher Bestockungen auf das Schwemmholzvorkommen in Wildbächen, Projektbericht Dezember, 2006.

Rickli, C., Böckli, M., Badoux, A., Rickenmann, D., Zurbrügg, S., Ruiz-Villanueva, V., and Stoffel, M.: Schwemmholztransport während des Hochwasserereignisses vom 24 Juli 2014 im Einzugsgebiet der Emme, Wasser Energie Luft, 108, 225–231, 2016 (in German).

Rickli, C., Badoux, A., Rickenmann, D., Steeb, N., and Waldner, P.: Large wood potential, piece characteristics, and flood effects in Swiss mountain streams, Phys. Geogr., 39, 542–564, https://doi.org/10.1080/02723646.2018.1456310, 2018.

Righini, M., Surian, N., Wohl, E., Marchi, L., Comiti, F., Amponsah, W., and Borga, M.: Geomorphic response to an extreme flood in two Mediterranean rivers (northeastern Sardinia, Italy): Analysis of controlling factors, Geomorphology, 290, 184–199, https://doi.org/10.1016/j.geomorph.2017.04.014, 2017.

Rigon, E., Comiti, F., and Lenzi, M. A.: Large wood storage in streams of the Eastern Italian Alps and the relevance of hillslope processes, Water Resour. Res., 48, W01518, https://doi.org/10.1029/2010WR009854, 2012.

Rinaldi, M., Surian, N., Comiti, F., and Bussettini, M.: A methodological framework for hydromorphological assessment, analysis and monitoring (IDRAIM) aimed at promoting integrated river management, Geomorphology, 251, 122–136, https://doi.org/10.1016/j.geomorph.2015.05.010, 2015.

Rinaldi, M., Amponsah, W., Benvenuti, M., Borga, M., Comiti, F., Lucía, A., Marchi, L., Nardi, L., Righini, M., and Surian, N.: An integrated approach for investigating geomorphic response to extreme events: methodological framework and application to the October 2011 flood in the Magra River catchment, Italy, Earth Surf. Proc. Land., 41, 835–846, https://doi.org/10.1002/esp.3902, 2016.

Roux, C., Alber, A., Bertrand, M., Vaudor, L., and Piégay, H.: "FluvialCorridor": A new ArcGIS toolbox package for multiscale riverscape exploration, Geomorphology, 242, 29–37, https://doi.org/10.1016/j.geomorph.2014.04.018, 2014.

Ruiz-Villanueva, V., Díez-Herrero, A., Ballesteros-Canovas, J. A., and Bodoque, J. M.: Potential large woody debris recruitment due to landslides, bank erosion and floods in mountain basins: a quantitative estimation approach, River Res. Appl., 30, 81–97, https://doi.org/10.1002/rra.2614, 2014.

Ruiz-Villanueva, V., Piégay, H., Gurnell, A. A., Marston, R. A., and Stoffel, M.: Recent advances quantifying the large wood dynamics in river basins: New methods and remaining challenges, Rev. Geophys., 54, 611–652, https://doi.org/10.1002/2015RG000514, 2016.

Savi, S., Schneuwly-Bollschweiler, M., Bommer-Denns, B., Stoffel, M., and Schlunegger, F.: Geomorphic coupling between hillslopes and channels in the Swiss Alps, Earth Surf. Proc. Land., 38, 959–969, https://doi.org/10.1002/esp.3342, 2013.

Seo, J. I. and Nakamura, F.: Scale-dependent controls upon the fluvial export of large wood from river catchments, Earth Surf. Proc. Land., 34, 16–25, https://doi.org/10.1002/esp.1765, 2009.

Impacts of a large flood along a mountain river basin: the importance of channel widening and estimating...

141

Seo, J. I., Nakamura, F., Akasaka, T., Ichiyanagi, H., and Chun, K. W.: Large wood export regulated by the pattern and intensity of precipitation along a latitudinal gradient in the Japanese archipelago, Water Resour. Res., 48, W03510, https://doi.org/10.1029/2011WR010880, 2012.

Sideris, I. V., Gabella, M., Sassi, M., and Germann, U.: The Com-biPrecip experience: development and operation of a real-time radar-raingauge combination scheme in Switzerland, Int. Symp. Weather Radar Hydrol., 2014.

Steeb, N., Rickenmann, D., Badoux, A., Rickli, C., and Waldner, P.: Large wood recruitment processes and transported volumes in Swiss mountain streams during the extreme flood of August 2005, Geomorphology, 279, 112–127, https://doi.org/10.1016/j.geomorph.2016.10.011, 2017.

Surian, N., Righini, M., Lucía, A., Nardi, L., Amponsah, W., Benvenuti, M., and Viero, A.: Channel response to extreme floods: Insights on controlling factors from six mountain rivers in northern Apennines, Italy, Geomorphology, 272, 78–91, https://doi.org/10.1016/j.geomorph.2016.02.002, 2016.

Thévenet, A., Citterio, A., and Piégay, H.: A new methodology for the assessment of large woody debris accumulations on highly modified rivers (example of two French piedmont rivers), Regulated Riv., 14, 467–483, https://doi.org/10.1002/(SICI)1099-1646(199810)14:6<467::AID-RRR514>3.0.CO;2-X, 1998.

Thompson, C. and Croke, J.: Geomorphic effects, flood power and channel competence of a catastrophic flood in confined and unconfined reaches of the upper Lockyer valley, south

east Queensland, Australia, Geomorphology, 197, 156–169, https://doi.org/10.1016/j.geomorph.2013.05.006, 2013.

Wei, T. and Simko V.: R package "corrplot": Visualization of a Correlation Matrix (Version 0.84), available at: https://github.com/taiyun/corrplot (last access: 21 July 2018), 2017.

Wells, S. G. and Harvey, A.: Sedimentologic and geomorphic variations in storm-generated alluvial fans, Howgill Fells, northwest England, Geol. Soc. Am. Bull., 98, 182–198, 1987.

Wohl, E.: Connectivity in rivers, Prog. Phys. Geog., 41, 345–362, https://doi.org/10.1177/0309133317714972, 2017.

Wohl, E., Cenderelli, D. A., Dwire, K. A., Ryan-Burkett, S. E., Young, M. K., and Fausch, K. D.: Large in-stream wood studies: a call for common metrics, Earth Surf. Proc. Land., 625, 618–625, https://doi.org/10.1002/esp.1966, 2010.

Wohl, E. E., Greenbaum, N., Schick, A. P., and Baker, V. R.: Controls on bedrock channel incision along Nahal Paran, Israel, Earth Surf. Proc. Land., 19, 1–13, 1994.

Wright, K.: corrgram: Plot a Correlogram, R package version 1.12, available at: https://CRAN.R-project.org/package=corrgram (last access: 21 July 2018), 2017.

Wyżga, B.: Methods for studying the response of flood flows to channel change, J. Hydrol., 198, 271–288, 1997.

Zurbrügg, S.: Hochwasseruntersuchung und Feldstudie zur Quantifizierung des Transports und der Ablagerung von Schwemmholz während einem extremen Ereignis: Die Überschwemmung vom Juli 2014 in der Emme (Schweiz), Bsc Thesis, University of Bern, Switzerland, 2015 (in German).

8

Morphological effects of vegetation on the tidal–fluvial transition in Holocene estuaries

Ivar R. Lokhorst[1], Lisanne Braat[1], Jasper R. F. W. Leuven[1], Anne W. Baar[1], Mijke van Oorschot[2],
Sanja Selaković[1], and Maarten G. Kleinhans[1]

[1]Faculty of Geosciences, Utrecht University, P.O. Box 80115, 3508 TC Utrecht, the Netherlands
[2]Department of Freshwater Ecology & Water Quality, Deltares, P.O. Box 177, 2600 MH Delft, the Netherlands

Correspondence: Maarten G. Kleinhans (m.g.kleinhans@uu.nl)

Abstract. Vegetation enhances bank stability and sedimentation to such an extent that it can modify river patterns, but how these processes manifest themselves in full-scale estuarine settings is poorly understood. On the one hand, tidal flats accrete faster in the presence of vegetation, reducing the flood storage and ebb dominance over time. On the other hand flow-focusing effects of a tidal floodplain elevated by mud and vegetation could lead to channel concentration and incision. Here we study isolated and combined effects of mud and tidal marsh vegetation on estuary dimensions. A 2-D hydromorphodynamic estuary model was developed, which was coupled to a vegetation model and used to simulate 100 years of morphological development. Vegetation settlement, growth and mortality were determined by the hydromorphodynamics. Eco-engineering effects of vegetation on the physical system are here limited to hydraulic resistance, which affects erosion and sedimentation pattern through the flow field. We investigated how vegetation, combined with mud, affects the average elevation of tidal flats and controls the system-scale planform. Modelling with vegetation only results in a pattern with the largest vegetation extent in the mixed-energy zone of the estuary, which is generally shallower. Here vegetation can cover more than 50 % of the estuary width while it remains below 10 %–20 % in the outer, tide-dominated zone. This modelled distribution of vegetation along the estuary shows general agreement with trends in natural estuaries observed by aerial image analysis. Without mud, the modelled vegetation has a limited effect on morphology, again peaking in the mixed-energy zone. Numerical modelling with mud only shows that the presence of mud leads to stabilisation and accretion of the intertidal area and a slight infill of the mixed-energy zone. Combined modelling of mud and vegetation leads to mutual enhancement with mud causing new colonisation areas and vegetation stabilising the mud. This occurs in particular in a zone previously described as the bedload convergence zone. While vegetation focusses the flow into the channels such that mud sedimentation in intertidal side channels is prevented on a timescale of decades, the filling of intertidal area and the resulting reduction in tidal prism may cause the infilling of estuaries over centuries.

1 Introduction

1.1 Problem definition

Estuaries are flanked by tidal marshes, which are unique ecosystems with a very high biomass that modify the local hydromorphodynamic conditions (Davidson et al., 1991; Meire et al., 2005; Friedrichs, 2010). Vegetation affects hydromorphodynamics in rivers (Corenblit et al., 2009;

Oorschot et al., 2015), and this effect on hydromorphodynamics has also been shown on the scale of individual tidal marshes (Bouma et al., 2005; D'Alpaos et al., 2006; Temmerman et al., 2007). The effect of vegetation on hydromorphodynamics in tidal marshes is therefore relatively well known on the individual plant or patch scale (Järvelä, 2002; Siniscalchi et al., 2012), while its effect on estuary-scale morphodynamics has barely been studied. Incorporating vegetation

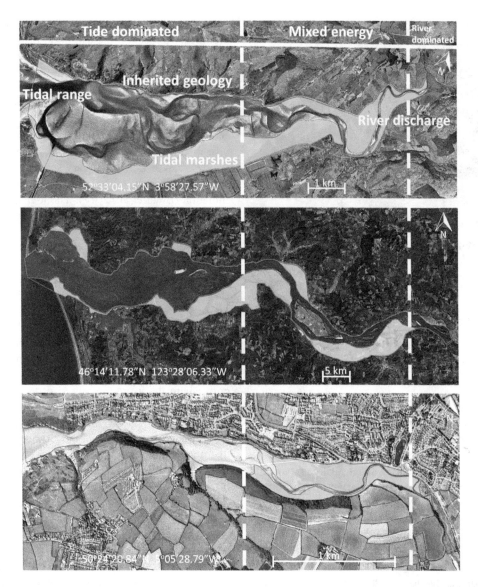

Figure 1. Active and vegetated parts of estuaries, showing proportionally more vegetated area in the upstream transition from single-thread river to multi-thread estuary. The estuaries are the Dyfi (UK), Columbia (USA) and Gannel (UK). The green areas are the vegetated parts of the estuary while the red lines project the morphologically active areas. Distinctions between dominant energy types are based on characteristic morphological features like tidal creeks, intertidal area, irregular shaped tidal bars and large meanders (Dalrymple et al., 1992).

in estuarine morphodynamic models is considered one of the three biggest challenges to overcome in modelling the long-term evolution of tidal networks (Coco et al., 2013). A comprehensive but qualitative model suggests that tidal marshes reach their largest extent in the mixed-energy zone of the estuary (Dalrymple et al., 1992). Here we investigate whether plant species can collectively have eco-engineering effects that are significant enough to modify entire estuarine landscapes. As we do not differentiate between different types of marshes, we will use a generic marsh species which will be referred to as either tidal marsh or marsh.

In rivers, riparian vegetation stabilises channels by reducing floodplain flow and adding bank strength to the flood-plains (Corenblit et al., 2009; Gurnell et al., 2012). These eco-engineering effects can be strong enough to cause the transition from braiding towards meandering or even sinuous rivers (Ferguson, 1987; Tal and Paola, 2007; Dijk et al., 2013; Oorschot et al., 2015). However, the presence of vegetation can also cause the bifurcation of channels by stabilising bar tips, causing flow resistance on point bars and diverging the flow from the channel onto the floodplain (Burge, 2005; Dijk et al., 2013). Furthermore this increased flow resistance causes flow to decelerate and water levels to rise, which may induce flooding events (Darby, 1999; Kleinhans et al., 2018). The presence of mud has a partly similar effect to vegetation because it can lead to the stabilisation of

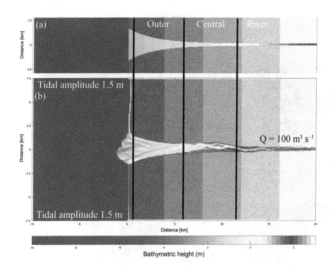

Figure 2. Initial model conditions. **(a)** The original initial bathymetry in Braat et al. (2017). **(b)** The bathymetry after 1000 years of simulation (Braat et al., 2017), which is the initial bathymetry for the present model runs. Bold lines indicate division between the outer, middle and river part of the estuary based on the decrease in flood velocity along the estuary.

systems as well, and mud has shown to preferentially accumulate in vegetated areas (Kleinhans et al., 2018). Based on these insights and general similarities between rivers and the tidal–fluvial transition, it is easily conceivable that similar biogeomorphological interactions shape upstream parts of estuaries. While salinity is an important variable determining which species prevail, here we focus on a single and often dominant tidal marsh vegetation species.

Tidal marsh vegetation flanks estuaries from the brackish zone to the mouth. Tidal marsh enhances sedimentation both through reduced flow velocities and through particle capture, somewhat comparable to what happens on river floodplains, but tidal marsh is not considered a particularly effective channel and bank stabiliser (Lee and Partridge, 1983; French, 1993; Allen, 1994; D'Alpaos et al., 2006; Bouma et al., 2007; Mudd et al., 2010). If the hydroperiod, the time that tidal marshes are submerged every day, gets longer the sediment supply to the marsh increases and therefore so does the sediment accretion. Several authors therefore found that tidal marshes are most productive at a certain rate of sea level rise (SLR) because this keeps the hydroperiod more or less constant as accretion rates balance with SLR (Redfield, 1972; Orson et al., 1985). However, tidal marshes may drown when the sea level rise rate is too large relative to the sediment supply, which leads to vegetation loss and therefore marsh drowning at an enhanced rate (Kirwan and Temmerman, 2009). In general, tidal marshes are thought to approach an equilibrium level relative to the sea level whether rising or not (Friedrichs and Perry, 2001; Marani et al., 2013).

For tidal marsh to accrete, the supply of mud is essential as the source of inorganic accumulation. This mud may have

a coastal or fluvial source, and the main source might have significant effects on the evolution of the estuary (de Haas et al., 2017). Although mud is transported in suspension and thus reaches higher, low-energetic elevations and areas more distal from the main channel, it is not unlimited. Suspended sediment rapidly settles in tidal marshes and therefore the concentration in the water quickly decreases with distance into the marsh (Townend et al., 2011). Nevertheless, cohesive mud is more difficult to erode than sand when it consolidates, so that on the estuary-scale mud leads to narrower systems with reduced bar dynamics through mudflat accumulation (Braat et al., 2017). The logical hypothesis is that the added effect of vegetation leads to even more accretion at the flanks of the estuary (Brew and Williams, 2010).

The availability of mud is partly determined by the changing hydrodynamic energy along the river continuum, especially in shallow, well-mixed estuaries that we focus on (Fig. 1) (Dalrymple et al., 1992). The tidal–fluvial transition appears to be a zone of sand and mud convergence, both of which are therefore conducive to tidal marsh establishment (Fig. 1). Alternatively, it could be the mixed-energy setting that is conducive to tidal marsh establishment, which, in turn, enhances sedimentation. A central zone of lower energy where the average grain size decreases has been observed where bedload converges (Johnson et al., 1982). Bedload convergence means that both the river and the sea transport more sediment towards this central zone in the estuary than they export, resulting in net accumulation. Dalrymple et al. (1992) suggested that this area of bedload convergence often coincides with the relatively largest tidal marsh extent (Fig. 1). Furthermore, in many estuaries a turbidity maximum zone (TMZ) occurs in the same mixed-energy zone of the estuary, which are characterised by elevated suspended sediment concentrations (e.g. Brenon and Le Hir, 1999). It is important to realise that the relative contribution of the tides, river and waves to the total hydrodynamic energy is gradually changing along the estuary (Dalrymple et al., 1992). We will use a rough classification of the estuary into an outer, central and river part, which is characterised by a dominance of tides, mixed importance of tides and river, and dominance of the river over hydrodynamics respectively.

Our hypothesis derives from a combination of three independent and complementary analyses. First, a reconstruction of the Holocene development of estuaries and tidal basins suggests that vegetation combined with mud tends to infilling of estuaries. Through a reduction in intertidal water storage at the system margins, due to vegetation-enhanced sedimentation, the tidal prism reduces and tends towards flood-dominant transport (Speer and Aubrey, 1985; Friedrichs and Perry, 2001; Friedrichs, 2010). Second, a large number of estuaries fill all space wider than that covered by an idealised convergent estuary with tidal bars (Leuven et al., 2017). This analysis excluded tidal marshes, but clearly a number of estuaries were larger in the past and have at least partly been filled by mudflats, tidal marsh or mangroves. A model study

Table 1. The main hydromorphological parameter settings.

Parameter	Value	Unit	Motivation
Time span model run	100	year	sufficient time to have changes on estuary scale
Hydrodynamic time step	0.2	min	to fulfill Courant number criteria
Morphological spin up time	24	h	two tidal cycles
Drying flooding depth	0.08	m	balance between capturing morphodynamics and time efficiency
Morphological acceleration factor	30	–	low value to allow vegetation processes
Active bed layer thickness	0.1	m	Braat et al. (2017)
Transverse bed slope parameter α	0.2	–	Braat et al. (2017)
Transverse bed slope parameter β	0.5	–	Braat et al. (2017)
Vegetation time step	21 900	min	to capture settling, growth and mortality

Table 2. Parameterisation of general characteristics of *Spartina anglica*.

Parameter	Unit	Value	Reference
Vegetation type	–	*Spartina anglica*	common European tidal marsh species
Maximum age	year	20	
Initial root length	m	0.02	based on *S. alterniflora* (Deng et al., 2009)
Initial shoot length	m	0.07	
Initial stem diameter	m	0.001	
Logarithmic growth factor root	–	0.19	based on *S. alterniflora* (Deng et al., 2009)
Logarithmic grow factor shoot	–	1	Nehring and Adsersen (2006)
Logarithmic growth factor stem diameter	–	0.005	
Timing of seed dispersal	Month	April	Nehring and Adsersen (2006)

by Braat et al. (2017) on the effects of mud on the system-scale development of estuaries over millennia showed that mud decreases the morphodynamics and decreases the total system width depending on mud concentration. All three approaches – geological, remote sensing and numerical – point to system-scale effects of mud and vegetation in estuaries.

Our aims are to determine the combined effects of mud and vegetation on estuarine planform and morphodynamics, specifically in the setting of a sandy estuary with mud input from the river. To this end we will use a numerical model for a century-scale simulation of flow, sediment transport, morphology and vegetation. We ignore the binding of sediment by roots because of the relatively shallow rooting and only explore the cohesive effects of mud, the floodplain-filling effects of mud and the flow resistance effects of vegetation. This allows us to apply an existing model for riparian vegetation to the tidal environment. Two questions of specific interest are how the zonation of vegetation, as found by Dalrymple et al. (1992), can be explained and what the morphological and hypsometric changes are as a result of the presence of vegetation.

2 Methods

To investigate whether the transition of dominantly fluvial energy to dominantly tidal energy is indeed the hotspot of sedimentation and tidal marsh formation, we combine a veg-

etation model with the morphological estuary model built in Delft3D by Braat et al. (2017), which includes cohesive sediment. Tidal marsh modelling is based on the recently developed riparian vegetation model by Oorschot et al. (2015). This model takes the vegetation cycle into account, which includes colonisation, growth and mortality due to flooding, uprooting, scour and high flow velocity. The processes of settlement, growth and mortality are similar for riparian and tidal marsh vegetation and the process of flow retardation due to flow obstruction remains a function of stem height, width and density. So, with a different parameterisation for plant growth, dimensions and mortality, we were able to realistically represent marsh vegetation with this model. We modelled the combined effects of mud and vegetation to investigate feedback mechanisms between these two and compare the model results with measurements in nine real estuaries.

The model consists of two interacting codes: the hydro-morphological modelling package Delft3D version 4.01.00 and our MATLAB-based vegetation module. The coupling is fast and the vegetation module slows down the model marginally, mainly due to file input and output. However, the need to compute at a very high temporal resolution leads to model runtimes for up to 2 months to simulate 100 years of development. To investigate the combined effects of mud and vegetation, an existing model schematisation was used that is loosely based on the Dyfi estuary in Wales (Braat et al., 2017). The large computation times of the interacting codes

Table 3. Parameterisation of life-stage-specific characteristics of *Spartina anglica*.

Parameter	Unit	*Spartina anglica*			Reference
		Ls 1	Ls2	Ls3	
Numbers of years in life stage	year	1	10	9	
Number of stems	stems m^{-2}	13 000	1500	600	Nehring and Adsersen (2006)
Area fraction (0–1)	–	0.05	0.5	0.8	
Drag coefficient	–	1	1	1	cylindrical stems
Desiccation threshold	days	360	360	360	no desiccation assumed
Desiccation slope	–	1	1	1	no desiccation assumed
Flooding threshold	days	20	40	40	
Flooding slope	–	0.75	0.75	0.75	
Flow velocity threshold	m s^{-1}	0.5	1	1	
Flow velocity slope	–	0.75	0.75	0.75	

Table 4. Channel area, vegetation area and estuary length derived from polygons digitised in Google Earth, accessed October 2017. The mixed-energy zone gives the approximate distance of the mixed-energy zone as a fraction of the distance from the estuary mouth.

Name	Location	Date aerial photography	Channel area (km^2)	Vegetation area (km^2)	Estuary length (km)	Mixed-energy zone
Columbia River	USA	31/12/2006	397.6	196.6	84.7	0.74
Dyfi estuary	UK	6/1/2009	11.9	6.7	11.9	0.63
Glaslyn estuary	UK	1/12/2006	9.9	4.2	11.3	0.56
Conwy estuary	UK	6/1/2009	5.3	3.1	16.0	0.78
Teign estuary	UK	1/12/2011	3.1	0.5	7.6	0.79
Gannel estuary	UK	12/31/2001	0.3	0.3	3.5	0.58
Clwyd estuary	UK	31/12/2006	0.3	0.6	4.7	0.74
Rodds Bay, Queensland	Australia	1/12/2006	10.1	6.5	10.2	0.86
Whitehaven beach	Australia	1/12/2011	2.3	3.4	6.8	0.80

necessitated that our model start from the well-developed morphology after 1000 years. To isolate the effect of vegetation in the simplest possible settings, we ignore salinity, waves and tidal components other than M2. The tidal marsh vegetation is represented by the settling, growth and mortality traits of *Spartina anglica* and the hydraulic resistance as a function of stem dimensions and density as detailed later. Although *Spartina anglica* is not the only pioneer species in these systems (e.g. *Salicornia*), the vegetation modelling here is simplified, given the large spatiotemporal scales and first application of a vegetation model. In our runs, the vegetation traits based on the commonly occurring *Spartina anglica* are to be seen as a generic tidal marsh plant species.

2.1 Hydromorphodynamic model

Delft3D is a widely tested open-source model that can calculate both sand and mud transport. The 2DH (depth-averaged) version was used with a parameterisation for bend flow effects on the direction of sediment transport. We used a rectangular grid, which affects the form of the equations given below. Here we will state the main equations used in Delft3D, which are either default or activated by choice. The only equations incorporated into our MATLAB model are related

to the settling, growth, mortality and bookkeeping of the vegetation.

The model is mainly based on two hydrodynamic equations, the first being the conservation of mass equation:

$$\frac{\partial h}{\partial t} + \frac{\partial hu}{\partial x} + \frac{\partial hv}{\partial y} = 0, \tag{1}$$

where h is the water depth, t is time, u is the flow velocity in the x direction and v is the flow velocity in the y direction. Equation (1) states that any change in water depth follows from a discharge gradient in the x direction (q_x) or a discharge gradient in the y direction (q_y) for a 2-D model. Momentum conservation is calculated as

$$\frac{\partial u}{\partial t} + u\frac{\partial u}{\partial x} + v\frac{\partial u}{\partial y} + g\frac{\partial z_w}{\partial x} + \frac{gu\sqrt{u^2 + v^2}}{C^2 h}$$
$$- V\left(\frac{\partial^2 u}{\partial x^2} + \frac{\partial^2 u}{\partial y^2}\right) + F_x = 0, \tag{2}$$

$$\frac{\partial v}{\partial t} + u\frac{\partial v}{\partial x} + v\frac{\partial v}{\partial y} + g\frac{\partial z_w}{\partial y} + \frac{gv\sqrt{u^2 + v^2}}{C^2 h}$$
$$- V\left(\frac{\partial^2 v}{\partial x^2} + \frac{\partial^2 v}{\partial y^2}\right) + F_y = 0, \tag{3}$$

where z_w is the water surface height, C is the Chézy roughness, which will be calculated by the vegetation model described below, V is the horizontal eddy viscosity and $F_{x,y}$ is the streamline curvature-driven acceleration term (Schuurman et al., 2013). These two equations describe the velocity variations in the $x-y$ plane in one grid cell over time under the influence of advection, eddy diffusivity, friction, changing water depth and streamline curvature. Sediment transport is calculated by separate equations for the different sediment constituents. Sand transport in the case of a non-cohesive bed is calculated with the Engelund–Hansen sediment transport predictor:

$$S = \frac{0.05\sqrt{u^2 + v^2}^5}{\sqrt{g}C^3 \frac{\rho_s - \rho_w}{\rho_w} D_{50}}, \tag{4}$$

where ρ_s the sediment density, ρ_w the water density and D_{50} the median grain size. The sediment transport of the mud fraction of the model is calculated by Partheniades–Krone equations (Partheniades, 1965) for erosion flux E_m,

$$E_m = M_m \left(\frac{\tau_{cw}}{\tau_{cr,e}} - 1 \right), \tag{5}$$

and for deposition flux D_m:

$$D_m = w_s c_b \left(1 - \frac{\tau_{cw}}{\tau_{cr,d}} \right) \tag{6}$$

for $\tau_{cw} > \tau_{cr,e}$, where τ_{cw} is the maximum bed shear stress due to currents, $\tau_{cr,e}$ is the critical erosion shear stress, M_m is an erosion parameter, w_s is the mud settling velocity and c_b the average sediment concentration in the near-bottom layer. Above a critical mud content threshold ($p_m > p_{m,cr}$), the sand and mud flux are proportional to their respective fractions in the sediment bed. Mud erosion is the same in the cohesive and non-cohesive regime, but the sand erosion becomes dependent on the mud entrainment in the cohesive regime, when the mud content in the bed exceeds 40 %. The transport of sand becomes fully dependent on the mud flux, as bedload transport is assumed to be zero in the cohesive regime. Once sediment is suspended following the Partheniades–Krone equation, it is transported by the advection–diffusion equations. A constant mud settling velocity of 2.5×10^{-4} m s^{-1} was assumed based on Braat et al. (2017).

A parameterisation is needed for helical flow due to streamline curvature in a depth-averaged simulation to create point bars in river bends and estuarine bars and is included as follows. The bedload transport direction ϕ_τ is given by the following equation:

$$\tan(\phi_\tau) = \frac{v - \alpha_I \frac{u}{U} I_s}{u - \alpha_I \frac{v}{U} I_s}, \tag{7}$$

where U is the depth-averaged flow velocity, I_s is the spiral flow intensity factor, here taken at unity, and α_I is given by

the following equation:

$$\alpha_I = \frac{2}{\kappa^2} \left(1 - \frac{1}{2} \frac{\sqrt{g}}{\kappa C} \right), \tag{8}$$

where κ is the von Kármán constant, taken as 0.41. Lastly, bed slope effects are included in the model to simulate a deviation in sediment transport direction from the shear stress direction due to grains moving downslope. The sediment transport in the x and y direction under influence of the bed slope effect is given by

$$q_x = q_s \left[\cos(\phi_\tau) - \frac{1}{f(\theta)} \frac{\partial z_b}{\partial x} \right], \tag{9}$$

$$q_y = q_s \left[\sin(\phi_\tau) - \frac{1}{f(\theta)} \frac{\partial z_b}{\partial y} \right], \tag{10}$$

where q_s is sediment transport, z_b is the bed height, and $f(\theta)$ is given by the following equation:

$$f(\theta) = \alpha \theta^\beta. \tag{11}$$

In this equation θ is the shields parameter and α and β are calibration parameters specified later.

2.2 Vegetation model

A model programmed in MATLAB was used to simulate the vegetation in the estuary (Oorschot et al., 2015). This model simulates vegetation colonisation, growth and mortality and translates this to hydraulic roughness used in Delft3D as based on the Baptist et al. (2007) equation:

$$C = \frac{1}{\sqrt{\frac{1}{C_b^2} + \frac{C_d n h_v}{2g}}} + \frac{\sqrt{g}}{\kappa} \ln \frac{h}{h_v}, \tag{12}$$

where C is the Chézy roughness value due to the bed and vegetation roughness ($\sqrt{m}/s$), C_b is the Chézy value for the bed without vegetation, C_d is the drag coefficient, n is the number of stems per square metre times the stem diameter, h_v the vegetation height and $\kappa = 0.41$ is the von Kármán constant. Vegetation of different ages and therefore with different characteristics can occur simultaneously in one grid cell up to a total fraction of 1. The Chézy value is calculated for each age class, and afterwards a total Chézy coefficient is calculated based on the fraction coverage of each age class.

The vegetation model divides the morphological year into 24 ecological time steps, which correspond with half a month of morphological development (Table 1). Following each ecological time step the hydromorphodynamic calculations are stopped and the bed level changes, water levels and flow velocities are exported from Delft3D to the vegetation model. A 2-week interval, during which vegetation properties are assumed constant, was chosen to capture the dominant vegetation development processes. Over a 2-week growth period, the species have no appreciable changes in size, and this time step balances with the computational cost that increases with a decreasing time step. The vegetation has both general and life-stage-specific characteristics (Tables 2 and 3). General characteristics are the seedling dimensions, i.e. shoot length and diameter and root length, maximum age, growth factors for logarithmic shoot, root and diameter development, and seed dispersal timing (Oorschot et al., 2015). Life-stage-specific characteristics are rules for mortality due to flooding and uprooting, number of stems per area, drag coefficient, and fraction of the grid cell surface covered with vegetation. All the variables in the Baptist et al. (2007) equation are thus accounted for. The new vegetation characteristics are then used to update the Chézy roughness field in Delft3D.

Colonisation takes place during the month of seed dispersal on every location where water has been present (Table 2). This means that all cells in the intertidal zone are colonised with *Spartina anglica* by the predefined colonisation density. Given that the tides in the model are simplified to M2, the supratidal zone where vegetation settles in nature can be seen as included as high intertidal. There is no seed dispersal module other than that we assume the seeds to spread through the water (hydrochorously), and neither do seeds end up above the water surface. This means that seedlings colonise lower intertidal areas, after which mortality determines which plants survive such that the lower intertidal zone is not occupied by plants during the flow modelling. We do not model rhizomic growth since this is a process occurring at a much smaller spatial scale than the grid cell size.

The vegetation follows a logarithmic growth function dependent on age, which limits their growth once they mature:

$$G = F_v \log(a),$$ (13)

in which G is the length or diameter of the shoot or root, F_v is a characteristic growth factor for the root or shoot, and a is the vegetation age in years. The initial dimensions of the seedlings are defined in the general characteristics, after which plant growth is calculated yearly following the equation.

Mortality is calculated yearly as a function of burial, uprooting, maximum flow velocities, flooding and ageing. Burial and uprooting are determined by comparison of the plant dimensions and bed level change. If the erosion in an ecological time step exceeds the length of the root, the plant is uprooted, and if the sedimentation exceeds the shoot length, it is considered buried, both leading to mortality (Oorschot et al., 2015). To calculate mortality due to flooding and flow velocity, the maximum, minimum and average water depth at each cell are determined during the tidal cycle. Because tidal marsh vegetation starts to occur above mean tide and usually quickly accretes to the high tide mark, the subsequent days that the cells are flooded during mean tide are recorded. For flow velocity, the maximum value during the tidal cycle in each cell is stored. Lastly, vegetation dies when its maximum age is reached.

A dose–effect relation (Oorschot et al., 2015) is applied to model gradual plant demise as the fraction of plants that do not survive the hydrodynamic pressure. Until a threshold is exceeded no mortality occurs, while above this threshold an increasing portion of the plants start dying with increasing stress. The threshold value and the slope of the stress–mortality relation are user-defined and can vary between the life stages of the plants (Table 3). Mortality was applied to each age class in all grid cells (Oorschot et al., 2015).

2.3 Model set-up

We set up four model scenarios based on our earlier work and about 30 preliminary test runs, where we balanced time efficiency and the processes that could be realistically represented (Braat et al., 2017; Oorschot et al., 2015).

The initial bathymetry is the final outcome of a model run that started from an idealised convergent shape (Braat et al., 2017). This avoids long computational time to develop sufficient bars and mudflats where vegetation can settle. The rectangular cell size varies from 50 m by 80 m in the estuary to 125 m by 230 m offshore. This is done to balance computational time and sufficient spatial resolution. A 0.2 min time step was used based on the Courant criterion. We applied a 1.5 m tidal amplitude defined by two harmonic water levels at the north and south coastal boundaries and a constant $100 \, \text{m}^{-3} \, \text{s}^{-1}$ discharge at the upstream river boundary. The bed is initially entirely composed of sand and has a sand supply equal to the transport capacity at the river boundary, which avoids sedimentation or erosion at the upstream boundary. Mud, on the other hand, is supplied as a constant concentration at the upstream boundary of $20 \, \text{mg} \, \text{L}^{-1}$, the same as in the run by Braat et al. (2017) that led to large-scale equilibrium of the estuary planform. This model was run for 1000 years without vegetation in Braat et al. (2017), and the final bathymetry was used as the initial condition for further simulations including vegetation (Fig. 2b). Note that this bathymetry was the result of calculations including mud. However, we only use the initial bathymetry and not the bed composition as our initial condition in order to isolate the effect of the addition of vegetation and mud through the upstream supply.

2.4 Parameters and scenarios

Several parameters for hydromorphodynamic processes, numerical processes and vegetation development were varied (Table 1) to study their effect on estuary developments. Model scenarios were run for 100 years, which is about the minimum time required for morphological changes at the system scale to occur due to vegetation and the practical maximum time given computational and input–output costs of about 2 months on a single node in a fast desktop computer (Table 1). A small morphological scale factor of 30 was used, since preliminary testing showed that this allowed vegetation

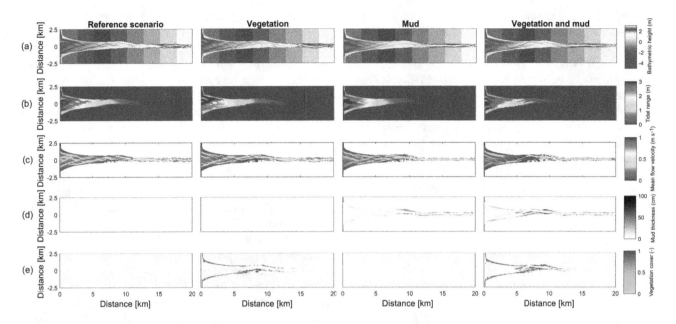

Figure 3. Results of the four scenarios after 100 years of simulation. **(a)** Morphology. Colours representing larger depths than −5 m were saturated to enhance contrast. **(b)** Tidal range. **(c)** Mean of absolute flow velocity during the tidal cycle. **(d)** Mud thickness in cm. **(e)** Vegetation cover at the surface, ranging from 0 to 1.

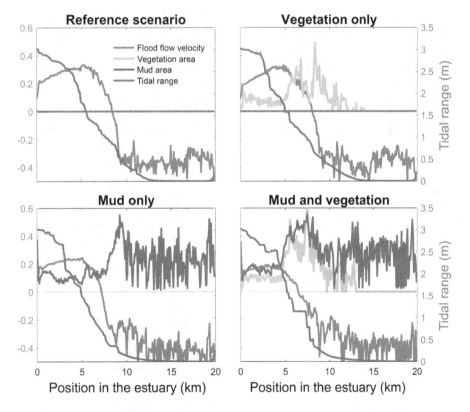

Figure 4. Tidal range, maximum flood flow velocity, vegetation cover and mud cover as a fraction of the estuary width plotted against landward distance from the coastline. In all four figures the left axis is used for three variables: width averaged flood velocity, mud cover and vegetation cover. The right axis is used for the maximum tidal range of the estuary cross section in all four subplots.

settlement, growth and mortality over a number of tidal cycles without significant morphological change. In contrast, for sandy estuaries without vegetation, values up to 1000 have been used (Van der Wegen and Roelvink, 2008). In the vegetation model a balance is required between morphological and hydrological timescales, since these both affect the development of the plants. If the morphology changes significantly faster than the hydrodynamics, plants are subject to large-scale burial and uprooting. A default Chézy value of 50 for bare sediment was chosen as in Braat et al. (2017). Vegetation traits of *Spartina anglica* were based on Nehring and Adsersen (2006) and Deng et al. (2009) (Tables 2, 3).

2.5 Data collection of real estuaries

For a first quantitative comparison of model results with real estuaries, we mapped along-channel variability of unvegetated channel width and width of the vegetated zone in nine natural estuaries. The real estuaries were selected from the dataset of Leuven et al. (2017) based on the presence of tidal marsh vegetation and include one system with mangrove species (Table 4).

The area of each estuary was visually classified as either unvegetated or vegetated in Google Earth. The unvegetated polygons come from the dataset by Leuven et al. (2017), and this analysis adds polygons of the vegetated area (Fig. 1). The vegetated area comprises the area that borders the active estuary and is covered with pioneering or fully grown tidal marsh vegetation. The presence of sinuous tidal creeks and vegetation other than, for instance, forest, were used as an indicator of present-day or recent tidal influence and older riparian vegetation was excluded. Tidal vegetation was distinguished by its different colour compared to surrounding forests and grass fields and by its clumpy and patchy structure. The elevation data in Google Earth were used as further evidence for the outer boundary of the tidal vegetation area to avoid steep gradients and cliffs at the transition from a supratidal elevation level to higher elevated areas bordering the estuary.

Subsequently, centre lines of the polygons were constructed along the channel, which allowed width measurements perpendicular to this centre line (following the approach of Leuven et al., 2017). This resulted in along-channel profiles of the active channel width, summed width of vegetation and estuary width, in which the estuary width is defined as the active channel width including bars plus the summed width of vegetation. The along-channel distance from the mouth was normalised with the length of the estuary. Estuary length is defined as the length from the mouth up to the point where the estuary width is equal within a few percent to the active channel width, in our case the upstream river. By this normalisation a direct comparison is possible between estuaries with different lengths and our modelled simulations. Through this normalisation it becomes possible to compare estuaries with different tidal–fluvial dominance.

Estuaries with a small river might have a smaller, more upstream, mixed-energy zone than estuaries with a larger river. As the mixed-energy zone is a somewhat objective designation because it is part of a continuum, we investigate vegetation cover as a function of the normalised position in the estuary and as a function of total energy. By doing this we do not delimit the mixed-energy zone but compare vegetation cover development with the development of the total energy along the estuary.

Estimates of local tidal prism and total energy were made for each of the real estuaries based on Leuven et al. (2017). Local tidal prism was estimated by multiplying the along-channel width profile with the tidal range profile and integrating over the distance upstream of a given point. The volume added by the river was characterised by river discharge multiplied by tidal period. We then calculated a characteristic velocity by dividing the local prism TP by the local active width W_a and half the tidal M2 period $T_{M2}/2$. As a proxy for the total flow energy, this velocity was taken to the power of 3 as this is also a common indicator of sediment movement (Aubrey and Speer, 1985), so that flow energy is here calculated as $2TP(W_a T_{M2})^{-3}$.

3 Results

3.1 Effects of mud and vegetation on the entire estuary

The mouth of the modelled estuary has a 3 m tidal range, which decreases gradually in the landward direction to disappear roughly 14 km into the estuary (Fig. 3). The flow velocity, on the other hand, increases in the outer part of the estuary because the convergence is stronger than the friction. Further in the estuary the convergence decreases and the increase in friction begins to dominate, which results in a decreasing flood velocity. Therefore, there is a peak in the flood flow velocity at roughly 5 km into the estuary (Fig. 4). The changes in tidal range along the estuary are thus similar to those in a hyposynchronous system while the changes in the current are similar to those in a hypersynchronous system (Fig. 4).

In the simulation without mud and vegetation, i.e. the reference scenario, channels and shoals are dynamic, but no system-scale changes occur as the initial system seems to be close to dynamic equilibrium. Only a slight change in hypsometry occurs: the intermediate heights are slightly eroded, while the higher parts accrete slightly (Fig. 5).

The simulation with vegetation only develops fringing marshes at the edges of the estuary. The marshes start from the estuary mouth up to the tidal limit, roughly 14 km upstream (Fig. 3). The relative width of the tidal marshes is fairly constant at ≈ 10 % of the estuary width in the outer zone. Between roughly 6 and 11 km, however, the relative width of the marshes suddenly increases. The relative width of the tidal marshes can go up to 60 % of the estuary width. This area coincides with the area where the flood velocity and river velocity start to decrease due to friction and estuary shape respectively (Fig. 4). Beyond 14 km there is no vegetation anymore; this is because this is beyond the tidal limit and therefore there is no drying and flooding area where seeds are distributed and seedlings survive. The morphology in the simulation with vegetation only shows little differences compared to the reference simulation. This indicates that the vegetation is unable to enhance sedimentation in the absence of suspended fine sediment and that it predominantly colonises locations that are not prone to erosion because there is no significant reduction in the erosion of the intertidal area (Fig. 5).

The simulation with mud only results in a fairly continuous mud cover along the entire estuary (Fig. 3). There are small amounts of mud which deposit on tidal bars, in the order of an accumulated 10 cm admixed in sand over 100 years, but the more pronounced accumulations occur on the edges of the system. Similar to the simulation with vegetation the relative mud abundance starts to increase landward of the maximum flood velocity, which occurs at roughly 6 km. The relatively large mud extent in the central zone of the estuary is due to the low flow velocities in this zone (Figs. 3, 4). Unlike the vegetation cover, however, the relative mud abundance does not decrease to zero at the tidal limit, but approaches a roughly constant value of approximately 30 % of the system width (Fig. 4). This is because the estuary is small in this area, as the river is only several cells wide, and not because there are large extensive mudflats. In terms of hypsometry the largest effect of mud is that the intermediate bed elevations increase slightly (Fig. 5). This shows that the higher elevations are nearly filled as much as possible, and that the estuary develops in a feedback of further filling and reduction in tidal prism.

The distribution of vegetation and mud in the combined simulation shows similar patterns to the simulations with either mud or vegetation only. There are some marshes and mud deposits in the outer estuary, but these become more pronounced towards the central zone (Fig. 3). There is a positive feedback between mud and vegetation. Not only do mud and vegetation occur in the same area, but their relative abundance also increases compared to simulations where one of them is absent (Figs. 3, 4). This is emphasised by the total mud and vegetation cover in the estuary, which are almost identical after 100 years (Fig. 7a). There is an especially strong feedback in the beginning of the simulation, when vegetation cover increases strongly, after which mud cover starts to increase faster (Fig. 7a). On top of that, the addition of vegetation to the simulation with mud further enhances the aggradation of the upper hypsometric heights and thus the intertidal area.

3.2 Effects of mud and vegetation in the mixed-energy zone

Vegetation presence affects the location and thickness of mud deposits mainly in the central estuary (Fig. 7b) and to a

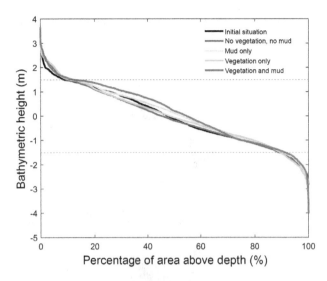

Figure 5. Hypsometry of the entire estuary after 100 years. Dashed lines indicate the tidal range at the seaward boundary. Around 70 % of the estuary area is intertidal in all scenarios, indicating that the model represents a shallow system. The hypsometry is determined over the surface occupied by the estuary of the initial condition, which excludes new areas formed by bank erosion that is modelled rather simplistically in Delft3D.

lesser degree in the outer area (Fig. 8). The vegetation cover develops faster than the mud cover but afterwards stimulates the mud sedimentation, which reaches a higher final area (Fig. 7). A major difference in hypsometry is, however, that the outer estuary has a concave profile while the central and river reach have a convex profile. This has direct consequences for the available area for vegetation. Because the effect of vegetation is largest in the central part of the estuary, a series of close-up images is provided (Fig. 9). The bathymetry of the reference simulation shows limited changes (Fig. 9a). Vegetation colonises the edges of the area in the simulation without mud but remains distal from the main ebb channel, and the bathymetry develops similar to that of the reference simulation (Fig. 9c). Larger differences occur in simulations where mud is present. When mud is added to the simulation, it first focusses the main ebb channel, but afterwards the entire area starts to gradually fill and becomes shallower (Fig. 9b).

The combined effect of vegetation and mud in the central estuary is to raise the intertidal areas and deepen the subtidal areas relative to the run with mud alone, but the overall depth compared to the control run and vegetation run is reduced. This means that the vegetation acts to focus flow into the channels, but the dominant effect is the filling of intertidal area that reduces the overall tidal prism over time. In the simulation with mud and vegetation, the deeper parts of the estuary no longer accrete. Instead the vegetation captures mud in the intertidal area, and the vegetation expands laterally towards the main channel while focusing the flow

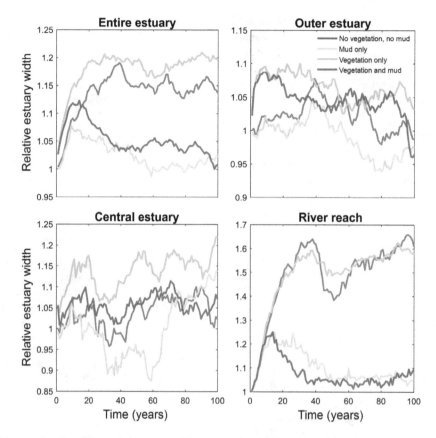

Figure 6. Estuary width over time for of the entire system and for zones along the estuary. Width is normalised by average initial width. See Fig. 2 for locations of zones.

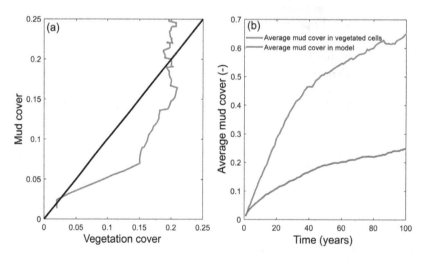

Figure 7. Interaction of mud and vegetation. **(a)** The development of the total mud and vegetation cover over time in the simulation where both are present, where the simulation begins in the origin of the plot. Black line indicates equality of mud and vegetation cover. **(b)** The average mud cover in vegetated cells and in the entire model, showing substantially higher cover in vegetated cells.

(Fig. 9d). Vegetation traps the mud in the higher intertidal areas and through this redistribution decreases the siltation of the deeper parts of the estuary. Simultaneously the accumulation of mud increases the bed level in the central part of the estuary, which enables the vegetation to laterally expand in

the direction of the channel. Because mud enables vegetation to expand laterally and because mud accumulation increases within vegetated areas, the total mud and vegetation cover increases when both are present (Fig. 10). Also, the vegetation causes the deposition of mud on bars in the middle of the es-

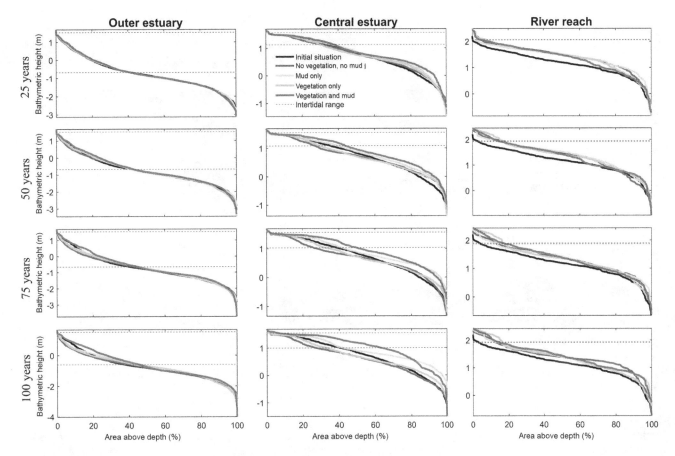

Figure 8. Development of hypsometry of three zones in the modelled estuaries. The outer estuary has a concave shape while the central and river areas have a convex shape. The middle part shows significant deposition compared to the outer estuary in simulations with mud and vegetation. Blue lines indicate initial minimum and maximum water surface elevation.

tuary (Fig. 9d) where mud barely occurs when vegetation is absent (Figs. 9c, 10).

The water elevation and mean flow velocity in the middle of the estuary were plotted over time to test the hypothesis that the system becomes flood dominant when vegetation (and mud) are present (Fig. 11). The system is ebb dominant from the start. The peak flow velocities occur roughly 1 h before low and high water, and thus the tidal velocity is slightly out of phase. The rise of the tide occurs somewhat faster than the fall of the tide. Normally this would result in higher flood velocities, but in the mixed-energy zone of the estuary they are compensated for by the river discharge. The tidal asymmetry does not change much over time for the four scenarios, but the tidal range decreases for the scenario with mud and vegetation and both simulations with vegetation cause a decreased average flow velocity (Fig. 11b). Furthermore, the effect of combined vegetation and mud is disproportionally larger than that of vegetation or mud alone, confirming the idea of interaction. Moreover, the effect of reduction in tidal

prism that determines overall flow energy dominates over the effect of reduction in intertidal area that determines the tendency of flood dominance.

3.3 Real estuaries

The model simulations showed that the relative vegetation abundance increases especially in the mixed-energy zone of the estuary. This is in close agreement with observations in nine real estuaries (Table 4). In real estuaries, vegetation increases in abundance from the estuary mouth towards a short distance before the tidal limit, while landward of the tidal limit the vegetation cover decreases quickly towards zero (Fig. 12). Similar to the modelled scenarios, the landward vegetation cover increase coincides with the decrease in the flow energy. The upper limit of the vegetation is slightly beyond the tidal limit, but this is probably because we included old marshes, which are rarely flooded.

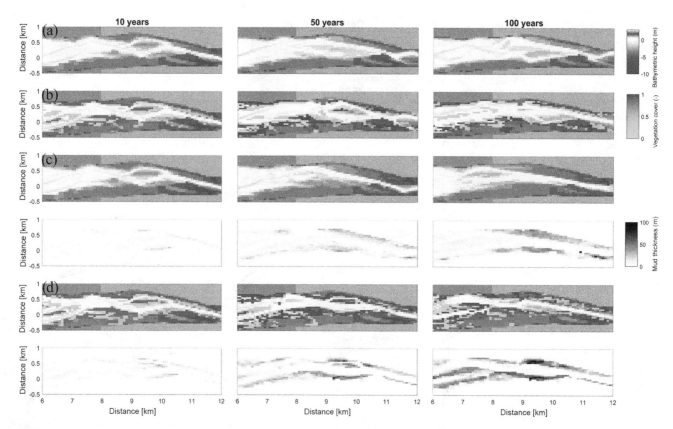

Figure 9. Development of the central zone of the estuary. (a) Simulation without mud and vegetation. (b) Simulation with only vegetation. (c) Simulation with only mud. (d) Simulation with both mud and vegetation. The mud maps belong to the simulation above it.

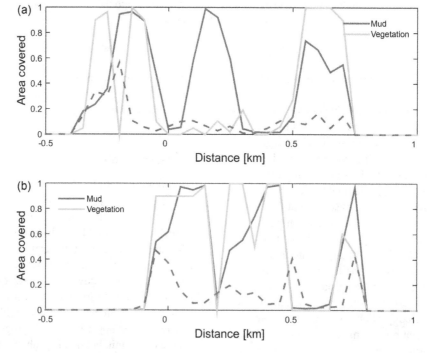

Figure 10. Positive feedback between vegetation and mud shown on cross sections through the estuary at (a) 6 km and (b) 8.5 km from the shoreline. The fraction of area covered by mud and by vegetation is plotted for the simulation with only mud (dashed line) and with both mud and vegetation (solid lines).

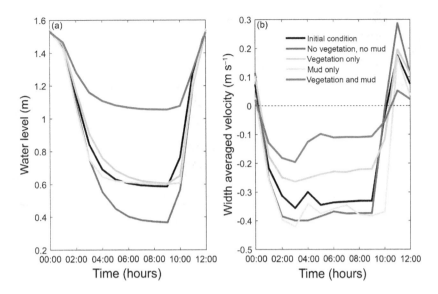

Figure 11. The final tidal cycle in the central estuary at 6 km from the mouth, showing the strongest reduction for the scenario with combined mud and vegetation. **(a)** Tidal water level. **(b)** Width-averaged flow velocities over the cycle.

4 Discussion

4.1 Marsh distribution

Modelled marshes reach their largest extent in the central estuary, where the tidal energy is the lowest in agreement with the qualitative model of Dalrymple et al. (1992). The tidal marsh expands mostly landward from the maximum flood current velocity. This is also where the bedload convergence zone begins and in natural estuaries where the turbidity maximum zone may occur (Fig. 13). The main reason for the increase in tidal marsh extent is the combination of flow velocities being low enough and the presence of suitable bed elevations. The establishment of tidal marshes requires a window of opportunity with a long enough mild hydrodynamic stress (Bouma et al., 2014). However, the modelled marshes develop primarily landward and not seaward of the maximum flood velocity, which shows that the hydrodynamics are not the only limiting factor. In reality, however, the hydrodynamic stresses will be larger in the outer part and wave magnitude is also more significant there (Dalrymple et al., 1992), and waves are a major limiting factor for seedling establishment in tidal marsh and mangrove landscapes (Balke et al., 2013). Waves would result in a further reduction in tidal marsh extent in the outer estuary but will have limited effect on the central part of the estuary and therefore strengthen the trends in our model.

4.2 Mixed-energy zone

The importance of sediment accumulation in the central part for tidal marsh development is shown in the scenario with mud and vegetation. This simulation shows a further extent of the marshes because mud preferably accumulates in the central part of the estuary, regardless of the fact that no preferential establishment of vegetation on a muddy substrate is included in the model. While it is known that suspended sediment is a requirement for tidal marshes to keep up with sea level rise (D'Alpaos et al., 2006, 2007; Murray et al., 2008 Fagherazzi et al., 2012), the present model results show that suspended sediment is also a requirement for significant lateral marsh progradation into the estuary. We show that the presence of vegetation increases the mud deposition in the *upper* intertidal area in agreement with observations (Larsen et al., 2007; Zong and Nepf, 2011; Follett and Nepf, 2012 but also that this reduces accumulation in the *lower* intertidal area. Once the vegetation starts to expand and approaches the main channel (Fig. 9), it starts to focus and concentrate the flow (Fig. 3). After vegetation settlement and stabilisation, vegetation causes flow focusing, similar to the fluvial environment (Tal and Paola, 2007; Dijk et al., 2013).

Despite the reduction in intertidal flood storage, the central zone barely becomes more flood dominant and the tidal limit shifts seaward. This is in contrast to expected tidal dynamics (Friedrichs, 2010), probably because the river in this part of the estuary already dominates over the tidal influence. The seaward shift of the tidal limit implies that the inundation time, and therefore stress, of the marshes decreases, explaining why vegetation density increases in the central estuary. Regardless, the river flow, if large enough to move sediment, will keep a channel open even if the floodplains fill up, such that an equilibrium tidal river may develop. This amounts to progradational filling of the estuary as observed in the Holocene (de Haas et al., 2017).

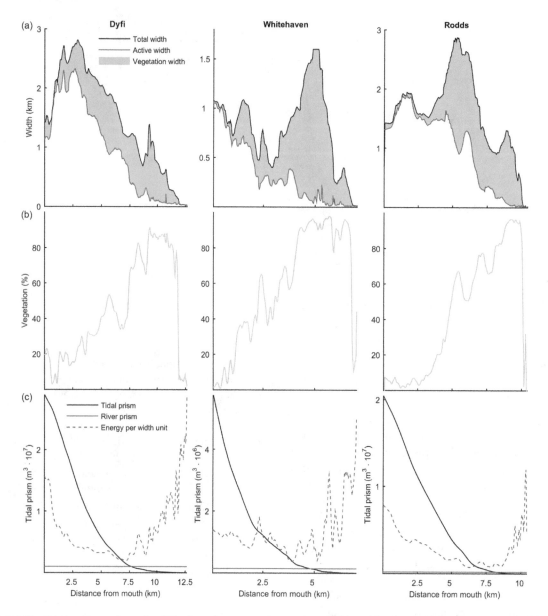

Figure 12. (a) The total, active and marsh width along three natural estuaries, partitioned by the method of Leuven et al. (2017). **(b)** The vegetated part as a percentage of the total width. **(c)** Tidal prism, discharge and energy taken as width-averaged tidal prism (for method, see Leuven et al., 2017).

4.3 Real estuaries

The general agreement between trends in real estuaries and the numerical model indicates that the overall pattern of tidal marsh and mudflats along the estuary is determined mainly by the tidal hydromorphodynamics and the interaction with mud and vegetation. Figure 14 shows the mean relative vegetation abundance for nine alluvial systems along the tidal–fluvial transition with pronounced marshes. The relative extent of the vegetation can be higher in real estuaries, which has three main causes. First, the modelled system started as a narrow convergent estuary while many real estuaries start from unfilled basins. This leads to the question of whether

the pattern of vegetation abundance and the tendency to accumulate sediment in the central estuary would have occurred for other initial conditions. The model results of Braat et al. (2017) show that mud generally settles in similar patterns over most of the modelled period and for most mud concentrations, suggesting that vegetation likewise would have formed similar patterns and central estuary sedimentation. Differences in patterns arise in conditions with very different boundary conditions as discussed below. Second, real estuaries are to a much larger degree infilling than our ebb-dominant system with little sediment import from the sea, and they had a much longer time to fill gradually. Third,

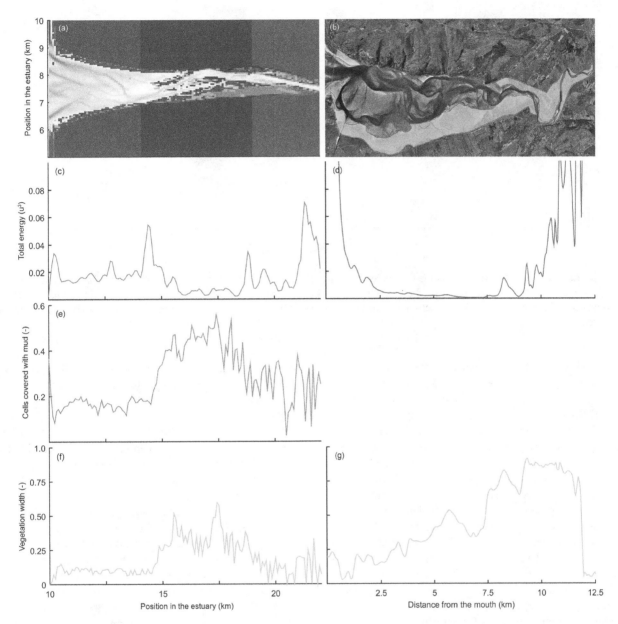

Figure 13. Comparison of mudflats and tidal marsh vegetation in a modelled (left) and natural (right) system. Here, velocity magnitude to the power of 3 is plotted as an indication for hydrodynamic energy. Panels **(a)** and **(b)** show the estuary bathymetry and vegetation, **(c)** and **(d)** show the total energy along the estuary, **(e)** shows the mud covered area along the estuary, and **(f)** and **(g)** show the relative vegetated width of the estuary.

many natural estuaries develop pronounced TMZs under influence of density-driven currents, tidal currents and river discharge. Such a TMZ would develop roughly in the mixed-energy zone, and a pronounced TMZ can be hypothesised to enhance accretion and tidal marsh expansion of the central part of the estuary that already occurs without a turbidity maximum zone (Braat et al., 2017).

Our model study simplifies real estuaries in several aspects. First, sediment supply coming from the sea could enhance tidal marsh establishment in the outer estuary. On the other hand, the presence of waves would reduce vegetation

survival mainly in the outer estuary where waves are most powerful. Third, the absence of multiple tidal components may reduce the ebb dominance and also limit vegetation development further upstream due to the absence of wetting and drying. Ebb dominance may arise due to the interaction of multiple tidal components which interact and result in a skewed velocity and thus ebb or flood dominance. In our model, there is only velocity asymmetry due to friction-induced lags as a function of tidal stage similar to the process described by Friedrichs (2010). The strongest driver of tidal asymmetry in the central zone is, however, the river

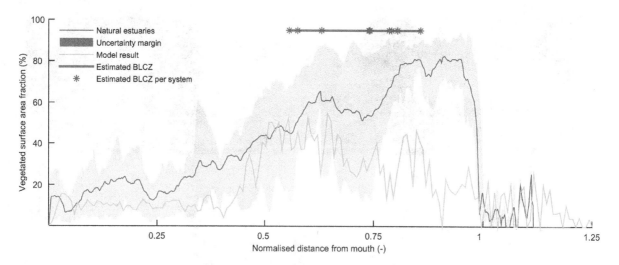

Figure 14. Relative vegetated width along the estuary averaged for nine natural estuaries compared to the simulation with mud and vegetation. Distance along the estuary is normalised by the approximate distance between coastline and tidal limit. The approximate location of the bedload convergence zone (BLCZ) is determined by the diminishing of the river energy. The uncertainty margin consists of the 20th and 80th percentile.

discharge. River discharge is known to affect velocity skewness and the timing of slack water and appears to be dominant in the central zone of the estuary (Nidzieko and Ralston, 2012). Fourth, the salinity gradient is ignored, the vegetation along the entire estuary is the same and there are no changes in how vegetation affects hydromorphodynamics along the estuary. While it is not yet known whether typical marsh species along the salinity gradient have different eco-engineering traits which significantly differently affect the long-term morphodynamics, our model is a new tool that, in further research, may lead to new insights in such patterns emerging along the estuary. Regardless, enhanced sedimentation would not change the conclusions, which is that the fundamental feedback mechanism between mud and vegetation affects the larger-scale estuary development: mud facilitates the expansion and survival of marshes while vegetation facilitates the capture of mud, especially in the mixed fluvial–tidal zone.

5 Conclusions

Numerical modelling of estuaries shows that vegetation follows mud accumulation patterns and simultaneously enhances mud accumulation rates. A positive feedback mechanism emerged in the model between the mud sedimentation and vegetation settlement. Mud sedimentation leads to higher elevated intertidal areas suitable for vegetation settling and development. The vegetation then increases local flow resistance which enhances the sedimentation of mud that would otherwise be resuspended again.

Through this biomorphological feedback loop vegetation has a strong effect on morphodynamics in the middle estuary while its effect in the outer estuary is marginal due to larger

flow energy. The relative extent of tidal marsh vegetation increases from the outer estuary towards the inner estuary and can increase from 10 % to 50 % of the estuary width or probably even more, which is in agreement with observations in real estuaries. In particular, the feedback enhances the sedimentary trend in what has been recognised in the literature as the bedload convergence zone in the mixed-energy tidal–fluvial transition. The main effect of the overall intertidal space filling is to reduce the tidal prism and progressively fill the estuary in agreement with observations of Holocene systems. The focusing of flow between flanking marsh vegetation has only a limited effect on channel depth, in contrast to observed effects in salt marsh channels and rivers. The reduction in flood storage has a negligible effect on the flood dominance of the estuary, in contrast to idealised modelling results in the literature, also because the river inflow more than balances the tidal velocity skewness. These results are mainly valid for shallow sandy estuaries.

The effect of vegetation alone on the hypsometry of the entire estuary is limited. This is mainly because its effect on the outer estuary is marginal, where it occupies only a small portion of the estuary surface. In the central part of the estuary, vegetation occupies a much larger fraction of the width so that its effects are most pronounced here. When mud is present and forms a new intertidal area, the vegetation expands towards the channel, which drives further accretion and forces the system into a single main channel. When mud is absent vegetation lacks an accreting effect because the sand does not reach the vegetated areas for lack of energy in the shallowest flows. This means that the greatest morphological effects of vegetation and mud emerge when they occur simultaneously as they have mutual positive feedbacks. The combined presence of mud and vegetation leads to the focusing of flow and channel incision on a decadal timescale but may lead to the infilling of the estuary on a centennial timescale due to the accumulation of the intertidal area and the consequent reduction in the tidal prism.

Author contributions. The authors contributed in the following proportions to conception and design, data collection, modelling, analysis and conclusions, and manuscript preparation: IRL (40 %, 50 %, 70 %, 60 %, 70 %), LB (10 %, 0 %, 10 %, 0 %, 0 %), JRFWL (0 %, 50v, 0 %, 0 %, 0 %), AWB (0 %, 0 %, 0 %, 0 %, 10 %), MvO (0 %, 0 %, 10 %, 0 %, 0 %), SS (20 %, 0 %, 10 %, 20 %, 0 %), MGK (30 %, 0 %, 0 %, 20 %, 10 %).

Competing interests. The authors declare that they have no conflict of interest.

Acknowledgements. Ivar R. Lokhorst, Sanja Selaković and Maarten G. Kleinhans were supported by the European Research Council (ERC Consolidator agreement 647570) to PI Maarten G. Kleinhans. Lisanne Braat, Jasper R. F. W. Leuven, Anne W. Baar and Maarten G. Kleinhans were supported by the Dutch Technology Foundation STW (part of the Netherlands Organisation for Scientific Research, grant Vici 016.140.316/13710) to PI Maarten G. Kleinhans. Mijke van Oorschot was supported by REFORM (FP7 grant agreement). We would like to thank Eli Lazarus and the anonymous reviewers for their contributions to improving the paper. Model support by Deltares is gratefully acknowledged. The modelling was a continuation of the MSc thesis of Ivar R. Lokhorst supervised by Sanja Selaković and Maarten G. Kleinhans.

References

Allen, J.: A continuity-based sedimentological model for temperate-zone tidal salt marshes, J. Geol. Soc., 151, 41–49, 1994.

Aubrey, D. and Speer, P.: A study of non-linear tidal propagation in shallow inlet/estuarine systems Part I: Observations, Estuar. Coast. Shelf Sci., 21, 185–205, 1985.

Balke, T., Webb, E. L., den Elzen, E., Galli, D., Herman, P. M., and Bouma, T. J.: Seedling establishment in a dynamic sedimentary environment: a conceptual framework using mangroves, J. Appl. Ecol., 50, 740–747, 2013.

Baptist, M., Babovic, V., Rodríguez Uthurburu, J., Keijzer, M., Uittenbogaard, R., Mynett, A., and Verwey, A.: On inducing equations for vegetation resistance, J. Hydraul. Res., 45, 435–450, 2007.

Bouma, T., Vries, M. D., Low, E., Kusters, L., Herman, P., Tanczos, I., Temmerman, S., Hesselink, A., Meire, P., and Van Regenmortel, S.: Flow hydrodynamics on a mudflat and in salt marsh vegetation: identifying general relationships for habitat characterisations, Hydrobiologia, 540, 259–274, 2005.

Bouma, T., Van Duren, L., Temmerman, S., Claverie, T., Blanco-Garcia, A., Ysebaert, T., and Herman, P.: Spatial flow and sedimentation patterns within patches of epibenthic structures: Combining field, flume and modelling experiments, Cont. Shelf Res., 27, 1020–1045, 2007.

Bouma, T. J., van Belzen, J., Balke, T., Zhu, Z., Airoldi, L., Blight, A. J., Davies, A. J., Galvan, C., Hawkins, S. J., Hoggart, S. P., Lara, J. L., Losada, I. J., Maza, M., Ondiviela, B., Skov, M. W., Strain, E. M., Thompson, R. C., Yang, S., Zanuttigh, B., Zhang, L., and Herman, P. M. J.: Identifying knowledge gaps hampering application of intertidal habitats in coastal protection: Opportunities & steps to take, Coast. Eng., 87, 147–157, https://doi.org/10.1016/j.coastaleng.2013.11.014, 2014.

Braat, L., van Kessel, T., Leuven, J. R. F. W., and Kleinhans, M. G.: Effects of mud supply on large-scale estuary morphology and development over centuries to millennia, Earth Surf. Dynam., 5, 617–652, https://doi.org/10.5194/esurf-5-617-2017, 2017.

Brenon, I. and Le Hir, P.: Modelling the turbidity maximum in the Seine estuary (France): identification of formation processes, Estuar. Coast. Shelf Sci., 49, 525–544, 1999.

Brew, D. S. and Williams, P. B.: Predicting the impact of large-scale tidal wetland restoration on morphodynamics and habitat evolution in south San Francisco Bay, California, J. Coast. Res., 26, 912–924, 2010.

Burge, L. M.: Wandering Miramichi rivers, New Brunswick, Canada, Geomorphology, 69, 253–274, 2005.

Coco, G., Zhou, Z., van Maanen, B., Olabarrieta, M., Tinoco, R., and Townend, I.: Morphodynamics of tidal networks: advances and challenges, Mar. Geol., 346, 1–16, 2013.

Corenblit, D., Steiger, J., Gurnell, A. M., and Naiman, R. J.: Plants intertwine fluvial landform dynamics with ecological succession and natural selection: a niche construction perspective for riparian systems, Global Ecol. Biogeogr., 18, 507–520, 2009.

D'Alpaos, A., Lanzoni, S., Mudd, S. M., and Fagherazzi, S.: Modeling the influence of hydroperiod and vegetation on the cross-sectional formation of tidal channels, Estuar. Coast. Shelf Sci., 69, 311–324, 2006.

D'Alpaos, A., Lanzoni, S., Marani, M., and Rinaldo, A.: Landscape evolution in tidal embayments: modeling the interplay of erosion, sedimentation, and vegetation dynamics, J. Geophys. Res.-Earth, 112, https://doi.org/10.1029/2006JF000537, 2007.

Dalrymple, R. W., Zaitlin, B. A., and Boyd, R.: Estuarine facies models: conceptual basis and stratigraphic implications: perspective, J. Sediment. Res., 62, 1130–1146, https://doi.org/10.1306/D4267A69-2B26-11D7-8648000102C1865D, 1992.

Darby, S. E.: Effect of riparian vegetation on flow resistance and flood potential, J. Hydraul. Eng., 125, 443–454, 1999.

Davidson, N., Laffoley, D. D., Doody, J., Way, L., Gordon, J., Key, R. E., Drake, C., Pienkowski, M., Mitchell, R., and Duff, K.: Nature conservation and estuaries in Great Britain, Nature Conservancy Council, Peterborough, 1–76, 1991.

de Haas, T., Pierik, H., van der Spek, A., Cohen, K., van Maanen, B., and Kleinhans, M.: Holocene evolution of tidal systems in the Netherlands: effects of rivers, coastal boundary conditions, eco-engineering species, inherited relief and human interference, Earth-Sci. Rev., 177, 139–163, https://doi.org/10.1016/j.earscirev.2017.10.006, 2017.

Deng, Z., Deng, Z., An, S., Wang, Z., Liu, Y., Ouyang, Y., Zhou, C., Zhi, Y., and Li, H.: Habitat choice and seed–seedling conflict of

Spartina alterniflora on the coast of China, Hydrobiologia, 630, 287–297, 2009.

Dijk, W., Teske, R., Lageweg, W., and Kleinhans, M.: Effects of vegetation distribution on experimental river channel dynamics, Water Resour. Res., 49, 7558–7574, 2013.

Fagherazzi, S., Kirwan, M. L., Mudd, S. M., Guntenspergen, G. R., Temmerman, S., D'Alpaos, A., Koppel, J., Rybczyk, J. M., Reyes, E., Craft, C., and Clough, J.: Numerical models of salt marsh evolution: Ecological, geomorphic, and climatic factors, Rev. Geophys., 50, https://doi.org/10.1029/2011RG000359, 2012.

Ferguson, R.: Hydraulic and sedimentary controls of channel pattern, in: River channels: environment and process, edited by: Richards, K., Inst. British Geographers Special Publication 18, 129–158, Blackwell, Oxford, UK, 1987.

Follett, E. M. and Nepf, H. M.: Sediment patterns near a model patch of reedy emergent vegetation, Geomorphology, 179, 141–151, 2012.

French, J. R.: Numerical simulation of vertical marsh growth and adjustment to accelerated sea-level rise, north Norfolk, UK, Earth Surf. Proc. Land., 18, 63–81, 1993.

Friedrichs, C. T.: Barotropic tides in channelized estuaries, Contemporary Issues in Estuarine Physics, 27–61, 2010.

Friedrichs, C. T. and Perry, J. E.: Tidal salt marsh morphodynamics: a synthesis, J. Coast. Res., 27, 7–37, 2001.

Gurnell, A. M., Bertoldi, W., and Corenblit, D.: Changing river channels: The roles of hydrological processes, plants and pioneer fluvial landforms in humid temperate, mixed load, gravel bed rivers, Earth-Sci. Rev., 111, 129–141, 2012.

Järvelä, J.: Flow resistance of flexible and stiff vegetation: a flume study with natural plants, J. Hydrol., 269, 44–54, 2002.

Johnson, M., Kenyon, N., Belderson, R., and Stride, A.: Offshore Tidal Sands, Processes and Deposits, chap. Sand transport, Chapman and Hall, London, 1982.

Kirwan, M. and Temmerman, S.: Coastal marsh response to historical and future sea-level acceleration, Quaternary Sci. Rev., 28, 1801–1808, 2009.

Kleinhans, M., de Vries, B., and van Oorschot, M.: Effects of mud on the morphodynamic development of a modelled river, Earth Surf. Proc. Land., https://doi.org/10.1002/esp.4437, 2018.

Larsen, L. G., Harvey, J. W., and Crimaldi, J. P.: A delicate balance: ecohydrological feedbacks governing landscape morphology in a lotic peatland, Ecol. Monogr., 77, 591–614, 2007.

Lee, W. G. and Partridge, T. R.: Rates of spread of Spartina anglica and sediment accretion in the New River Estuary, Invercargill, New Zealand, New Zeal. J. Bot., 21, 231–236, 1983.

Leuven, J., Haas, T., Braat, L., and Kleinhans, M.: Topographic forcing of tidal sand bar patterns for irregular estuary planforms, Earth Surf. Proc. Land., https://doi.org/10.1002/esp.4166, 2017.

Marani, M., Da Lio, C., and D'Alpaos, A.: Vegetation engineers marsh morphology through multiple competing stable states, P. Natl. Acad. Sci. USA, 110, 3259–3263, 2013.

Meire, P., Ysebaert, T., Van Damme, S., Van den Bergh, E., Maris, T., and Struyf, E.: The Scheldt estuary: a description of a changing ecosystem, Hydrobiologia, 540, 1–11, 2005.

Mudd, S. M., D'Alpaos, A., and Morris, J. T.: How does vegetation affect sedimentation on tidal marshes? Investigating particle capture and hydrodynamic controls on biologically mediated sedimentation, J. Geophys. Res.-Earth, 115, https://doi.org/10.1029/2009JF001566, 2010.

Murray, A., Knaapen, M., Tal, M., and Kirwan, M.: Biomorphodynamics: Physical-biological feedbacks that shape landscapes, Water Resour. Res., 44, https://doi.org/10.1029/2007WR006410, 2008.

Nehring, S. and Adersen, H.: NOBANIS–Invasive Alien Species Fact Sheet Spartina anglica, From: Online Database of the North European and Baltic Network on Invasive Alien Species–NOBANIS, available at: https://www.artportalen.se/nobanis (last access: 1 January 2018), 2006.

Nidzieko, N. J. and Ralston, D. K.: Tidal asymmetry and velocity skew over tidal flats and shallow channels within a macrotidal river delta, J. Geophys. Res.-Oceans, 117, C03001, https://doi.org/10.1029/2011JC007384, 2012.

Oorschot, M. V., Kleinhans, M., Geerling, G., and Middelkoop, H.: Distinct patterns of interaction between vegetation and morphodynamics, Earth Surf. Proc. Land., https://doi.org/10.1002/esp.3864, 2015.

Orson, R., Panageotou, W., and Leatherman, S. P.: Response of tidal salt marshes of the US Atlantic and Gulf coasts to rising sea levels, J. Coast. Res., 1, 29–37, 1985.

Partheniades, E.: Erosion and deposition of cohesive soils, J. Hydr. Div.-ASCE, 91, 105–139, 1965.

Redfield, A. C.: Development of a New England salt marsh, Ecol. Monogr., 42, 201–237, 1972.

Schuurman, F., Marra, W. A., and Kleinhans, M. G.: Physics-based modeling of large braided sand-bed rivers: Bar pattern formation, dynamics, and sensitivity, J. Geophys. Res.-Earth, 118, 2509–2527, 2013.

Siniscalchi, F., Nikora, V. I., and Aberle, J.: Plant patch hydrodynamics in streams: Mean flow, turbulence, and drag forces, Water Resour. Res., 48, https://doi.org/10.1029/2011WR011050, 2012.

Speer, P. and Aubrey, D.: A study of non-linear tidal propagation in shallow inlet/estuarine systems Part II: Theory, Estuar. Coast. Shelf Sci., 21, 207–224, 1985.

Tal, M. and Paola, C.: Dynamic single-thread channels maintained by the interaction of flow and vegetation, Geology, 35, 347–350, 2007.

Temmerman, S., Bouma, T., Van de Koppel, J., Van der Wal, D., De Vries, M., and Herman, P.: Vegetation causes channel erosion in a tidal landscape, Geology, 35, 631–634, 2007.

Townend, I., Fletcher, C., Knappen, M., and Rossington, K.: A review of salt marsh dynamics, Water Environ. J., 25, 477–488, 2011.

Van der Wegen, M. and Roelvink, J.: Long-term morphodynamic evolution of a tidal embayment using a two-dimensional, process-based model, J. Geophys. Res.-Oceans, 113, https://doi.org/10.1029/2006JC003983, 2008.

Zong, L. and Nepf, H.: Spatial distribution of deposition within a patch of vegetation, Water Resour. Res., 47, https://doi.org/10.1029/2010WR009516, 2011.

9

Directional dependency and coastal framework geology: implications for barrier island resilience

Phillipe A. Wernette[1,2,a], **Chris Houser**[1], **Bradley A. Weymer**[3], **Mark E. Everett**[4], **Michael P. Bishop**[2], and **Bobby Reece**[4]

[1]Department of Earth and Environmental Sciences, University of Windsor, Windsor, Ontario, N9B 3P4, Canada
[2]Department of Geography, Texas A&M University, College Station, Texas, 77843, USA
[3]GEOMAR Helmholtz Centre for Ocean Research Kiel, 24148 Kiel, Germany
[4]Department of Geology and Geophysics, Texas A&M University, College Station, Texas, 77843, USA
[a]now at: Department of Earth and Environmental Sciences, University of Windsor, 401 Sunset Ave., Windsor, Ontario, N9B 3P4, Canada

Correspondence: Phillipe A. Wernette (wernette@uwindsor.ca)

Abstract. Barrier island transgression is influenced by the alongshore variation in beach and dune morphology, which determines the amount of sediment moved landward through wash-over. While several studies have demonstrated how variations in dune morphology affect island response to storms, the reasons for that variation and the implications for island management remain unclear. This paper builds on previous research by demonstrating that paleo-channels in the irregular framework geology can have a directional influence on alongshore beach and dune morphology. The influence of relict paleo-channels on beach and dune morphology on Padre Island National Seashore, Texas, was quantified by isolating the long-range dependence (LRD) parameter in autoregressive fractionally integrated moving average (ARFIMA) models, originally developed for stock market economic forecasting. ARFIMA models were fit across ∼ 250 unique spatial scales and a moving window approach was used to examine how LRD varied with computational scale and location along the island. The resulting LRD matrices were plotted by latitude to place the results in the context of previously identified variations in the framework geology. Results indicate that the LRD is not constant alongshore for all surface morphometrics. Many flares in the LRD plots correlate to relict infilled paleo-channels, indicating that the framework geology has a significant influence on the morphology of Padre Island National Seashore (PAIS). Barrier island surface morphology LRD is strongest at large paleo-channels and decreases to the north. The spatial patterns in LRD surface morphometrics and framework geology variations demonstrate that the influence of paleo-channels can be asymmetric (i.e., affecting beach–dune morphology preferentially in one direction alongshore) where the alongshore sediment transport gradient was unidirectional during island development. The asymmetric influence of framework geology on coastal morphology has long-term implications for coastal management activities because it dictates the long-term behavior of a barrier island. Coastal management projects should first seek to assess the framework geology and understand how it influences coastal processes in order to more effectively balance long-term natural variability with short-term societal pressure.

1 Introduction

Since modern barrier island morphology is the product of past and present coastal processes acting over preexisting morphologies, effective barrier island management requires a comprehensive knowledge of how an island has evolved to its current state in order to understand how it may change in the future. Continued sea level rise and future climatic uncertainty represent significant concerns about the resiliency of barrier islands and threats to many coastal communities (U.S. Environmental Protection Agency, 2016). Barrier island geomorphology can exhibit considerable variability alongshore, leading to varying responses to storm activity that ultimately determine the response of the island to sea level rise. Understanding the source of variability in beach and dune morphology can provide insight into how the barrier island is likely to change in response to future storms and sea level rise.

Storm waves interact with the variable morphology of the nearshore, beach, and dunes to determine how vulnerability varies along a barrier island. To some degree, variations in the nearshore, beach, and dune morphology are influenced by the framework geology (Hapke et al., 2010, 2016; Houser et al., 2008, 2018a; Houser, 2012; Riggs et al., 1995). In this paper, the term "framework geology" is defined as any subsurface variation in geologic structure, where variability in geologic structure can result from variations in sediment type (i.e., sand vs. silt), differences in compaction, or significant changes in the subsurface organic content or mineralogy. This term encompasses the subsurface and bathymetric geologic structure (onshore and offshore), which may include rhythmic bar and swale structures (Houser and Mathew, 2011; Houser, 2012), shoreface attached sand ridges (SASRs) overlying offshore glacial outwash headlands (Hapke et al., 2010; Schwab et al., 2013), or buried infilled paleo-channels (Anderson et al., 2016; Browder and McNinch, 2006; Fisk, 1959; McNinch, 2004; Schupp et al., 2006; Simms et al., 2010). Since the framework geology can provide insight into historical patterns of island transgression (Hapke et al., 2016; Houser, 2012; Houser et al., 2015; Lentz et al., 2013), it is vital to better understand how the framework geology influences variability in modern beach and dune morphology (Cooper et al., 2018). Despite its importance, framework geology remains absent from contemporary barrier island change models that treat the geology as being uniform alongshore (Goldstein and Moore, 2016; Goldstein et al., 2017; Gutierrez et al., 2015; Moore et al., 2010; Murray et al., 2015; Plant and Stockdon, 2012; Wilson et al., 2015). Sections of a barrier island that experience greater wash-over will experience a net loss of sediment landward and localized erosion, but the dissipative nature of shoreline change (see Lazarus et al., 2011) means that those losses are distributed alongshore. In this respect, the variation in beach and dune morphology alongshore forced by the framework geology can influence the rate of historical shoreline retreat and island transgression and needs to be considered in models of barrier island response to sea level rise.

The influence of framework geology on barrier island morphology is well documented by work along the New York, Florida, and North Carolina coasts. Submerged glacial outwash headlands along Fire Island, NY, are reflected in the nearshore bathymetry as a series of shore-oblique ridges and swales (Hapke et al., 2010; Schwab et al., 2013). The nearshore bathymetry impacts sediment transport gradients along the island, which has implications for beach and dune response and recovery following a storm (Brenner et al., 2018). Using sediment cores in conjunction with ground-penetrating radar (GPR) and seismic surveys, Houser (2012) demonstrated that variations in shoreline change patterns, beach width, and dune height corresponded to ridges and swales at Pensacola, FL. Shoreline position was more stable along the ridges, resulting in a wider beach which provided more sediment for onshore winds to create higher and more persistent dunes (Houser, 2012). Paleo-channels dissecting the southeastern US Atlantic coast also align with hotspots of shoreline change (Lazarus et al., 2011; Schupp et al., 2006). However, Lazarus et al. (2011; p.1) argued that "shoreline change at small spatial scales (less than kilometers) does not represent a peak in the shoreline change signal and that [shoreline] change at larger spatial scales dominates the [shoreline change] signal". This implies that variations in the framework geology, such as paleo-channels, do not influence long-term shoreline change, but, as noted, shoreline change is influenced by the alongshore variation in beach and dune morphology. The dissipative behavior of shoreline change does not negate the importance of framework geology. While alongshore variation in dune morphology is also influenced by the distribution of vegetation in both space and time (Goldstein et al., 2017; Lazarus et al., 2011; Lazarus, 2016), the self-organized behavior of the dune morphology is ultimately set up by the framework geology (see Houser, 2012; Stallins and Parker, 2003; Weymer et al., 2015b).

The purpose of this paper is to test the hypothesis that relict infilled paleo-channels in the framework geology of a barrier island play a significant role and have an asymmetric influence on the alongshore variation in beach and dune morphology at a range of alongshore length scales. Based on the combination of a variable framework geology and a dominant alongshore current, it is feasible that the framework geology may influence barrier island geomorphology at discrete spatial scales and that this influence may be asymmetric. Central to this hypothesis is the idea that the modern island morphology itself is scale-dependent, which has been proposed and supported by previous studies (Houser, 2012; Houser et al., 2015; Lazarus et al., 2011; Lazarus and Armstrong, 2015; Lazarus, 2016). Padre Island National Seashore (PAIS) on North Padre Island, Texas, represents an ideal location to test this hypothesis because previous studies have documented significant variability in the subsurface framework geology (Fig. 1; Anderson et al., 2016; Fisk, 1959;

Wernette et al., 2018; Weymer et al., 2018; Weymer, 2012, 2016), and there is substantial alongshore variation in beach and dune morphology. Given that the dominant current along the central Texas coast flowed from north to south during the Holocene (Sionneau et al., 2008), it follows that the dominant alongshore sediment transport gradient during that time also flowed from north to south. It is feasible that paleo-channels along PAIS would have had interacted with the southerly alongshore current and sediment transport to asymmetrically influence barrier island geomorphology during island transgression. In this scenario, areas up-drift of a paleo-channel would be distinctly different from areas down-drift of the paleo-channel because the channel acts as a unidirectional sediment sink in the coastal sediment budget during island development.

2 Methods

2.1 Regional setting

Padre Island National Seashore encompasses a large portion of North Padre Island, the longest continuous barrier island in the world. Located along the south Texas, USA, coast, PAIS represents an ideal location to quantify the alongshore influence of framework geology on barrier island geomorphology because of the multiple previously identified paleo-channels dissecting the island (Fig. 1; Anderson et al., 2016; Fisk, 1959; Simms et al., 2007). Similarly, the modern surface morphology varies alongshore. Central PAIS is characterized by large, relatively continuous dunes, compared to the elongated parabolic dunes along northern PAIS and the heavily scarped and dissected dunes in southern PAIS. Padre Island is separated from the mainland by Laguna Madre, Baffin Bay, and the Intracoastal Waterway (ICW), which was dredged during the 1950s.

Multiple paleo-channels dissect the framework geology of central PAIS and Laguna Madre (Fig. 1; Fisk, 1959). These channels were suggested to have been incised into the Pleistocene paleo-surface and infilled during Holocene transgression. The prevailing theory of formation of PAIS is that the island was initially a series of disconnected barrier islands during the Last Glacial Maximum (LGM; ~ 18 ka), when a series of channels were incised into the paleo-topographic surface (Weise and White, 1980). Rapid sea level transgression during the late Pleistocene and Holocene drowned the relict dunes and submerged other dunes located approximately 80 km inland from the LGM shoreline, resulting in disconnected offshore shoals in the current location of PAIS. The disconnected shoals coalesced around 2.8 ka because sand from the relict Pleistocene dunes (~ 80 km offshore from the LGM shoreline) and sediment discharged from rivers were reworked via alongshore currents, resulting in a continuous subaqueous shoal. Eventually, sediment from offshore relict dunes and increased river discharge supplied enough sediment to the shoals that they aggraded vertically,

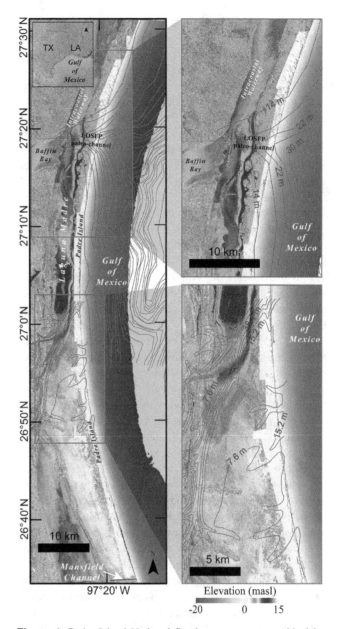

Figure 1. Padre Island National Seashore represents an ideal location to test for a directional influence of the framework geology because of the variability in the surface morphology, represented by the topobathy digital elevation model (DEM), and the underlying framework geology, represented by the Pleistocene paleo-surface contour lines from Fisk (1959) and MIS II paleo-surface contour lines from Anderson et al. (2016).

becoming subaerially exposed in the same location as the modern barrier island (Weise and White, 1980).

A series of studies in the Gulf of Mexico have focused on extracting a buried ravinement surface, also referred to as the marine isotope stage (MIS) II paleo-surface and buried Pleistocene surface, including the area offshore of PAIS (Fig. 1; Anderson et al., 2016; Fisk, 1959; Simms et al., 2010). Maps of the MIS II surface indicate that PAIS is dissected by at least two substantial paleo-channels. One large channel dissects PAIS at an oblique angle near "the hole" in Laguna Madre, an area immediately landward of PAIS characterized by consistently deeper water (Fisk, 1959). Based on knickpoints in the MIS II paleo-surface, this large channel appears to meander from a northeasterly orientation to an easterly orientation as it crosses PAIS, eventually flowing into a large paleo-channel adjacent to Baffin Bay. The large paleo-channel forming Baffin Bay is the combined ancestral Los Olmos, San Fernando, and Patronila (LOSP) Creeks, which was drowned during sea level transgression and eventually filled with sediment (Simms et al., 2010). Complexities in the framework geology and modern island geomorphology, coupled with the fact that PAIS framework geology has already been mapped, make PAIS an ideal location to examine how framework geology influences barrier island geomorphology.

Previous studies of PAIS have utilized geophysical surveys and sediment cores to document variation in the depth to a buried Pleistocene paleo-surface (Anderson et al., 2016; Fisk, 1959; Wernette et al., 2018; Weymer et al., 2016). Weymer et al. (2016) confirmed paleo-channels in the buried Pleistocene paleo-surface using a 100 km alongshore electromagnetic induction (EMI) survey, where areas of lower apparent conductivity are indicative of a deeper buried surface based on the difference in conductivity between overlying Holocene sand and the buried silty clay Pleistocene paleo-surface. Areas where the subsurface apparent conductivity decreased alongshore coincided with paleo-channels which had been previously mapped. Wavelet decomposition of the alongshore EMI survey and offshore bathymetry serve as proxies for the onshore and offshore framework geology, respectively. When analyzed and interpreted in conjunction with alongshore beach and dune morphometrics, these metrics reveal that larger beach and dune systems are located within the previously mapped paleo-channels (Wernette et al., 2018). The current paper expands on previous research by adapting economic forecast models to determine how paleo-channels in the framework geology have influenced beach and dune evolution and whether this influence is directional and scale-dependent. Identifying these spatial lags, their spatial scale(s), and their lag direction(s) is the first step toward integrating this information into morphodynamic prediction models.

2.2 Data sources and validation

Examining the relationships between surface and subsurface barrier island geomorphology requires continuous alongshore data for surface morphology and subsurface framework geology. Barrier island surface morphometrics (i.e., beach width, beach volume, dune toe elevation, dune crest elevation, dune height, dune volume, island width, and island volume) were extracted every 1 m along the entire length of PAIS using an automated multi-scale approach (Wernette et al., 2016). This approach is advantageous because it is less

subjective and more efficient than conventional approaches to extracting island morphology. Long-term shoreline change (1950–2007) was used in this analysis because fine-scale and shorter-term changes are unlikely to persist given that there has not been ample storm activity to continually force shorter-term variations in shoreline change at PAIS (Houser et al., 2018b). Offshore bathymetric depth profiles were extracted every 1 m from a National Geophysical Data Center (NGDC) coastal relief model (CRM; Fig. 1).

Dune height is an important morphometric to examine the influence of framework geology on barrier island morphology, since initial patterns in dune height and dune crest elevation can persist through time (Houser, 2012; Lazarus, 2016; Weymer et al., 2015b) and determine the response of a barrier island to storms (Sallenger, 2000). Areas of high dunes are more likely to limit wash-over and inundation during a storm, and instead sediment is likely to be partially eroded from the dune and deposited on the beach and nearshore (Houser, 2012; Sallenger, 2000). Following the storm, sediment deposited in the nearshore is available for beach recovery through nearshore bar migration and welding. Onshore winds can transport sediment inland (i.e., from the beach to dune) following a storm, promoting dune recovery and development. Conversely, areas with lower or no dunes are more likely to be over-washed or completely inundated, resulting in the net landward transportation of sediment to the back-barrier. Since dune sand is not deposited in the nearshore or along the beach during the storm, sediment is not available for nearshore, beach, and, eventually, dune recovery. In this way, variations in dune height and dune crest elevation are likely to persist through time by directly affecting patterns of over-wash and represent a control on patterns of coastal resiliency and shoreline change. Identifying processes that set up modern patterns in dune morphology provides valuable insight into how the barrier island formed and how it continues to be influenced by the framework geology. Since dune height and development are partially a function of beach width, it follows that beach width is a valuable morphometric to evaluate for patterns of long-range dependence (LRD) and short-range dependence (SRD).

Information about the subsurface framework geology of the coast was derived from a ~ 100 km alongshore EMI survey (Wernette et al., 2018; Weymer et al., 2016). EMI works by inducing a primary electromagnetic field in the subsurface half-space and measuring the deformation (i.e., response) of a secondary current. From the secondary field deformation, it is possible to compute the apparent conductivity of the half-space at a specific frequency. While the apparent conductivity is influenced by a multitude of factors (Huang and Won, 2000; Huang, 2005), recent fieldwork suggests that hydrology has a minimal influence on the subsurface conductivity at PAIS at broad geographic scales, relative to the influence of stratigraphic and lithologic variation. A series of piezometer shore-normal transects were collected in fall 2016, which indicated that sand was dry within the first 2 m of the surface along the back beach. Since the EMI surveys were collected along the back beach, the piezometer measurements support the use of EMI as a proxy for the subsurface framework geology. Previous research used EMI surveys to confirm the location of several paleo-channels and to begin to quantify their influence on coastal geomorphology EMI surveys (Wernette et al., 2018; Weymer, 2016), while the current paper aims to determine the alongshore influence (direction and scale) of the paleo-channels.

2.3   Statistical modeling of spatial series

Previous research demonstrates that island morphology and framework geology can be spatially variable at multiple scales alongshore (Hapke et al., 2016; Lentz and Hapke, 2011; Schwab et al., 2013; Wernette et al., 2018; Weymer, 2012, 2016; Weymer et al., 2015a); however, previous approaches utilized models unable to identify spatial lags that may occur given alongshore sediment transport gradients. Since the goal of this paper is to evaluate SRD and LRD of island morphology and framework geology and to test whether there is directional dependence in island morphology, the current study requires a statistical model capable of accounting for SRD and LRD. Short-range dependence includes localized relationships in the data series, such as an autoregressive or moving average; there, LRD is the dependence of values on all other data values within the data series, irrespective of trend or window size. While fractal Gaussian noise (fGn) and fractal Brownian motion (fBm) models can model the SRD, both are unable to model the LRD of a series because both models are limited to two parameters (fGn: range and standard deviation; fBm: variance and scaling). Therefore, we used an autoregressive fractionally integrated moving average (ARFIMA) model to capture the LRD of a data series.

ARFIMA models may be considered a special case of autoregressive moving average (ARMA) models that have been most widely applied in predicting financial market behavior; however, it is possible to analyze spatial data series by substituting space for time. The most significant advantage of ARFIMA models over ARMA, fGn, and fBm models is their potential to account for autoregressive (AR) relationships, LRD, and moving average (MA) relationships simultaneously through fitting p, d, and q parameters, respectively. Many ARFIMA models utilize all three parameters simultaneously to describe a data series, although it is possible to isolate the influence of AR, LRD, or MA within the data in order to better understand more specifically how the data are structured (Fig. 2). By isolating one of the three parameters, it is possible to distinguish the degree to which LRD influences a data series, independent of any SRD influence. This ability to distinguish and isolate LRD from SRD is unique and represents the most significant reason that ARFIMA models were used to test for directional dependencies in coastal geomorphology.

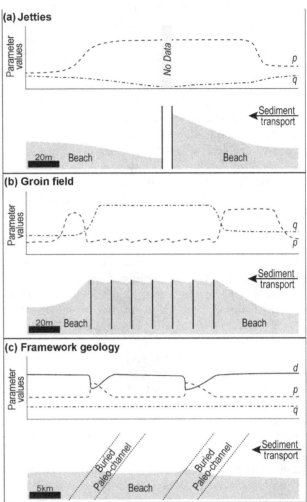

NOTE Scale in panel (c) is different than panels (a) and (b).

Figure 2. Sample beach–dune alongshore data series and ARFIMA model parameters ($p = $ AR; $d = $ LRD; and $q = $ MA) for three coastal geomorphology scenarios. **(a)** Jetties trap sediment on the beach up-drift side and starve the down-drift beach of sediment (see Ocean City, Maryland, USA), resulting in increased AR values on either side of the jetty. **(b)** Groin fields can trap sediment between the groins within the field, while starving the down-drift beach of sediment. In this case, beach volume at a particular location within the field can be modeled as the MA of adjacent beach volume measurements. Outside of the groin field, beach volume may increase/decrease, resulting in increased AR values and decreased MA values. **(c)** Framework geology, such as infilled paleo-channels, influences coastal geomorphology on broader spatial scales (see "oblique sandbars" in the Outer Banks, North Carolina, USA; McNinch, 2004) and is much more likely to appear in the LRD values. While coastal morphology at broad scales is influenced by the entire data series, sediment transport gradients can be influenced by more localized processes, resulting in an inverted trend with the AR component. The degree to which a particular point is influenced by the entire data series at a particular scale can be modeled and plotted using the LRD parameter.

The p and q parameters provide information about SRD structures within the data series, representing AR and MA, respectively. Data series modeled with high p values are those where the data value at a particular location is dependent on the trend in nearby values. For example, large jetties or groins can affect the overall alongshore sediment transport (Fig. 2a and b), trapping sediment on the up-drift side of the structure and starving down-drift areas of beach sediment. Alongshore beach–dune metrics, such as beach volume, provide valuable information about the alongshore influence of the coastal engineering structures. Using an ARMA model to characterize the data series, we would find that p values are very high adjacent to the jetties and decrease moving away from the structure (Fig. 2a). This simple AR relationship between the structures and beach volume is effectively represented by the p parameter because this relationship is relatively localized to either side of the structures and the data series does not extend for several kilometers alongshore. Moving beyond the accumulated sediment on the up-drift side or shadow on the down-drift side of the jetties, p parameter values decrease. It is important to note that the p parameter is useful for modeling localized AR relationships; however, given a more complex and/or substantially larger data series, the p parameter is less likely to capture directional trends simply due to the increased "noise" inherent in larger data series. In other words, the AR relationships may become obfuscated with increasingly large and/or complex data series.

Data series modeled with high q values also exhibit strong local dependence, although the data value at a particular location is dependent not on localized directional trends but on the average of nearby values (i.e., moving average). For example, assuming a groin field is effectively able to trap sediment and build a stable beach, the influence of these structures on beach volume can be effectively captured by the q parameter (Fig. 2b). The q parameter values for beach volume are much higher within the groin field than outside of the field because the beach volume is being influenced by sediment trapped up-drift and down-drift of a specific point. Similar to the p parameter, it is important to note that the effectiveness of using q parameter values to identify MA relationships decreases with increasing data series complexity and/or length. MA relationships are less evident in larger or very complex series simply because of the noise within the larger data series.

Unlike ARMA models which only utilize the p and q values, ARFIMA models include an additional d parameter that can vary fractionally and provides information about the degree to which values within the series are dependent on all other values in the series, not simply localized effects (i.e., moving average and autoregressive). This d parameter makes ARFIMA particularly well suited for modeling series with broad-scale dependencies (Fig. 2c). In the case of coastal geomorphology, d parameter values may be particularly useful for identifying the influence of very broad-scale influenc-

ing factors, such as paleo-channels in the framework geology (Weymer et al., 2018; Weymer, 2016).

ARFIMA modeling in the geosciences remains relatively unexplored, despite its potential for better understanding spatial and temporal patterns of variability in complex datasets. While previous research demonstrated that ARFIMA modeling can provide insight into long-range dependence patterns in alongshore barrier island surface and subsurface morphology at discrete scales (Weymer et al., 2018; Weymer, 2016), the current paper expands the ARFIMA approach to analyze alongshore morphometrics at all scales along the entire length of spatial data series. In other words, while previous research discretized a data series into arbitrary alongshore lengths and locations to characterize LRD along PAIS, the current paper assesses LRD at all alongshore length scales along the entire length of PAIS. In this sense, the current paper presents a new approach to assessing how LRD changes alongshore and interprets these changes with respect to coastal processes and barrier island evolution. While wavelet decomposition can provide insight into relationships between two variables in the same location (Wernette et al., 2018), utilizing ARFIMA as a sliding window across multiple spatial scales can shed light on relationships that exhibit a lag in one or both directions. The advantage of this new approach is its application to examine alongshore influences of various natural and anthropogenic features (e.g., jetties, seawalls, groin fields, paleo-channels, and/or headlands) and identify their effective zone(s) of influence on coastal processes and geomorphology.

In this paper, the effects of LRD within each spatial data series was isolated using a 0, d, 0 ARFIMA model. Each ARFIMA model was fit using the *fracdiff* package (Fraley et al., 2012, in R Core Team, 2016), where the p and q parameters were set equal to 0. Setting both p and q parameters to 0 eliminates the short-range autoregressive and moving average terms from the fitted models. Each surface, subsurface, and bathymetric spatial data series contains 96 991 measurements in total. Each spatial series was divided into $\sim$ 250 unique computational windows, corresponding to alongshore length scales, ranging from two observations (2 m alongshore length scale) to the entire 96 991 observations (96 991 m alongshore length scale). While the number of computational windows can be decreased, or increased, it is important to note that the ARFIMA modeling process is computationally intensive, requiring days to complete an analysis of a single spatial data series on a high-performance desktop computer. Increasing the number of computational windows would provide more detailed information about the structure of the dataset but would significantly increase the computing power required to fit the models. Decreasing the number of computational scales would decrease the computing power required and speed up the computations; however, it would become more difficult to resolve the scales at which the structure breaks down. The range of computational windows could also be adjusted to a specific range, depending on the objectives of the research. At each scale the computational window is moved along the dataset and the appropriate d parameter is computed. The fitted d parameter is then assigned to the center of the window at the corresponding length scale. Repeating this process for each alongshore length scale yields a matrix of values, where the row corresponds to the alongshore length scale of the data subset used to compute the d parameter and the column represents the alongshore location of the center of the computational window. This matrix can be plotted similar to a wavelet plot to examine spatial patterns of LRD throughout the entire dataset at all length scales.

2.4 Interpreting LRD plots

Figure 3 represents a sample LRD plot using a 10 km alongshore portion of PAIS dune height, where the x axis represents the alongshore position or space (in meters) and the y axis represents the alongshore spatial scale (in meters). Plots are oriented by latitude on the x axis, from south (left) to north (right). In this paper, all plots utilize a color ramp from blue to red, where blue hues represent smaller d parameter values and red hues represent larger d parameter values. Given this color scheme, locations or segments of the data lacking LRD are likely to appear as "flares" or flames. Each of the flares, such as the flare at location A, represent the scale and areas of the dataset where LRD begins to break down in favor of SRD. LRD dominates at a particular location at a broad spatial scale (indicated by red hues) and becomes less influential as the spatial scale becomes increasingly finer (indicated by the transition from red to yellow to blue hues). In the case of the flare at location A (Fig. 3) we can see that the dune height series exhibits strong LRD at scales broader than ~ 20 km alongshore. This suggests that dune height at location A is related to adjacent values down to ~ 10 km on both sides of A. Morphology at scales finer than ~ 20 m is more locally dependent. In this respect, ARFIMA represents an approach to determine the limiting scale to self-similarity.

Depending on the structure of the morphology and/or geology, it is feasible that the LRD may not appear to be symmetrical. Long-range dependence is asymmetric at location B, where the LRD begins to break down more rapidly to the right side of the plot than the left. While the physical interpretation of a LRD plot depends on the variable, asymmetric flares can be broadly interpreted as areas where the variable is more locally dependent on the surrounding values at the scales and in the direction that the flare is oriented. In the case of flare B, dune height is more dependent on adjacent values to the north up to ~ 39 km alongshore. Asymmetries in the LRD plots can provide valuable information about the underlying structure influencing the variable of interest.

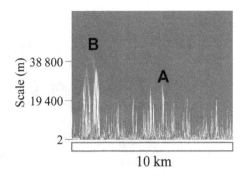

Figure 3. Example LRD plot using alongshore dune height at PAIS. The y axis represents the alongshore length scale (in meters), and the x axis represents the alongshore location. LRD is persistent at greater alongshore length scales at location B than location A. Additionally, location B is asymmetric, which may suggest a directional dependence in the data series.

3 Results

Subsurface apparent conductivity exhibits substantial LRD along the entire length of PAIS (Fig. 4a). Patterns in the subsurface framework geology LRD plot demonstrate that the framework geology is self-similar at broader scales and that this structure varies alongshore at finer alongshore length scales which correspond to the scale of the previously identified paleo-channels. The large LRD values at broad spatial scales (Fig. 4a) demonstrate that the paleo-topographic structure dominated by broad-scale coastal curvature over very broad spatial scales. Since the framework geology reflects the paleo-topography and the modern barrier island surface is dissipative at very broad scales, based on large LRD values at broad scales in the modern barrier island morphology, it follows that the framework geology is dissipative. The substantial LRD along much of the island supports previous work by Weymer (2016) and Weymer et al. (2018), which demonstrated that subsurface framework geology exhibits LRD at discrete locations and alongshore length scales.

The shoreline change LRD plot exhibits the greatest LRD values (i.e., highest LRD values across all broad spatial scales) along the length of PAIS, as indicated by the dominance of red hues in Fig. 4b. Most flares present in the shoreline change LRD are at relatively fine spatial scales, shorter than a few kilometers. Peaks in the shoreline change LRD plot are very narrow, which we interpret to mean that the long-term shoreline change is dominantly dissipative with only minor undulations due to localized coastal processes, consistent with the findings of Lazarus et al. (2011), who demonstrated that broad-scale and long-term shoreline change is dissipative. Waves impacting the coast can erode sediment from one area and transport it to another area, resulting in undulations in the shoreline orientation. Since long-term shoreline change is the result of cumulative daily wave processes eroding undulations in the shoreline shape

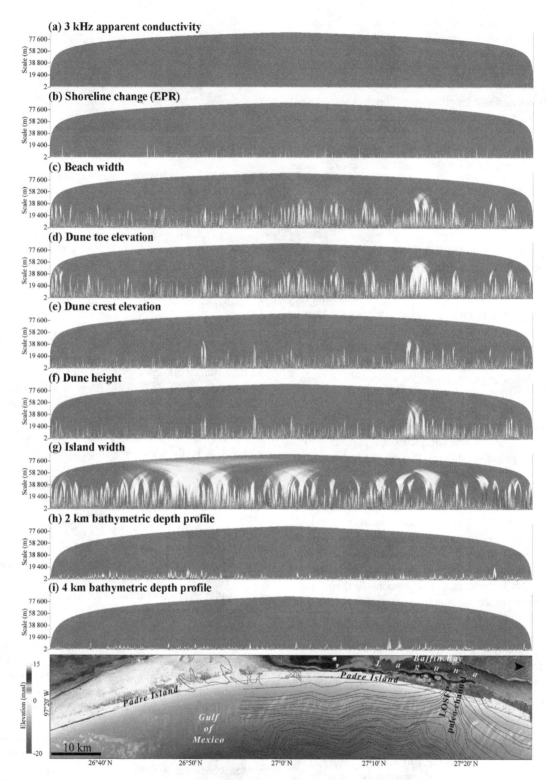

Figure 4. Long-range dependence plots of alongshore morphometrics: **(a)** 3 kHz apparent conductivity, **(b)** shoreline change rate (end-point rate), **(c)** beach width, **(d)** dune toe elevation, **(e)** dune crest elevation, **(f)** dune height, **(g)** island width, **(h)** bathymetric depth profile at 2 km offshore, and **(i)** bathymetric depth profile at 4 km offshore. All LRD plots are aligned with the map below, based on latitude. Previously documented variability in the framework geology is indicated by the contour lines representing the Pleistocene (i.e., MIS II) paleo-surface (Anderson et al., 2016; Fisk, 1959).

and dissipating any short-term undulations, fine-scale variations in the nearshore bathymetry, such as nearshore bars and troughs, can affect patterns of erosion and deposition along the coast over longer periods of time (Hapke et al., 2016). Therefore, it follows that the long-term shoreline change LRD plot would exhibit a large amount of LRD.

Beach width LRD is more variable than shoreline change (Fig. 4c), with the least amount of variability concentrated in approximately the southern third of the island. These flares are likely present because transverse ridges in the nearshore bathymetry affect localized wave refraction patterns, thereby influencing fine-scale patterns in beach morphology. Patterns in the beach morphology in southern PAIS are likely more localized because the incoming wave energy is refracted around the transverse ridges, which impacts sediment transport gradients along this part of the island. Any variations in beach morphology are more locally influenced by relatively closely spaced transverse ridges ($\sim$ 0.8 to 1.5 km alongshore spacing), resulting in broad-scale LRD along southern PAIS.

The central third of PAIS beach width is characterized by several significant flares in LRD, with many of the strongest flares adjacent to infilled paleo-channels previously identified by Fisk (1959) (Figs. 4c and 5a). The scale at which LRD transitions to SRD is at the broadest alongshore length scales proximal to Baffin Bay, and this threshold decreases in scale to the north (Figs. 4c and 6a). Given a dominant southerly alongshore current during island development in the Holocene (Anderson et al., 2016; Sionneau et al., 2008) and corresponding southerly sediment transport gradient, patterns in the beach morphology LRD plot suggest that the paleo-channels are asymmetrically influencing beach morphology. Simms et al. (2010) presented seismic profiles extending from north to south across the ancestral LOSP Creeks, which exhibit a series of onlapping reflectors on the northern edge of the seismic profiles. These onlapping reflectors are indicative of deposition on the northern edge of the paleo-channel and support the hypothesis that alongshore spit development occurred within the LOSP Creeks paleo-channel. The beach north of the large paleo-channel identified by Fisk (1959) would have been nourished by sediment discharged from the ancestral LOSP Creeks, now forming Baffin Bay. Similarly, the beach north of the ancestral LOSP Creeks paleo-channel may have been nourished by sediment from the ancestral Nueces River. In this way, beach morphology up-drift of the large paleo-channels would impact beach morphology within and south of the large paleo-channels.

Alongshore LRD in the dune crest elevation and dune height varies similarly to beach width LRD along PAIS (Figs. 4e, 4f, 5b, 5c, 6b, and 6c). The southern third of PAIS is characterized by LRD–SRD transitioning at finer alongshore length scales than the northern two-thirds of the island, as indicated by the flares in the dune height LRD plot (Fig. 4e and f). The most significant flares are proximal to the ancestral LOSP Creeks paleo-channels dissecting central PAIS and the ancestral Nueces River paleo-channel extending into

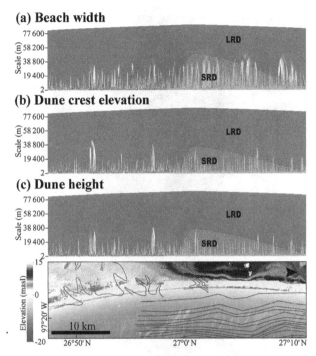

Figure 5. LRD plots of **(a)** beach width, **(b)** dune crest elevation, and **(c)** dune height for central PAIS, where Fisk (1959) identified a series of relict infilled paleo-channels dissecting the island. The scale at which LRD breaks down in favor of SRD is greatest at the southern edge of large paleo-channels, and this scale gradually decreases to the north. Smaller paleo-channels do not appear to be as influential in the modern beach and dune morphology, suggesting that small channels may not have as significant an influence as larger channels.

Baffin Bay (Fig. 6). Given that the dominant alongshore sediment transport gradient is from north to south and that the beach morphology exhibits an asymmetric LRD to the north of the large paleo-channels, it follows that LRD and SRD patterns in dune morphology would exhibit similar asymmetry to beach morphology.

The transition from dune height LRD to SRD occurs at the largest scale, i.e., approximately at 35 km alongshore length scales (Figs. 4f and 6c). This maximum occurs at the southern edge of the ancestral LOSP Creeks paleo-channel, adjacent to Baffin Bay (Fig. 6c). The alongshore length scale can be interpreted as the alongshore distance that the paleo-channel affected wave refraction patterns and sediment distribution along the beach, ultimately affecting sediment supply to develop larger dunes. It follows that paleo-channel influence on dune crest elevation and dune height would be asymmetric, with greater LRD to the north of the paleo-channels, assuming paleo-channels inhibited southern alongshore sediment transport and starved the beach down-drift. The wide beach up-drift of a paleo-channel represents a larger sediment supply and greater fetch for aeolian transport and dune growth and is consistent with peaks in dune height identified by Wernette et al. (2018).

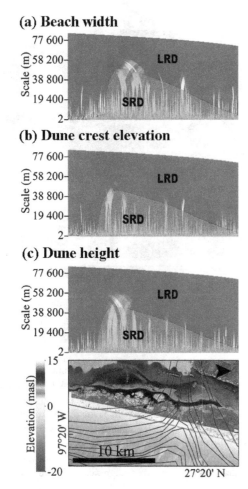

(a) Beach width

(b) Dune crest elevation

(c) Dune height

Figure 6. LRD plots of **(a)** beach width, **(b)** dune crest elevation, and **(c)** dune height for PAIS adjacent to the ancestral LOSP Creeks, forming the modern Baffin Bay. LRD breaks down in favor of SRD at the largest scales at the southern edge of the previously identified paleo-channel. The scale at which LRD breaks down to SRD decreases gradually to the north of the channel, suggesting that the paleo-channel asymmetrically influenced beach and dune morphology.

Island width exhibits the greatest alongshore variability in LRD of all island and framework geology morphometrics (Fig. 4g) and demonstrates that island width is dependent on broad- and fine-scale patterns of change. Areas of low dunes are likely to be overtopped during a storm, transporting sediment to the landward margin of the island. Waves and currents along the landward margin of the island erode the wash-over fans and redistribute sediment along the island. In this sense, the island width at one location is directly influenced by sedimentation patterns along the adjacent parts of the island. Undulations in the Gulf of Mexico shoreline are smoothed out over the long term, thereby reducing the likelihood that patterns in island width are solely caused by shoreline change patterns. This repeat wash-over, followed by sediment redistribution along the back-barrier shoreline, repre-

sents the mechanism by which barrier islands can transgress landward and keep up with sea level rise.

Bathymetric depth profiles at 2 and 4 km offshore exhibit substantial LRD at broad scales, but this breaks down at scales finer than ∼ 15 km alongshore (Fig. 4h and i). Long-range dependence breaks down at larger alongshore length scales in the 2 km bathymetry, compared to the 4 km bathymetry. Since modern coastal processes continue to affect alongshore sediment transport, large undulations in the bathymetry are smoothed out over time by sediment redistributed along the coast. Finer-scale variations in the modern nearshore bathymetry occur at similar spatial scales as previously identified at PAIS (Wernette et al., 2018). The 2 km bathymetric profile LRD breaks down at broader spatial scales than the 4 km bathymetry (Fig. 4h and i). This suggests that localized variations in coastal processes manifest themselves in the nearshore bathymetry closer to the shoreline. Wave shoaling and breaking will erode and deposit sediment along the coast, impacting bathymetric structures closer to the shoreline.

4 Discussion

As noted, flares in the LRD plots are interpreted as areas where the morphometrics are more locally dependent on the adjacent values. Since flares in the LRD plots of surface morphometrics are most pronounced adjacent to the infilled paleo-channels and decrease to the north (Figs. 4, 5, and 6), this spatial correlation supports the hypothesis that the modern barrier island morphology was influenced by variations in the framework geology. Paleo-channels along PAIS range in scale, with the smallest channels only ∼ 13 m below the modern surface and the deepest and widest channels ∼ 50 to ∼ 64 m deep. Regardless of the paleo-channel dimensions, patterns in the LRD plots demonstrate that paleo-channels affect the nearshore bathymetry and modern island morphometrics asymmetrically and decrease in minimum alongshore scale to the north. Beach and dune morphology updrift of a paleo-channel directly affects sediment available for areas of the beach down-drift. In this way, larger paleo-channels (depth and width) will have a greater accommodation space and influence beach–dune morphology along a greater stretch of coast, while smaller paleo-channels have a more limited accommodation space and, therefore, influence a smaller stretch of adjacent coastal morphology. Given that a paleo-channel would have acted as a sediment sink for excess sediment transported alongshore during sea level transgression, it follows that LRD values would remain high at fine spatial scales up-drift of the paleo-channel locations (Figs. 5 and 6).

The current paper is in agreement with previous research that demonstrates barrier island morphology is dissipative at broad spatial scales (Wernette et al., 2018; Lazarus et al., 2011). Long-range dependence is significant at very broad spatial scales in all island morphometrics except for island width. Previous research also demonstrates that rhythmic undulations and isolated paleo-channels can influence short-term shoreline change patterns (Lazarus et al., 2011; McNinch, 2004; Schupp et al., 2006) and beach and dune morphology (Houser et al., 2008; Houser and Barrett, 2010). This paper presents new information supporting the hypothesis that paleo-channels in the framework geology interact with alongshore currents to drive asymmetries in barrier island geomorphology and that the scale of influence is ultimately limited. This asymmetry is likely caused by paleo-channels acting as sediment sinks for sediment transported south by a prevailing southerly alongshore current during barrier island formation.

The alongshore distance that variations in the framework geology influence beach and dune morphology is dependent on paleo-channel scale and orientation, relative to the average shoreline orientation. Long-range dependence plots of beach and dune morphometrics suggest that beach and dune morphology within the largest paleo-channel dissecting the island, the ancestral LOSP Creeks, was influenced by beach and dune morphology up to 25 km north of the channel edge (Figs. 4c–f, 5, and 6). The large paleo-channel identified by Fisk (1959) is slightly smaller in scale than the paleo-channel forming Baffin Bay; however, the large Fisk (1959) channel intersects the coast at an oblique angle. Since the channel dissects PAIS at an oblique angle, the influence of this channel is more apparent on beach morphology than dune morphology. An oblique channel would have required more sediment and take longer to fill than a shore-normal channel. Subsequently, a wide beach and dunes would begin to form in the shore-normal paleo-channel before the oblique paleo-channel. For an oblique paleo-channel the volume of sediment required to fill the channel from alongshore sediment transport and fluvial deposition from the mainland would likely have been insufficient to build a wide beach to supply sediment for significant dune growth.

Paleo-currents during the Holocene were predominantly from north to south (Sionneau et al., 2008), which would have set up a southerly alongshore sediment transport gradient. Sediment transported from north to south along the coast would have nourished beaches up-drift (i.e., north) of the channel. Consequently, nourished beaches up-drift of the paleo-channel had a greater sediment supply and increased fetch for aeolian transport inland to promote large dune development (Bauer and Davidson-Arnott, 2002; Bauer et al., 2009). While beach nourishment and dune growth continued up-drift of the channel, excess sediment entering the channel was deposited along the up-drift edge of the channel (Fig. 7). Deposition on the up-drift edge was caused by the increased accommodation space within the channel. Increasing the area that the alongshore current flows through (i.e., transitioning from a confined alongshore current to an open channel), while maintaining the alongshore current discharge, resulted in a decreased flow along the northern edge. Reducing along-

shore current velocity caused sands to be deposited along the northern edge of the channel (Fig. 7), while finer particles are transported farther into the channel and funneled offshore through the channel outlet. Given enough time and with continued sea level rise during the Holocene, this preferential deposition would have built a spit into the channel. Sediment trapped in the paleo-channel would be unavailable to the beach down-drift. The closest modern analogy to this alongshore sedimentation process is the formation and evolution of an alongshore spit eventually completely crossing the outflowing river channel, where the river is eventually cut off by the elongating spit. In this case, sediment is supplied to the up-drift beach and provides a sediment source for dunes to form.

Directional dependencies in beach and dune morphology, initially set up by the interaction of framework geology with a dominant southerly alongshore current, persist through time due to preferential wash-over reinforcing preexisting alongshore variation in dune height. Areas of the island with limited or no dune development are preferentially overtopped by elevated water levels during a storm. Conversely, areas with higher dunes resist storm wash-over/inundation and recover more rapidly following a storm. Alongshore variations in the barrier island morphometrics, such as dune height, persist through time because these patterns are reinforced by episodic wash-over of small dunes during storms.

The apparent disconnect between long-term shoreline change and framework geology is due to the cumulative influence of waves continuously interacting with the coast. This disconnect is further highlighted by the lack of storms impacting PAIS. Long-term shoreline change rate is the cumulative result of waves moving sediment on a daily basis, while short-term variations in shoreline position caused by storms are feasible. It is unlikely that short-term variations in PAIS shoreline position identified here are caused by storms because PAIS has not been significantly impacted by a storm since Hurricane Bret in 1999. Any short-term undulations in shoreline position are likely to disappear over longer timescales, especially since no storm has hit the island to cause significant localized shoreline erosion. Therefore, the long-term shoreline change rate LRD (Fig. 4b) is unlikely to exhibit substantial variation alongshore. Beach, dune, and island morphology do show significant variation in patterns of LRD along PAIS (Figs. 4c, 4d, 4e, 4f, 4g, 5, and 6) because the initial barrier island morphology was set up by the framework geology. If hurricanes had impacted PAIS more frequently, it is likely that the alongshore variations in dune morphology, which were initially set up by the paleo-channels, would have been reinforced. This is because areas set up as low dunes would be preferentially overwashed while areas of high dunes would be more resistant and resilient during and following a storm. Therefore, the impact of a hurricane would highlight alongshore variations in dune morphology set up by the paleo-channels. Predicting future changes to barrier island geomorphology requires

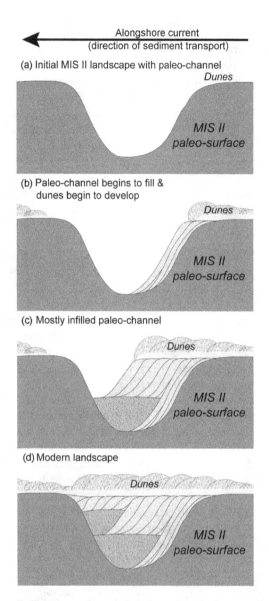

Figure 7. The stages of how paleo-channels in the framework geology affected barrier island development and evolution at PAIS, beginning with **(a)** initial paleo-channel incision (approximately MIS II). **(b)** As sea-level began to rise, sediment was transported south from river outlets to the north and was deposited along the northern edge of the paleo-channel. **(c)** Continued sea-level rise and sedimentation from the north, coupled with episodic fluvial channel fill, continue to fill the paleo-channel until **(d)** the paleo-channel is completely infilled and the island coalesces alongshore. From **(a)** to **(d)** beaches to the north of the active channel area are supplied with sediment from river outlets up-drift, which provide ample sediment supply for aeolian sediment transport and dune development.

a comprehensive knowledge of how the framework geology affected initial variation in the beach and dunes.

Understanding how the framework geology influences barrier island geomorphology has important implications for understanding how barrier islands are likely to respond to and recover following a storm or series of storms. While many

models of barrier island recovery focus on spatiotemporal models of change, Parmentier et al. (2017) demonstrated that spatial autocorrelation outperformed temporal autocorrelation (e.g., "space beats time", SBT) when predicting the recovery of vegetation following Hurricane Dean. Since vegetation recovery and dune geomorphic recovery are related (Houser et al., 2015), it follows that understanding spatial autocorrelation in beach and dune features is essential to predicting future changes to barrier island geomorphology. The current paper supports the conclusions of Parmentier et al. (2017) by demonstrating that spatial variations in the framework geology directly relate to alongshore variations in beach and dune morphology (Figs. 5 and 6). In the context of SBT theory, results of the current paper support the hypothesis that spatial variations in the framework geology (i.e., "space") control barrier island evolution (i.e., "time"). Accurately predicting future barrier island change is predicated on comprehensively understanding what processes influenced its initial formation and what processes continue to influence island morphology. Predicting coastal change without accounting for all factors affecting formation and evolution, such as directional dependencies due to framework geology, is more prone to uncertainty, which can have important managerial applications.

Given that framework geology influences beach and dune morphology along the coast, the methods and results of this paper represent an opportunity for managers to improve coastal engineering projects, such as beach nourishment. Sediment budget imbalances set up by the framework geology dictate the long-term barrier island trajectory. Utilizing ARFIMA models to evaluate the alongshore beach and dune morphology can provide valuable insight into how the coast is likely to change naturally in the future. To reduce waste by coastal nourishment, future projects should seek to first comprehensively understand how the paleo-topography of an area continues to affect coastal processes and morphology. By understanding the long-term influence of framework geology, coastal nourishment projects can more effectively balance how a project focuses on the near-future coastal morphology with long-term natural changes.

5 Conclusions

This paper quantitatively demonstrates that variation in the framework geology influences patterns of beach and dune morphology along a barrier island. Understanding what controls beach and dune morphology and barrier island development is integral to predicting future changes to barrier island geomorphology and island transgression caused by storms and sea level rise. Storm impact and barrier island transgression patterns are controlled by beach slope, dune height, and wave run-up. Given a persistent alongshore sediment gradient during the Holocene, paleo-channels in the framework geology at PAIS likely acted as sediment sinks during island development. While wide beaches and, subsequently, large dunes are nourished with sediment up-drift of the channel, excess sediment can become trapped in the channel. These channels trap sediment, starving sediment from down-drift portions of the coast. The result of this asymmetry in sediment supply is that large dunes occur up-drift of the paleo-channel and small dunes occur down-drift of the paleo-channel. Effectively managing a barrier island underlain by a variable framework geology should seek to balance short-term societal pressures in the context of long-term natural change (i.e., framework geology).

Author contributions. Because the work is so collaborative, all co-authors contributed equally. There is no author solely involved in any one phase of the research, rather, everybody was involved throughout all phases of the work.

Competing interests. The authors declare that they have no conflict of interest.

Acknowledgements. This material is based upon work supported by the National Science Foundation under grant no. 1634077. Publication supported in part by an Institutional Grant (NA14AR4170102) to the Texas Sea Grant College Program from the National Sea Grant Office, National Oceanic and Atmospheric Administration, U.S. Department of Commerce and the National Science Foundation (DDRI grant number 1634077). This project was supported by a Natural Science and Engineering Research Council of Canada (NSERC) Discovery Grant to Chris Houser. Undergraduate student research assistants aided in fieldwork, with funding through a High-Impact Learning Experiences grant from the Texas A&M University College of Geosciences.

References

Anderson, J. B., Wallace, D. J., Simms, A. R., Rodriguez, A. B., Weight, R. W. R., and Taha, Z. P.: Recycling sediments between source and sink during a eustatic cycle: Systems of late Quaternary northwestern Gulf of Mexico Basin, Earth-Sci. Rev., 153, 111–138, https://doi.org/10.1016/j.earscirev.2015.10.014, 2016.

Bauer, B. O. and Davidson-Arnott, R. G. D.: A general framework for modeling sediment supply to coastal dunes including wind angle, beach geometry, and fetch effects, Geomorphology, 49, 89–108, 2002.

Bauer, B. O., Davidson-Arnott, R. G. D., Hesp, P. A., Namikas, S. L., Ollerhead, J., and Walker, I. J.: Aeolian sediment transport on a beach: Surface moisture, wind fetch, and mean transport, Geomorphology, 105, 106–116, https://doi.org/10.1016/j.geomorph.2008.02.016, 2009.

Brenner, O. T., Lentz, E. E., Hapke, C. J., Henderson, R. E., Wilson, K. E., and Nelson, T. R.: Characterizing storm response and recovery using the beach change envelope: Fire Island, New York, Geomorphology, 300, 189–202, https://doi.org/10.1016/j.geomorph.2017.08.004, 2018.

Browder, A. G. and McNinch, J. E.: Linking framework geology and nearshore morphology: Correlation of paleo-channels with shore-oblique sandbars and gravel outcrops, Mar. Geol., 231, 141–162, https://doi.org/10.1016/j.margeo.2006.06.006, 2006.

Cooper, J. A. G., Green, A. N., and Loureiro, C.: Geological constraints on mesoscale coastal barrier behaviour, Global Planet. Change, 168, 15–34, https://doi.org/10.1016/j.gloplacha.2018.06.006, 2018.

Fisk, H. N.: Padre Island and Lagunas Madre Flats, coastal south Texas, Second Coastal Geography Conference, 103–151, 1959.

Goldstein, E. B. and Moore, L. J.: Stability and bistability in a one-dimensional model of coastal foredune height, J. Geophys. Res.-Earth, 121, 964–977, https://doi.org/10.1002/2015jf003783, 2016.

Goldstein, E. B., Moore, L. J., and Durán Vinent, O.: Lateral vegetation growth rates exert control on coastal foredune "hummockiness" and coalescing time, Earth Surf. Dynam., 5, 417–427, https://doi.org/10.5194/esurf-5-417-2017, 2017.

Gutierrez, B. T., Plant, N. G., Thieler, E. R., and Turecek, A.: Using a Bayesian network to predict barrier island geomorphologic characteristics, J. Geophys. Res.-Earth, 120, 2452–2475, https://doi.org/10.1002/2015JF003671, 2015.

Hapke, C. J., Lentz, E. E., Gayes, P. T., McCoy, C. A., Hehre, R., Schwab, W. C., and Williams, S. J.: A Review of Sediment Budget Imbalances along Fire Island, New York: Can Nearshore Geologic Framework and Patterns of Shoreline Change Explain the Deficit?, J. Coast. Res., 263, 510–522, https://doi.org/10.2112/08-1140.1, 2010.

Hapke, C. J., Plant, N. G., Henderson, R. E., Schwab, W. C., and Nelson, T. R.: Decoupling processes and scales of shoreline morphodynamics, Mar. Geol., 381, 42–53, https://doi.org/10.1016/j.margeo.2016.08.008, 2016.

Houser, C.: Feedback between ridge and swale bathymetry and barrier island storm response and transgression, Geomorphology, 173–174, 1–16, https://doi.org/10.1016/j.geomorph.2012.05.021, 2012.

Houser, C. and Barrett, G.: Divergent behavior of the swash zone in response to different foreshore slopes and nearshore states, Mar. Geol., 271, 106–118, https://doi.org/10.1016/j.margeo.2010.01.015, 2010.

Houser, C. and Mathew, S.: Alongshore variation in foredune height in response to transport potential and sediment supply: South Padre Island, Texas, Geomorphology, 125, 62–72, https://doi.org/10.1016/j.geomorph.2010.07.028, 2011.

Houser, C., Hapke, C., and Hamilton, S.: Controls on coastal dune morphology, shoreline erosion and barrier island response to extreme storms, Geomorphology, 100, 223–240, https://doi.org/10.1016/j.geomorph.2007.12.007, 2008.

Houser, C., Wernette, P., Rentschlar, E., Jones, H., Hammond, B., and Trimble, S.: Post-storm beach and dune recovery: Implications for barrier island resilience, Geomorphology, 234, 54–63, https://doi.org/10.1016/j.geomorph.2014.12.044, 2015.

Houser, C., Barrineau, P., Hammond, B., Saari, B., Rentschlar, E., Trimble, S., Wernette, P. A., Weymer, B. A., and Young, S.: Role of the foredune in controlling barrier island response to sea level rise, in: Barrier Dynamics and Response to Climate Change, edited by: Moore, L. J. and Murray, A. S., Springer, 2018a.

Houser, C., Wernette, P., and Weymer, B. A.: Scale-dependent behavior of the foredune: Implications for barrier island response to storms and sea-level rise, Geomorphology, 303, 362–374, https://doi.org/10.1016/j.geomorph.2017.12.011, 2018b.

Huang, H.: Depth of investigation for small broadband electromagnetic sensrs, Geophysics, 70, G135–G142, https://doi.org/10.1190/1.2122412, 2005.

Huang, H. and Won, I. J.: Conductivity and susceptibility mapping using broadband electromagnetic sensors, J. Environ. Eng. Geoph., 5, 31–41, 2000.

Lazarus, E., Ashton, A., Murray, A. B., Tebbens, S., and Burroughs, S.: Cumulative versus transient shoreline change: Dependencies on temporal and spatial scale, J. Geophys. Res., 116, F02014, https://doi.org/10.1029/2010jf001835, 2011.

Lazarus, E. D.: Scaling laws for coastal overwash morphology, Geophys. Res. Lett., 43, 12113–112119, https://doi.org/10.1002/2016gl071213, 2016.

Lazarus, E. D. and Armstrong, S.: Self-organized pattern formation in coastal barrier washover deposits, Geology, 43, 363–366, https://doi.org/10.1130/g36329.1, 2015.

Lentz, E. E. and Hapke, C. J.: Geologic framework influences on the geomorphology of an anthropogenically modified barrier island: Assessment of dune/beach changes at Fire Island, New York, Geomorphology, 126, 82–96, https://doi.org/10.1016/j.geomorph.2010.10.032, 2011.

Lentz, E. E., Hapke, C. J., Stockdon, H. F., and Hehre, R. E.: Improving understanding of near-term barrier island evolution through multi-decadal assessment of morphologic change, Mar. Geol., 337, 125–139, https://doi.org/10.1016/j.margeo.2013.02.004, 2013.

McNinch, J. E.: Geologic control in the nearshore: shore-oblique sandbars and shoreline erosional hotspots, Mid-Atlantic Bight, USA, Mar. Geol., 211, 121–141, https://doi.org/10.1016/j.margeo.2004.07.006, 2004.

Moore, L. J., List, J. H., Williams, S. J., and Stolper, D.: Complexities in barrier island response to sea level rise: Insights from numerical model experiments, North Carolina Outer Banks, J. Geophys. Res., 115, F03004, https://doi.org/10.1029/2009jf001299, 2010.

Murray, A. S., Lazarus, E., Moore, L., Lightfoot, J., Ashton, A. D., McNamara, D. E., and Ells, K.: Decadal Scale Shoreline Change Arises from Large-Scale Interactions, While Small-Scale Changes Are Forgotten: Observational Evidence, Proceedings of the Coastal Sediments, San Diego, USA, 2015.

Parmentier, B., Millones, M., Griffith, D. A., Hamilton, S. E., Chun, Y., and McFall, S.: When Space Beats Time: A Proof of Concept with Hurricane Dean, Springer, 207–215, https://doi.org/10.1007/978-3-319-22786-3_19, 2017.

Plant, N. G. and Stockdon, H. F.: Probabilistic prediction of barrier-island response to hurricanes, J. Geophys. Res.-Earth, 117, F03015, https://doi.org/10.1029/2011jf002326, 2012.

Riggs, S. R., Cleary, W. J., and Snyder, S. W.: Influence of inherited geologic framework on barrier shoreface morphology and dynamics, Mar. Geol., 126, 213–234, 1995.

Sallenger, A. H.: Storm impact scale for barrier islands, J. Coast. Res., 16, 890–895, 2000.

Schupp, C. A., McNinch, J. E., and List, J. H.: Nearshore shore-oblique bars, gravel outcrops, and their correlation to shoreline change, Mar. Geol., 233, 63–79, https://doi.org/10.1016/j.margeo.2006.08.007, 2006.

Schwab, W. C., Baldwin, W. E., Hapke, C. J., Lentz, E. E., Gayes, P. T., Denny, J. F., List, J. H., and Warner, J. C.: Geologic Evidence for Onshore Sediment Transport from the Inner Continental Shelf: Fire Island, New York, J. Coast. Res., 288, 526–544, https://doi.org/10.2112/jcoastres-d-12-00160.1, 2013.

Simms, A. R., Lambeck, K., Purcell, A., Anderson, J. B., and Rodriguez, A. B.: Sea-level history of the Gulf of Mexico since the Last Glacial Maximum with implications for the melting history of the Laurentide Ice Sheet, Quaternary Sci. Rev., 26, 920–940, https://doi.org/10.1016/j.quascirev.2007.01.001, 2007.

Simms, A. R., Aryal, N., Miller, L., and Yokoyama, Y.: The incised valley of Baffin Bay, Texas: a tale of two climates, Sedimentology, 57, 642–669, https://doi.org/10.1111/j.1365-3091.2009.01111.x, 2010.

Sionneau, T., Bout-Roumazeilles, V., Biscaye, P. E., Van Vliet-Lanoe, B., and Bory, A.: Clay mineral distributions in and around the Mississippi River watershed and Northern Gulf of Mexico: sources and transport patterns, Quaternary Scie. Rev., 27, 1740–1751, https://doi.org/10.1016/j.quascirev.2008.07.001, 2008.

Stallins, J. A. and Parker, A. J.: The Influence of Complex Systems Interactions on Barrier Island Dune Vegetation Pattern and Process, Ann. Assoc. Am. Geogr., 93, 13–29, https://doi.org/10.1111/1467-8306.93102, 2003.

U.S. Environmental Protection Agency: Climate change indicators in the United States, 2016, 2016.

Weise, B. R. and White, W. A.: Padre Island National Seashore: A guide to the geology, natural environments, and history of a Texas barrier island, Texas Bureau of Economic Geology, 94, 1980.

Wernette, P., Houser, C., and Bishop, M. P.: An automated approach for extracting Barrier Island morphology from digital elevation models, Geomorphology, 262, 1–7, https://doi.org/10.1016/j.geomorph.2016.02.024, 2016.

Wernette, P. A., Houser, C., Weymer, B. A., Bishop, M. P., Everett, M. E., and Reece, R.: Influence of a spatially complex framework geology on island geomorphology, Mar. Geol., 398, 151–162, 2018.

Weymer, B., Everett, M. E., Houser, C., Wernette, P., and Barrineau, P.: Differentiating tidal and groundwater dynamics from barrier island framework geology: Testing the utility of portable multifrequency electromagnetic induction profilers, Geophysics, 81, E347–E361, https://doi.org/10.1190/GEO2015-0286.1, 2016.

Weymer, B. A.: A geologic characterization of the alongshore variability in beach-dune morphology: Padre Island National Seashore, Texas, Master of Science, Geology and Geophysics, Texas A&M University, 104 pp., 2012.

Weymer, B. A.: An investigation of the role of framework geology on modern barrier island transgression, Doctor of Philosophy, Geology and Geophysics, Texas A&M University, 2016.

Weymer, B. A., Everett, M. E., de Smet, T. S., and Houser, C.: Review of electromagnetic induction for mapping barrier island framework geology, Sedimentary Geol., 321, 11–24, https://doi.org/10.1016/j.sedgeo.2015.03.005, 2015a.

Weymer, B. A., Houser, C., and Giardino, J. R.: Post-storm Evolution of Beach-Dune Morphology: Padre Island National Seashore, Texas, J. Coast. Res., 31, 634–644, https://doi.org/10.2112/Jcoastres-D-13-00020.1, 2015b.

Weymer, B. A., Wernette, P., Everett, M. E., and Houser, C.: Statistical modeling of the long-range-dependent structure of barrier island framework geology and surface geomorphology, Earth Surf. Dynam., 6, 431–450, https://doi.org/10.5194/esurf-6-431-2018, 2018.

Wilson, K. E., Adams, P. N., Hapke, C. J., Lentz, E. E., and Brenner, O.: Application of Bayesian Networks to hindcast barrier island morphodynamics, Coast. Eng., 102, 30–43, https://doi.org/10.1016/j.coastaleng.2015.04.006, 2015.

10

Mechanical state of gravel soil in mobilization of rainfall-induced landslides in the Wenchuan seismic area, Sichuan province, China

Liping Liao[1,2,3], Yunchuan Yang[1,2,3], Zhiquan Yang[4], Yingyan Zhu[5], Jin Hu[5], and D. H. Steve Zou[6]

[1]College of Civil Engineering and Architecture, Guangxi University, Nanning 530004, China
[2]Key Laboratory of Disaster Prevention and Structural Safety of
Ministry of Education, Guangxi University, Nanning 530004, China
[3]Guangxi Key Laboratory of Disaster Prevention and Engineering Safety,
Guangxi University, Nanning 530004, China
[4]Faculty of Land Resource Engineering, Kunming University of Science
and Technology, Kunming 650500, China
[5]Institute of Mountain Hazards and Environment, Chinese Academy of Sciences
and Ministry of Water Conservancy, Chengdu 610041, China
[6]Department of Civil and Resource Engineering, Dalhousie University, Halifax, NS, B3H4K5, Canada

Correspondence: Yingyan Zhu (zh_y_y_imde@163.com)

Abstract. Gravel soils generated by the Wenchuan earthquake have undergone natural consolidation for the past decade. However, geological hazards, such as slope failures with ensuing landslides, have continued to pose great threats to the region. In this paper, artificial model tests were used to observe the changes of soil moisture content and pore water pressure, as well as macroscopic and microscopic phenomena of gravel soil. In addition, a mathematical formula of the critical state was derived from the triaxial test data. Finally, the mechanical states of gravel soil were determined. The results had five aspects. (1) The time and mode of the occurrence of landslides were closely related to the initial dry density. The process of initiation was accompanied by changes in density and void ratio. (2) The migration of fine particles and the rearrangement of coarse–fine particles contributed to the reorganization of the microscopic structure, which might be the main reason for the variation of dry density and void ratio. (3) If the confining pressure were the same, the void ratios of soils with constant particle composition would approach approximately critical values. (4) Mechanical state of gravel soil can be determined by the relative position between state parameter (e, p') and e_c–p' planar critical state line, where e is the void ratio, e_c is the critical void ratio and p' is the mean effective stress. (5) In the process of landslide initiation, dilatation and contraction were two types of gravel soil state, but dilatation was dominant. This paper provided insight into interpreting landslide initiation from the perspective of critical state soil mechanics.

1 Introduction

In 2008, the gravel soil generated by the Wenchuan earthquake produced a large amount of loose deposits (Tang and Liang, 2008; Xie et al., 2009). These deposits had features such as wide grading, weak consolidation and low density. They were located on both sides of roads and gullies and led to the formation of soil slopes (Cui et al., 2010; Qu et al., 2012; Zhu et al., 2011). Although gravel soils have been subjected to natural consolidation processes for nearly a decade, geological hazards, such as slope failures with ensuing landslides, are readily motivated in rainy season. At present, geohazards still pose the great threats to the region (Chen et al., 2017; Cui et al., 2013; Yin et al., 2016).

The variation of mechanical state, such as the transformation from a relatively stable state to a critical state, has been commonly used to analyze the initiation of landslides (Iverson et al., 2010, 2000; Liang et al., 2017; Sassa, 1984; Schulz et al., 2009). Therefore, a deep understanding of the soil state is the scientific basis for the study of landslide occurrence (Chen et al., 2017). Generally, the critical void ratio is an important parameter to determine the state of soil quantitatively (Been and Jefferies, 1985; Schofield and Wroth, 1968). The theoretical research has its origins in Reynold's work in 1885. He defined the characteristic of the volumetric deformation of granular materials due to shear strain as dilatation (Reynolds, 1885). Casagrande (1936) pointed out that loose soil contracted, and dense soil dilated to the same critical void ratio in the drained shearing test. He drew the F line to distinguish the dilative zone and the contractive zone. The F line's horizontal and vertical coordinate is effective normal stress and void ratio. Since the 1980s, critical state soil mechanics has received extensive attention (Fleming et al., 1989; Gabet and Mudd, 2006; Iverson et al., 2000). Some of the observed landslides, such as the Salmon Creek landslide in Marin County (Fleming et al., 1989), Slumgullion landslide in Colorado (Schulz et al., 2009), and Guangming New Distinct landslide in Shenzhen (Liang et al., 2017), could be approximately explained by this theory. Based on the F line drawn by Casagrande (1936), Fleming et al. (1989) found that the increase in pore water pressure contributed to the dilation and caused the debris flow characterized by the intermittent movement. Iverson et al. (1997, 2000) pointed out that porosity played an important role in the occurrence of landslides; in the soil shearing process, the density of loose sand increased, and the density of dense sand decreased to the same critical density. The formula of the void ratio was derived, which was the function of the mean effective stress (Gabet and Mudd 2006). Schulz et al. (2009) found out that the dilative strengthening might control the velocity of a moving landslide through the hourly continuous measurement of displacement of landslides. Liang et al. (2017) found that the initial solid volume fraction affected the soil state of the granular–fluid mixture. Other scholars also found that, in the shearing process, dilation or contraction was existing in

residual soil, loess and coarse-grained soil (Dai et al., 2000, 1999a, b; Liu et al., 2012; Zhang et al., 2010).

The above researchers provided the meaningful insights in order to explain the occurrence of landslides and drew the instructive conclusion that the initial density or porosity can affect the mechanical state of soil (Iverson et al., 2000) and the formation of landslides (McKenna et al., 2011). However, most of them focused on qualitative results and lacked mutual verification between indoor tests and model tests. In addition, for the gravel soil generated by seismic activity, the study on its mechanical state is lacking. Some scientific issues need to be resolved. For example, what are the differences and similarities of landslide occurrence? Why does the void ratio or the density change? Is the mechanical state a contraction or dilation? The purpose of this paper is to resolve the above issues through artificial flume model tests and triaxial tests. Firstly, the macroscopic phenomena were observed and summarized. Secondly, the variations of soil moisture content and pore water pressure were analyzed. Thirdly, the microscopic property of soil was obtained. Fourthly, the mathematical expression of critical state of soil was proposed. Finally, the mechanical state of gravel soil was determined by the relative position between state parameter (e, p') and $e_c - p'$ planar critical state line.

2 Field site and method

2.1 Field site

Niujuan valley is located in Yingxiu, Wenchuan County, Sichuan province, which was the epicenter of the 12 May 2008 Wenchuan earthquake in China (Fig. 1). The main valley of the basin has an area of $10.46\,\mathrm{km^2}$ and a length of 5.8 km. The highest elevation is 2693 m, and the largest relative elevation is 1833 m. The gradient ratio of the valley bed is 32.7 %–52.5 % (Tang and Liang, 2008; Xie et al., 2009). Six small ditches are distributed in the basin. Most of the valley is covered with the abundant gravel soil. Extremely complex terrain and adequate rainfall triggers the frequent landslides and the large-scale debris flows. Thus, this valley is the most typical basin in the seismic area. Its excellent landslide-formative environment can provide comprehensive reference models and abundant soil samples for artificial flume model tests.

2.2 Soil tests and quantitative analysis

2.2.1 Artificial flume model test

Based on the field surveys along Duwen highway, Niujuan valley and the literature review (Chen et al., 2010; Fang et al., 2012; Tang et al., 2011; Yu et al., 2010), most of the rainfall-induced landslides are shallow. The range of the slope angle is 25–40°, and its average value is 27°. The rainfall intensity triggering the landslide is $10-70\,\mathrm{mm\,h^{-1}}$. As shown in

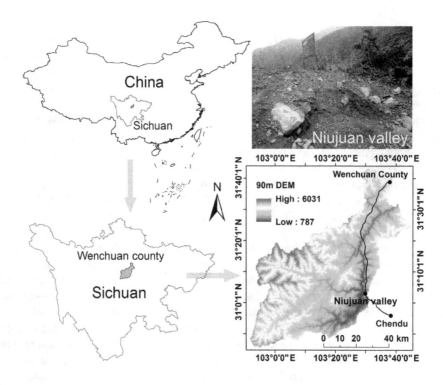

Figure 1. Study area.

Fig. 2a, the length, width and height of the flume model are 300, 100 and 100 cm.

The gravel soil samples are from Niujuan valley. The specific gravity is 2.69. The range of dry density is 1.48–2.36 g cm^{-3}; in addition, the minimum and the maximum void ratios are 0.14 and 0.82. Figure 2b shows that the cumulative content of gravel (diameter < 2 mm), and silt and clay (diameter < 0.075 mm) are 30.74 % and 2.78 %. The content of silt and clay plays an important role in the mobilization of landslides and debris flows (Chen et al., 2010). Four initial dry densities are designed as 1.50, 1.60, 1.70 and 1.80 g cm^{-3}. According to previous investigations, the water content mainly changes within a depth of less than 50 cm, and its average value varies from 6 % to 8 %, while water content below 50 cm basically keeps stable. Therefore, the total thickness of the soil model is 60 cm. In order to achieve a predetermined initial dry density, the soils of the models are divided into four layers, and each layer is compacted. The thickness of each layer is 20, 15, 15 and 10 cm (Fig. 2a). Due to the experimental error, the actual initial dry density (IDD) is 1.54, 1.63, 1.72 and 1.81 g cm^{-3} (Table 1).

Artificial rainfall system, designed by the Institute of Soil and Water Conservation, Chinese Academy of Sciences, is comprised of two spray nozzles, a submersible pump, water box and a bracket. The range of nozzle sizes is 5–12 mm; thus, the different rainfall intensity can be simulated. The rain intensity triggering the large-scale debris flow on 21 August 2011 was 56.5 mm h^{-1}, which is the designed rainfall for the test. The real rainfall intensity was 47–50.2 mm h^{-1}

because the model test was disturbed by the direction of wind. Three groups of sensors, including the micro-pore pressure sensors (the model is TS-HM91) and moisture sensors (the model is SM300), are placed between two layers of the soil to measure the volume water content and the pore water pressure (Fig. 2a). A data-acquisition system (the model is DL2e) is used to collect the data; it can scan 30 channels within the same second. A camera is used to record the macroscopic process of the entire experiment.

2.2.2 Triaxial test

Tests are performed by using a dynamic apparatus at the Institute of Mountain Hazards and Environment, Chinese Academy of Sciences. The diameter and the height of sample are 15 and 30 cm (Fig. 3). The test is the saturated and consolidated drainage shear test at a shear rate of 0.8 mm min^{-1}, which is comprised of two sets: the initial dry density of 1.94 and 2.00 g cm^{-3}. The confining pressure σ_3 is 50, 100 and 150 Kpa.

2.2.3 Quantitative analysis method

Quantitative analysis is mainly based on artificial flume model test and triaxial test. Firstly, the state parameters of soil are represented by the void ratio e and the mean effective stress p', which are from the model test. In the model test, at least three soil samples are collected by a soil sampler at the same depth of lines 1, 2 and 3 and are used to calculate their natural density ρ, mass moisture content ω and dry

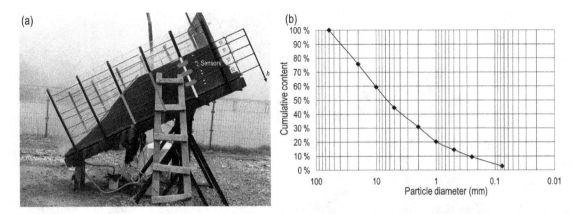

Figure 2. Test model and grain composition of gravel soil. **(a)** Artificial flume model (the position of sampling: red line – 1, pink line – 2, white line – 3). **(b)** Grain composition of gravel soil.

Table 1. Sets of artificial flume model test.

| | Factor | | | |
Number	Initial volume moisture content (%)	Slope angle (°)	Rainfall intensity) (mm h^{-1})	Initial dry density (g cm^{-3})
1				1.54
2				1.63
3	6–8	27	47–50.2	1.72
4				1.81

Figure 3. Dynamic triaxial apparatus.

density ρ_d. Later, e can be calculated by the following formula: $e = G_s/\rho_d - 1$ (where G_s is the specific gravity). The cumulative content of coarse P_5 (particle diameter > 5 mm), gravel (particle diameter < 2 mm) P_2, and silt and clay (particle diameter < 0.075 mm) $P_{0.075}$ is obtained from the particle grading tests. p' can be calculated by the following formula: $p' = (\sigma_x + \sigma_y + \sigma_z)/3$, where $\sigma_z = \gamma h$ and $\sigma_x = \sigma_y = K_a\gamma h$.

h is the vertical distance between a certain point inside the slope and the surface of the slope; β is the slope angle. γ is the unit weight of soil. K_a is the lateral pressure coefficient, which can be calculated by formula (1) (Chen et al., 2012). ϕ is the internal friction angle of soil. In this paper, $\beta = 27°$, $\phi = 33°$.

$$K_a = \cos\beta \frac{\cos\beta - \sqrt{\cos^2\beta - \cos^2\varphi}}{\cos\beta + \sqrt{\cos^2\beta - \cos^2\varphi}} \qquad (1)$$

Secondly, the critical state line (CSL) is derived from the triaxial test. Finally, based on the critical state soil mechanics, according to the relative position of the state parameter (e, p') at the CSL, the mechanical state of the soil can be estimated. When the soil state (e, p') is located at the upper right of the CSL, the soil is contracted. When the soil state (e, p') is located at the lower left of the CSL, the soil is dilated (Casagrande, 1936; Schofield and Wroth, 1968).

3 Results

3.1 Macroscopic phenomena of experiment

According to the camera recording, when IDD is 1.54–1.72 g cm^{-3} except 1.81 g cm^{-3}, the landslide can be triggered by rainfall. The processes of the occurrences of landslides have their similarities and differences. The similarity

Figure 4. Similarity of process of landslide initiation. (**a**) Shallow soil is compacted. (**b**) Muddy water is generated.

Figure 5. Process of landslide initiation (IDD of 1.54–1.63 g cm^{-3}). (**a**) Soil of the superficial layer slowly slides. (**b**) A small-scale slip occurs. (**c**) The overall slide is motivated.

is that at the beginning of rainfall, the shallow soil is compacted by seepage force and soil weight (Fig. 4a). In addition, during the rainfall duration, surface runoff cannot be observed, whereas muddy water appears and overflows the slope foot (Fig. 4b). This phenomenon indicates that the entire rainfall can seep into the internal soil, followed by the formation of subsurface flow. At this moment, the fine particles along the percolation paths begin to move in translation and rotation under the action of gravity (Gao et al., 2011; Igwe, 2015) and cause a re-distribution of the microstructure of soil (Chen et al., 2004; Zhuang et al., 2015). These moving fine particles will fill the interval space of porosity, even block the downstream channels of the seepage path (Fang et al., 2012; McKenna et al., 2011), which can lead to a decrease in void ratio and an increase in the pore water pressure (Gao et al., 2011).

The difference in experiments is time and mode of the occurrence of landslides. When IDD is 1.54–1.63 g cm^{-3}, the total time of landslide occurrence is 30–40 min, including the time of partial sliding and overall sliding. The processes of landslide occurrence involve three steps. Firstly, the partial soil of the superficial layer slowly slides in the shape of mudflow when rainfall duration is about 8 min (Fig. 5a). Secondly, small-scale slips occur in a layered manner (Fig. 5b). Thirdly, the overall sliding is motivated when the rainfall du-

ration is about 33 min (Fig. 5c). The above processes represent the mode of landslide is the progressive failure. This mode reflects four mechanisms. Firstly, in the early stage of rainfall, the shearing strength of shallow soil decreases and partial sliding occurs due to the rapid infiltration of rainfall. Secondly, partial sliding takes away the saturated soil, which causes the internal soil exposed on the surface. Thirdly, the exposed soil slides again, which can change the geometrical shape of the slope and prompt the shearing force increase. Fourthly, when the increase in the shearing force destroys the balance of the slope, overall sliding will occur.

When IDD is 1.72 g cm^{-3}, the total time of landslide occurrence is 18 min. Landslide formation process is divided into three steps. Firstly, the shear opening gradually occurs accompanied by the visible cracks developing in the slope foot (Fig. 6a). Secondly, surface cracks begin to develop on the slope top (Fig. 6b). Finally, the landslide is initiated and is accompanied by the instantaneous propagation of cracks (Fig. 6c–d), which takes 5 s. The above steps imply the mode of landslide is the tractive failure. The mechanism includes three aspects. Firstly, an increase in soil weight causes an increase in shearing force, which breaks the equilibrium state of slope, so cracks can develop in the slope foot and cause the shear opening. Secondly, the instability of the slope continues to deteriorate, which leads to new cracks located at the

Figure 6. Process of landslide initiation (IDD of 1.72 g cm^{-3}). (**a**) Shearing opening appears in slope foot. (**b**) Cracks develop on the slope top. (**c**) Crack propagation. (**d**) Landslide is triggered.

Figure 7. Process of experiment (IDD of 1.81 g cm^{-3}). (**a**) Shearing opening appears at the slope foot. (**b**) Muddy water flows. (**c**) Fine particles disappear and coarse particles are exposed.

top of the slope. Thirdly, the overall sliding is triggered by crack extension.

When IDD is 1.81 g cm^{-3}, the shearing opening appears at the slope foot (Fig. 7a). Next, the muddy water can flow from the slope foot (Fig. 7b). Even though on the slope surface fine particles disappear and coarse particles are exposed, rainfall could not trigger a landslide (Fig. 7c). One reason is that the fine particles within the surface soil move with the water seepage. After the fine particles of the shallow soil are all migrated, the soil skeleton begins to consist of coarse particles. This skeleton can provide some smooth paths for the subsurface runoff. The other reason is that when the soil is in a dense state, the change of volume moisture content is limited due to the low permeability. Even if the soil shows a small shearing strain, the loss of pore water pressure is diffi-

cult to recover in time due to the lack of rainfall infiltration. Therefore, the shearing strength can remain unchanged.

Macroscopic phenomena of experiments imply that the initial dry density can influence the time and mode of landslide occurrence. It coincides with the existing research (Iverson et al., 2000). As the IDD increases from 1.54 to 1.72 g cm^{-3}, the failure mode of soil changes from progressive sliding to traction sliding. When IDD is less than 1.63 g cm^{-3}, partial sliding is a dominant phenomenon that affects the entire deformation failure. When IDD is 1.72 g cm^{-3}, shear opening and cracks are responsible for deformation failure. Although the total time of overall sliding of loose soil is longer than that of relatively dense soil, the time of partial sliding is shorter. This difference may be associated with failure modes, relative timescales of shearing strength loss and changes of pore water pressure.

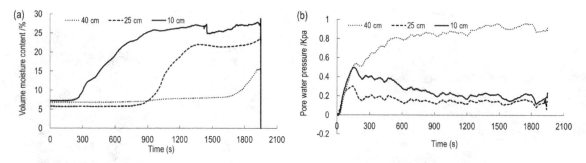

Figure 8. Volume moisture content and pore water pressure when IDD is $1.54\,\mathrm{g\,cm^{-3}}$.

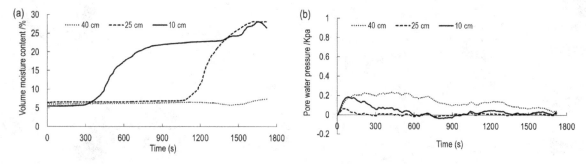

Figure 9. Volume moisture content and pore water pressure when IDD is $1.63\,\mathrm{g\,cm^{-3}}$.

3.2 Volume moisture content (VMC) and pore water pressure (PWP)

The maximum x label in Figs. 8–10 represents the total time for the occurrence of the landslide. This value is also the rainfall duration. In order to compare with Figs. 8–10, the maximum x label in Fig. 11 is 1800 s. As shown in Figs. 8 to 11, the first change is VMC of the depth of 10 cm, followed by VMC of the depth of 25 and 40 cm. This change order of VMC is related to the process of rainfall penetration. Specifically, rainfall penetration is from shallow soil to deep soil. Therefore, the VMC of 10 cm can increase first. The variation of VMC at the depth of 10–25 cm exhibits a similar tendency. The tendency consists of three phases. Since the beginning of rainfall, VMC has been in a constant state. When the rainfall seeps into soil, VMC increases rapidly and eventually grows steadily. The time when VMC of the depth of 10 cm begins to increase is 203, 292, 313 s for 1.54–$1.72\,\mathrm{g\,cm^{-3}}$. This result indicates these three densities have different permeability, the higher density, the lower hydraulic conductivity and the longer time of penetration. The time when VMC of the depth of 25 cm begins to increase is about 900 s for 1.54–$1.72\,\mathrm{g\,cm^{-3}}$.

When IDD is 1.54 and $1.63\,\mathrm{g\,cm^{-3}}$, VMC at a depth of 40 cm initially remains stable and eventually shows an increasing trend. Change trend of $1.54\,\mathrm{g\,cm^{-3}}$ is more obvious than that of $1.63\,\mathrm{g\,cm^{-3}}$. When IDD is $1.72\,\mathrm{g\,cm^{-3}}$, VMC at a depth of 40 cm is almost constant. The reason is that when a landslide occurs, rain stops; at this time, no abundant wa-

ter can penetrate to this depth. When IDD is $1.81\,\mathrm{g\,cm^{-3}}$, if the rainfall duration is less than 1300 s, VMC of 40 cm remains stable. When the duration is about 1300 s, compared to Figs. 8–10, VMC of 40 cm starts to increase. This difference between Fig. 11 and other three figures may be attributed to the following aspect. As mentioned in Sect. 3.1, the landslide cannot be triggered by rainfall. Therefore, there is sufficient time for rainfall to penetrate to a depth of 40 cm, although the hydraulic conductivity is low. However, when the rainfall time is greater than 1800 s, VMC of 10–40 cm keeps constant. This means, due to the accumulation of fine particles, there may be an impermeable layer in the depth of 0–10 cm. This layer can prevent the rain from penetrating to depths deeper than 10 cm. When rainfall continues, rainfall can be converted into the subsurface runoff, flowing out of the soil skeleton that consists of coarse particles.

As shown in Figs. 8 to 11, PWP at a depth of 10–25 cm has a similar tendency. This tendency consists of a sharp increase at first, a rapid decrease and a continuous dynamic fluctuation. However, the variation of PWP is inconsistent with the variation of VMC. Before VMC increases, PWP with the depth of 10–25 cm has experienced the sharp increase and decrease. Soil inhomogeneity may contribute to this inconsistency. As mentioned in Sect. 3.1, at the beginning of experiment, the surface layer less than 10 cm is compacted by seepage force and soil weight. The compaction and penetration process leads to the increase in the force acting on the subsoil, which causes the increase in PWP. During the saturation process of the surface layer, the fine particles

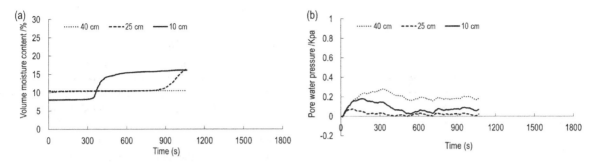

Figure 10. Volume moisture content and pore water pressure when IDD is $1.72 \, \mathrm{g \, cm^{-3}}$.

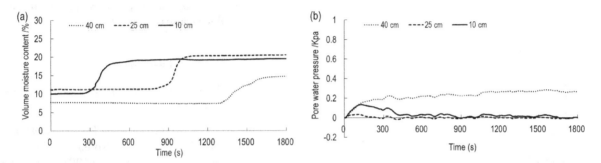

Figure 11. Volume moisture content and pore water pressure when IDD is $1.81 \, \mathrm{g \, cm^{-3}}$.

of this layer are taken away and fill the porosity of the subsoil, which prompt PWP to the peak value quickly. When the surface soil slowly moves or cracks begin to develop in the slope foot, the internal deformation due to dilation will occur, which causes PWP releases. When VMC increases, PWP has a dynamic fluctuation. This fluctuation may be attributed to the rearrangement of the soil skeleton.

The curve of PWP with a depth of 40 cm is drawn above that of 10–25 cm. The variation has no significant increase or decrease but exhibits a smooth fluctuation. During the whole rainfall duration, the corresponding VMC shows that the soil is not saturated. Therefore, the pore pressure of 40 cm is dominated by air pressure.

3.3   Microscopic property of gravel soil

As shown in Table 2, when IDD is from 1.54 to $1.72 \, \mathrm{g \, cm^{-3}}$, the natural density and the dry density with the depth of 5–20 cm are larger than those before the tests, and the void ratio is less than that before the tests. Of these three lines, the line 1 has the greatest change rate in density. When IDD is $1.63 \, \mathrm{cm^3}$, the density of 40 cm is less than the value before the test. When IDD is $1.81 \, \mathrm{g \, cm^{-3}}$, the densities with the depth of 5–10 cm are increased compared to those before the test.

As shown in Sect. 2.2.1, P_5 before the test is 55.32 %. Therefore, coarse particles and fine particles interact to form the soil skeleton, which affects changes in dry density (Guo, 1998) and landslide characteristics (Li et al., 2014). In this

paper, the particle content before and after the test is compared to understand the change in the void ratio. As shown in Table 3, when IDD is 1.54 and $1.63 \, \mathrm{g \, cm^{-3}}$, the loss of $P_{0.075}$ of the shallow soil of line 3 is the largest, followed by that of line 1. The result indicates that the fine particles of surface soil at the slope top begin to move along the direction of gravity firstly. When subsurface runoff occurs, these particles begin to move to the slope foot. This process causes two results. One is that the porosity of the position related to particle migration increases. The other is that the porosity filled by fine particles decreases (Fang et al., 2012; McKenna et al., 2011), which is the seepage-compacting effect (Jiang et al., 2013). As a result, the shallow soil of the slope top is looser than that of the slope foot. The loss of $P_{0.075}$ ($\Delta P_{0.075}$, which is negative) at the slope top decreases significantly with depth. Specifically, it is about -1.26 % at the depth of 40 cm. It implies that the depth of rainfall infiltration is about 40 cm. In the case of IDD of $1.72\text{–}1.81 \, \mathrm{g \, cm^{-3}}$, the variation of $P_{0.075}$ of the slope top changes from negative to positive accompanied by an increase in depth. This trend indicates that the fine particles may concentrate at the depth of 5–25 cm. The depth range of particle concentration is 10–25, 5–10 cm for 1.72 and $1.81 \, \mathrm{g \, cm^{-3}}$.

On the slope top, P_5 at a depth of 5 cm changes from positive to negative with an increase IDD, whose range is from -5.11 % to 10.25 %. The reason is that the loss of fine particles contributes to the relative increase in the content of coarse particles. Both $P_{0.075}$ on the slope top and $P_{0.075}$ on the slope foot decrease. The range of $\Delta P_{0.075}$ is from

Table 2. Density and void ratio of gravel soil with initial dry density 1.54–1.81 g cm^{-3}.

Number	Initial dry density (g cm^{-3})	Line number	Soil depth h (cm)	Natural density ρ (g cm^{-3})	Mass moisture content ω (%)	Dry density ρ_d (g cm^{-3})	Void ratio $e = G_s/\rho_d - 1$	$\sigma_z = \gamma h$ (Kpa)	$\sigma_x = \sigma_y = K_a\gamma h$ (Kpa)	$p' = (\sigma_x + \sigma_y + \sigma_z)/3$ (Kpa)
1	1.54	3	5	2.08±0.05	9.35±0.85	1.90±0.04	0.39±0.03	1.04	0.59	0.74
		3	28	1.93±0.03	8.61±1.16	1.77±0.02	0.49±0.02	5.39	3.07	3.84
		2	33	2.07±0.05	9.15±0.15	1.89±0.04	0.40±0.03	6.82	3.88	4.86
		1	21	2.10±0.05	9.63±1.01	1.91±0.05	0.39±0.04	4.40	2.51	3.14
2	1.63	3	5	2.19±0.01	13.36±0.09	1.98±0.01	0.34±0.01	0.44	0.25	0.31
		3	40	1.67±0.03	6.15±0.17	1.58±0.02	0.68±0.02	6.68	3.80	4.76
		2	20	2.09±0.04	10.18±0.21	1.90±0.04	0.39±0.03	4.19	2.38	2.99
		1	13	2.23±0.04	10.84±0.83	2.01±0.02	0.32±0.02	2.90	1.65	2.07
3	1.72	3	10	2.22±0.02	8.45±0.72	2.05±0.02	0.30±0.01	2.22	1.26	1.58
		3	25	2.34±0.04	8.59±0.261	2.16±0.05	0.23±0.03	5.86	3.33	4.17
		1	10	2.30±0.01	9.26±0.42	2.10±0.01	0.26±0.01	2.30	1.31	1.64
4	1.81	3	5	2.14±0.04	9.57±0.75	1.95±0.04	0.36±0.03	1.28	0.73	0.91
		3	10	2.26±0.01	8.16±0.39	2.09±0.02	0.27±0.01	2.26	1.28	1.61

$-25.83\,\%$ to $-76.24\,\%$ and from $-49.09\,\%$ to $-72.36\,\%$ accordingly. The relationship between $\Delta P_{0.075}$ and ρ_d is shown in Fig. 12. The regression equation is as follows: $\Delta P_{0.075} = 1.2632\rho_d - 2.6464$, $\Delta P_{0.075} = 1.709\rho_d - 3.4391$, and R^2 is 0.8827 and 0.8199, respectively. The result indicates that $\Delta P_{0.075}$ has a significant correlation with ρ_d. Specifically, the greater initial dry density causes the smaller loss of $P_{0.075}$. When IDD is 1.53 g cm^{-3}, P_2 decreases and its change value is $-0.16\,\%$. When IDD is 1.63–1.81 g cm^{-3}, P_2 increases, and the range of the change is 2.15 %–15.45 %. The reason for the loss of $P_{0.075}$ and P_2 is that the fine particles are taken away by subsurface runoff. The reason for the increase in P_2 may be that the particles larger than 2 mm roll downward, which causes a relative increase in P_2.

3.4 Critical state of gravel soil

3.4.1 Definition of critical state and calculation of critical void ratio

Under the action of continuous shear load, the state of soil is critical when principal stress q and volume strain ε_v tends to be stable (Casagrande, 1936; Liu et al., 2011; Roscoe et al., 1963; Schofield and Wroth, 1968). In the triaxial shear tests, when the axial strain reaches 16 %, the deviation stress is stable, and the absolute value of the ratio of ε_v to the present ε_v is less than 0.01; at this time, the soil enters the critical state (Liu et al., 2012). Formula (2) indicates that there is a certain relationship between the current void ratio e and ε_v, wherein e_0 is the initial void ratio (Xu et al., 2009). Thus, the critical void ratio e_c also can be calculated by formula (2).

$$e = (1 + e_0)\exp(-\varepsilon_v) - 1 \tag{2}$$

3.4.2 The critical state line in the e_c–p' plane

Table 4 shows e_c, q and p' for two initial dry densities: 1.94 and 2.00 g cm^{-3}. As shown in Table 4, when the confining pressure is same, two densities have approximately similar e_c value. This result has the consistent principle with existing research (Gabet and Mudd, 2006; Iverson et al., 2000). The principle is that the soil with the same granular composition can obtain the approximate critical void ratio under uniform stress conditions (Casagrande, 1936; Roscoe et al., 1963; Schofield and Wroth, 1968).

The fitting curve of e_c and $\ln p'$ is shown in Fig. 13a. The correlation coefficient is 0.8566, which indicates a statistically significant relationship between e_c and p'. According to the normalized residual probability, the P value of 0.964 is greater than the selected significance level, which indicates that the residuals follow a normal distribution. Therefore, the mathematical expression of e_c–$\ln p'$ of gravel soil is as follows:

$$e_c = 0.5241 - 0.04304\ln p'. \tag{3}$$

Table 3. Variation of coarse- and fine-particle contents.

Number	Initial dry density ($\mathrm{g\,cm^{-3}}$)	Line number	Soil depth h (cm)	P_5	ΔP_5	$P_{0.075}$	$\Delta P_{0.075}$	P_2	ΔP_2
1	1.54	3	5	61.00 %	10.25 %	0.66 %	−76.24 %	30.69 %	−0.16 %
		3	28	55.91 %	1.05 %	2.01 %	−27.90 %	34.36 %	11.76 %
		1	21	58.98 %	6.60 %	0.77 %	−72.36 %	31.07 %	1.05 %
2	1.63	3	5	58.69 %	6.09 %	0.91 %	−67.23 %	31.40 %	2.15 %
		3	40	57.98 %	4.80 %	2.75 %	−1.26 %	31.69 %	3.07 %
		1	13	67.66 %	22.30 %	1.26 %	−54.81 %	26.23 %	−14.68 %
3	1.72	3	5	55.98 %	1.18 %	1.03 %	−62.98 %	32.70 %	6.38 %
		3	10	54.01 %	2.37 %	1.78 %	−36.14 %	33.94 %	10.40 %
		3	25	55.32 %	0 %	3.17 %	13.85 %	34.05 %	10.75 %
		1	10	56.15 %	1.5 %	1.42 %	−49.09 %	33.67 %	9.53 %
4	1.81	3	5	52.50 %	−5.11 %	2.06 %	−25.83 %	35.49 %	15.45 %
		3	10	52.55 %	−5.01 %	2.86 %	2.68 %	33.91 %	10.30 %

Note: the positive value of the change represents an increase while the negative value represents a decrease.

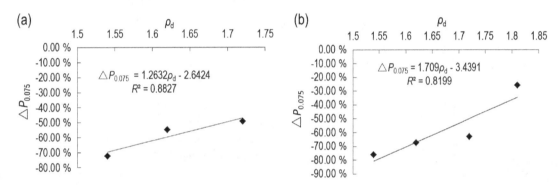

Figure 12. Relationship between $\Delta P_{0.075}$ and ρ_d. (**a**) Slope foot. (**b**) Slope top.

Table 4. Critical void ratio e_c of gravel soil.

Confining pressure σ_3 (Kpa)	Initial dry density ($\mathrm{g\,cm^{-3}}$)	e_c	q (Kpa)	p' (Kpa)
50	1.94	0.32	93.41	95.98
	2.00	0.34	69.50	84.65
100	1.94	0.30	227.43	213.80
	2.00	0.30	159.14	178.13
150	1.94	0.27	324.79	312.39
	2.00	0.29	181.12	239.86

The fitting cure of e_c and $\ln p'$ represents the critical state of soil. It can divide the graphical space into two states. The space above this curve is the contractive zone, and the space below this curve is the dilative zone. If the state parameter (e, p') is determined, the soil state can be judged by this line (Gabet and Mudd, 2006; Iverson et al., 2000).

3.4.3　The critical state line in the q–p' plane

The fitting curve of q and the p' is shown in Fig. 13b. The correlation coefficient is 0.9465, which indicates a statistically significant relationship between q and p'. The mathematical expression of q–p' is as follows:

$$q = 0.6641(p')^{1.063}. \tag{4}$$

4　Discussion

The relative position of the state parameter (e, p') at the critical state line is shown in Fig. 14. The critical states are from Table 4 and represented by filled circles, and the state parameters of four densities are from Table 2 and represented by the hollow rectangle, cross-shape, hollow triangle and circle. Figure 14 shows that when IDD is 1.54 and 1.63 $\mathrm{g\,cm^{-3}}$, contraction occurs at 28 and 40 cm of line 3. In addition, dilation appears in the remaining positions. These positions include the surface layer of line 3 with the depth of 5–10 cm,

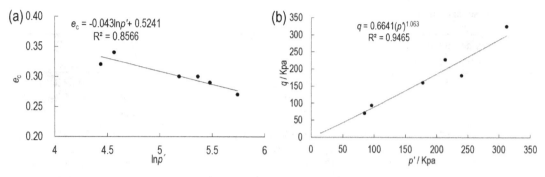

Figure 13. Critical state line of gravel soil. **(a)** e_{c}–$\ln p'$. **(b)** q–p'.

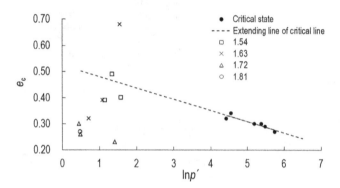

Figure 14. The states of gravel soil.

the depth of 20–33 cm of line 2 and the depth of 10–21 cm of line 1. The results show that dilation and contraction are two types of the mechanical state of gravel soil when the landslide initiates. Dilation is the primary type.

In this research, at the beginning of rainfall, the shallow soil is compacted by seepage force and soil weight. The consequent contraction causes the increase in pore water pressure. However, the process of the rapid rise of PWP is short. After PWP reaches the peak, PWP begins to release. The reason is that the surface soil slowly moves or cracks begin to develop in the slope foot, which causes the sliding force increase. Subsequently, the effective stress decreases and the shearing deformation occurs. At this moment, the loss of shearing strength because of strain softening can be restored due to the decrease in PWP. Soil deformation will stop eventually. If there is the sufficient water penetration, pore water pressure can recover, and the soil deformation can continue. It can be seen that the loss and recovery of PWP are the reasons for the dynamic fluctuations of PWP. When soil is dense (relative density $D_{\mathrm{r}} > 2/3$) and the infiltration rate is less than the rainfall intensity, the soil will not reach the critical state. At this point, the slope can remain stable. The macroscopic phenomenon of soil deformation is mainly local deformation, such as circumferential cracks and partial collapse. If the infiltration rate is greater than the rainfall intensity, the abundant rainfall can break the mechanical balance of slope. However, its process still takes a long time. Therefore, the macroscopic deformation is progressive, such as frequent sliding. When the soil is in a medium dense state ($1/3 < D_{\mathrm{r}} \leq 2/3$), the loss of the pore water pressure due to dilation will be recovered, and the shearing deformation will continue. At this moment, the macroscopic deformation will be a sudden failure (Dai et al., 2000, 1999b).

5 Conclusions

The initial dry density can influence the time and mode of landslide occurrence. When IDD is 1.54–1.72 g cm^{-3}, the failure mode of soil changes from progressive sliding to traction sliding. When IDD is less than 1.63 g cm^{-3}, partial sliding is a dominant phenomenon that affects the entire deformation failure. When IDD is 1.72 g cm^{-3}, shear opening and cracks are responsible for deformation failure. Although the total time of overall sliding of loose soil is longer than that of relatively dense soil, the time of partial sliding is shorter.

During the experiments, the first change is VMC of the depth of 10 cm, followed by VMC of the depth of 25 and 40 cm. The variation of PWP is inconsistent with the variation of VMC.

The occurrence of landslides is accompanied by change in density and void ratio. The slope foot has the greatest change rate in density. The migration of fine particle and the rearrangement of coarse–fine particle contributed to the reorganization of the microscopic structure, which might be the main reason for the variation of density and void ratio.

The mathematical expression of the critical state line of gravel soil is $e_{\mathrm{c}} = 0.5241 - 0.04304\ln p'$. Mechanical state of gravel soil can be determined by the relative position between the state parameter (e, p') and the critical state line. Dilation and contraction are two types of soil state when the landslide initiates. Dilation is the primary type.

Author contributions. LL carried out the artificial model tests and the triaxial tests, analyzed the experimental data and wrote the manuscript. YY participated in the implementation of the triaxial tests. ZY and JH participated in the implementation of the tests and analyzed part of data. YZ provided the guidance on the tests as well as the content and the structure of the manuscript. DHSZ proposed the suggestions on the tests.

Competing interests. The authors declare that they have no conflict of interest.

Acknowledgements. This study was funded by the National Natural Science Foundation of China (nos. 41071058, 41402272, 51609041), Disaster Prevention and Mitigation and Engineering Safety Key Laboratory Project of Guangxi Province (no. 2016ZDX09) and the Natural Scientific Project of Guangxi Zhuang Autonomous Region (no. 2018GXNSFAA138187).

References

Been, K. and Jefferies, M. G.: A state parameter for sands, Geotechnique, 35, 99–112, 1985.

Casagrande, A: Characteristics of cohesionless soils affecting the stability of slopes and earth fills, Journal of the Boston Society of Civil Engineers, 23, 13–32, 1936.

Chen, N. S., Cui, P., Wang, X. Y., and Di, B. F.: Testing study on strength reduction of gravelly soil in triggering area of debris flow under earthquake, Chinese Journal of Rock Mechanics and Engineering, 23, 2743–2747, 2004 (in Chinese).

Chen, N. S., Zhou, W., Yang, C. L., Hu, G. S., Gao, Y. C., and Han, D.: The processes and mechanism of failure and debris flow initiation for gravel soil with different clay content, Geomorphology, 121, 222–230, https://doi.org/10.1016/j.geomorph.2010.04.017, 2010.

Chen, N. S., Zhu, Y. H., Huang, Q., Lqbal, J., Deng, M. F., and He, N.: Mechanisms involved in triggering debris flows within a cohesive gravel soil mass on a slope: a case in SW China, J. Mt. Sci., 14, 611–620, https://doi.org/10.1007/s11629-016-3882-x, 2017.

Chen, Z. Y., Zhou, J. X., and Wang, H. J.: Soil Mechanics, 19th Edn., Tsinghua University Press, Beijing, 2012.

Cui, P., Zhuang, J. Q., Chen, X. C., Zhang, J. Q., and Zhou, X. J.: Characteristics and countermeasures of debris flow in Wenchuan area after the earthquake, Journal of Sichuan University (Engineering Science Edition), 42, 10–19, 2010 (in Chinese).

Cui, P., Xiang, L. Z., and Zou, Q.: Risk assessment of highways affected by debris flows in Wenchuan earthquake area, J. Mt. Sci., 10, 173–189, https://doi.org/10.1007/s11629-013-2575-y, 2013.

Dai, F. C., Lee, C. F., and Wang, S. J.: Analysis of rainstorm-induced slide-debris flows on natural terrain of Lantau Island, Hong Kong, Eng. Geol., 51, 279–290, https://doi.org/10.1016/s0013-7952(98)00047-7, 1999a.

Dai, F. C., Lee, C. F., Wang, S. J., and Feng, Y. Y.: Stress-strain behaviour of a loosely compacted volcanic-derived soil and its significance to rainfall-induced fill slope failures, Eng. Geol., 53, 359–370, https://doi.org/10.1016/s0013-7952(99)00016-2, 1999b.

Dai, F. C., Chen, S. Y., and Li, Z. F.: Analysis of landslide initiative mechanism based on stress-strain behavior of soil, Chinese Journal of Geotechnical Engineering, 22, 127–130, 2000 (in Chinese).

Fang, H., Cui, P., Pei, L. Z., and Zhou, X. J.: Model testing on rainfall-induced landslide of loose soil in Wenchuan earthquake region, Nat. Hazards Earth Syst. Sci., 12, 527–533, https://doi.org/10.5194/nhess-12-527-2012, 2012.

Fleming, R. W., Ellen, S. D., and Algus, M. A.: Transformation of dilative and contractive landslide debris into debris flows-An example from marin County, California, Eng. Geol., 27, 201–223, 1989.

Gabet, E. J. and Mudd, S. M.: The mobilization of debris flows from shallow landslides, Geomorphology, 74, 207–218, https://doi.org/10.1016/j.geomorph.2005.08.013, 2006.

Gao, B., Zhou, J., and Zhang, J.: Macro-meso analysis of water-soil interaction mechanism of debris flow starting process, Chinese Journal of Rock Mechanics and Engineering, 30, 2567–2573, 2011 (in Chinese).

Guo, Q. Q.: Engineering features and utilization of coarse-grained soil, 1st Edn., The Yellow River Water Conservancy Press, Zhengzhou, 1998 (in Chinese).

Igwe, O.: The compressibility and shear characteristics of soils associated with landslides in geologically different localities – case examples from Nigeria, Arab. J. Geosci., 8, 6075–6084, https://doi.org/10.1007/s12517-014-1616-3, 2015.

Iverson, N. R., Mann, J. E., and Iverson, R. M.: Effects of soil aggregates on debris-flow mobilization: Results from ring-shear experiments, Eng. Geol., 114, 84–92, https://doi.org/10.1016/j.enggeo.2010.04.006, 2010.

Iverson, R. M., Reid, M. E., and Lahusen, R. G.: Debris-flow mobilization from landslides, Annu. Rev. Earth Pl. Sc., 25, 85–138, 1997.

Iverson, R. M., Reid, M. E., Iverson, N. R., LaHusen, R. G., and Logan, M.: Acute sensitivity of landslide rates to initial soil porosity, Science, 290, 513–516, https://doi.org/10.1126/science.290.5491.513, 2000.

Jiang, Z. M., Wang, W., Feng, S. R., and Zhong, H. Y.: Experimental of study on the relevance between stress state and seepage failure of sandy-gravel soil, Shuili Xuebao, 44, 1498–1505, 2013 (in Chinese).

Li, Y., Liu, J. J., Su, F. H., Xie, J., and Wang, B. L.: Relationship between grain composition and debris flow characteristics: a case study of the Jiangjia Gully in China, Landslides, 12, 19–28, https://doi.org/10.1007/s10346-014-0475-z, 2014.

Liang, H., He, S. M., Lei, X. Q., Bi, Y. Z., Liu, W., and Ouyang, C. J.: Dynamic process simulation of construction solid waste (CSW) landfill landslide based on SPH considering dilatancy effects, B. Eng. Geol. Environ., 2, 1–15, https://doi.org/10.1007/s10064-017-1129-x, 2017.

Liu, E. L., Chen, S. S., Li, G. Y., and Zhong, Q. M.: Critical state of rockfill materials and a constitutive model considering grain crushing, Rock and Soil Mechanics, 32, 148–154, 2011 (in Chinese).

Liu, E. L., Qin, Y. L., Chen, S. S., and Li, G. Y.: Investigation on critical state of rockfill materials, Shuili Xuebao, 43, 505–511, 2012 (in Chinese).

McKenna, J. P., Santi, P. M., Amblard, X., and Negri, J.: Effects of soil-engineering properties on the failure mode of shallow landslides, Landslides, 9, 215–228, https://doi.org/10.1007/s10346-011-0295-3, 2011.

Qu, Y. P., Tang, C., Wang, J. L., Tang, H. X., Liu, Y., Chen, H. L., and Huang, W.: Debris flow initiation mechanisms in strong earthquake area, Mountain Research, 30, 336–341, 2012 (in Chinese).

Reynolds, O.: On the dilatancy of media composed of rigid particles in contact, with experimental illustrations, Philos. Mag., 20, 469–481, 1885.

Roscoe, K. H., Schofield, A. N., and Thuraijajah, A.: Yielding of clays in states wetter than critical, Geotechnique, 13, 211–240, 1963.

Sassa, K.: The mechanism to initiate debris flows as undrained shear of loose sediments, Internationales Symposium Interpraevent, 73–87, 1984.

Schofield, A. N. and Wroth, C. P.: Critical state soil mechanics, University of Cambridge, 1968.

Schulz, W. H., McKenna, J. P., Kibler, J. D., and Biavati, G.: Relations between hydrology and velocity of a continuously moving landslide – evidence of pore-pressure feedback regulating landslide motion?, Landslides, 6, 181–190, https://doi.org/10.1007/s10346-009-0157-4, 2009.

Tang, C. and Liang, J. T.: Characteristics of debris flows in Beichuan epicenter of the Wenchuan earthquake triggered by rainstorm on september 24, 2008, Journal of Engineering Geology, 16, 751–758, 2008 (in Chinese).

Tang, C., Li, W. L., Ding, J., and Huang, X. C.: Field Investigation and Research on Giant Debris Flow on August 14, 2010 in Yingxiu Town, Epicenter of Wenchuan Earthquake, Earthscience-Journal of China University of Geosciences, 36, 172–180, 2011 (in Chinese).

Xie, H., Zhong, D. L., Jiao, Z., and Zhang, J. S.: Debris flow in Wenchuan quake-hit area in 2008, Mountain Research, 27, 501–509, 2009 (in Chinese).

Xu, S. H., Zheng, G., and Xu, G. L.: Critical state constitutive model of sand with shear hardening, Chinese Journal of Geotechnical Engineering, 31, 953–958, 2009 (in Chinese).

Yin, Y. P., Cheng, Y. L., Liang, J. T., and Wang, W. P.: Heavy-rainfall-induced catastrophic rockslide-debris flow at Sanxicun, Dujiangyan, after the Wenchuan Ms 8.0 earthquake, Landslides, 13, 9–23, https://doi.org/10.1007/s10346-015-0554-9, 2016.

Yu, B., Ma, Y., and Wu, Y. F.: Investigation of serve debris flow hazards in Wenjia gully of Sichuan province after the Wenchuan earthquake, Journal of Engineering Geology, 18, 827–836, 2010 (in Chinese).

Zhang, M., Hu, R. L., and Yin, Y. P.: Study of transform mechanism of landslide-debris flow with ring shear test, Chinese Journal of Rock Mechanics and Engineering, 29, 822–832, 2010 (in Chinese).

Zhu, J., Ding, J., and Liang, J. T.: Influences of the Wenchuan Earthquake on sediment supply of debris flows, J. Mt. Sci., 8, 270–277, https://doi.org/10.1007/s11629-011-2114-7, 2011.

Zhuang, J. Q., Cui, P., Hu, K. H., and Chen, X. Q.: Fine particle size moving and its effective on debris flow initiation, Mountain Research, 33, 713–720, 2015 (in Chinese).

11

Optimising 4-D surface change detection: an approach for capturing rockfall magnitude–frequency

Jack G. Williams[1], **Nick J. Rosser**[1], **Richard J. Hardy**[1], **Matthew J. Brain**[1], and **Ashraf A. Afana**[2]

[1]Department of Geography, Durham University, Lower Mountjoy, South Road, Durham, UK
[2]National Trust, Kemble Drive, Swindon, UK

Correspondence: Jack G. Williams (j.g.williams@durham.ac.uk)

Abstract. We present a monitoring technique tailored to analysing change from near-continuously collected, high-resolution 3-D data. Our aim is to fully characterise geomorphological change typified by an event magnitude–frequency relationship that adheres to an inverse power law or similar. While recent advances in monitoring have enabled changes in volume across more than 7 orders of magnitude to be captured, event frequency is commonly assumed to be interchangeable with the time-averaged event numbers between successive surveys. Where events coincide, or coalesce, or where the mechanisms driving change are not spatially independent, apparent event frequency must be partially determined by survey interval.

The data reported have been obtained from a permanently installed terrestrial laser scanner, which permits an increased frequency of surveys. Surveying from a single position raises challenges, given the single viewpoint onto a complex surface and the need for computational efficiency associated with handling a large time series of 3-D data. A workflow is presented that optimises the detection of change by filtering and aligning scans to improve repeatability. An adaptation of the M3C2 algorithm is used to detect 3-D change to overcome data inconsistencies between scans. Individual rockfall geometries are then extracted and the associated volumetric errors modelled. The utility of this approach is demonstrated using a dataset of $\sim 9 \times 10^3$ surveys acquired at $\sim 1\,\mathrm{h}$ intervals over 10 months. The magnitude–frequency distribution of rockfall volumes generated is shown to be sensitive to monitoring frequency. Using a 1 h interval between surveys, rather than 30 days, the volume contribution from small ($< 0.1\,\mathrm{m}^3$) rockfalls increases from 67 to 98 % of the total, and the number of individual rockfalls observed increases by over 3 orders of magnitude. High-frequency monitoring therefore holds considerable implications for magnitude–frequency derivatives, such as hazard return intervals and erosion rates. As such, while high-frequency monitoring has potential to describe short-term controls on geomorphological change and more realistic magnitude–frequency relationships, the assessment of longer-term erosion rates may be more suited to less-frequent data collection with lower accumulative errors.

1 Introduction

1.1 Size distribution of geomorphic events

The processes that erode landscapes involve a broad range of event sizes, the distribution of which is commonly characterised using magnitude–frequency curves. Wolman and Miller (1960) proposed that the frequency of events that denude the Earth's surface is log-normally distributed, and that their geomorphic effectiveness (the product of magnitude and frequency) is greatest for the frequent, moderately sized events. This concept has been widely applied to study both the geomorphic efficacy of rivers (Wolman and Gerson, 1978; Hooke, 1980; Nash, 1994; Gintz et al., 1996) and the characteristics of landslides (Hovius et al., 1997, 2000; Dussauge-Peisser et al., 2002; Turcotte et al., 2002; Dussauge et al., 2003; Malamud et al., 2004; Guthrie and Evans, 2007; Li et al., 2016) using inverse power-law distributions or similar.

The exponent of the inverse power law describes the proportional contribution of increasingly small events, with larger exponents representing an increase in the proportion of small events in the inventory. However, many landslide volume distributions have been characterised by a decrease in the frequency density of the smallest events in log magnitude–log frequency space, known as a "rollover" (Malamud et al., 2004). At this point, the inverse power law breaks down, and so alternative distributions such as the double Pareto (Stark and Hovius, 2001; Guzzetti et al., 2002) or inverse gamma (Malamud et al., 2004; Guzzetti et al., 2005) have been drawn upon to model observations. Explanations for this rollover have been widely considered and include mechanical differences and physically based minimum possible event sizes (Pelletier et al., 1997; Guzzetti et al., 2002; Guthrie and Evans, 2004) or censoring of the smallest events by the resolution or frequency of monitoring (Lim et al., 2010). For rockfalls, Malamud et al. (2004) hypothesised that a rollover may not occur due to rock mass fragmentation.

1.2 Temporal resolution of monitoring

The duration of monitoring relative to the return period of all possible event sizes determines the likelihood of detecting changes that are representative of how a landform evolves over longer timescales. The creation of a representative inventory is also a function of the smallest event size that can be detected and the temporal frequency of monitoring compared to the rate at which such small events occur. Abellán et al. (2014) suggested that the spatial resolution of rockfall monitoring should be sufficient to discretise the smallest events in a magnitude–frequency distribution and that the recording frequency should fall below the timescale on which superimposition and coalescence may occur. In practice, defining this timescale a priori is challenging and requires the ability to monitor the rock face over a sustained period in (near) real time. For rockfalls, high-resolution monitoring also shows evolution of failures through time, with event sequences and patterns related to the incremental growth of scars (Rosser et al., 2007, 2013; Stock et al., 2012; Kromer et al., 2015a; Rohmer and Dewez, 2015; Royán et al., 2015). Barlow et al. (2012) showed that a monitoring interval of 19 months underestimated the frequency distribution of small rockfall events, which coalesced into or were superimposed by larger rockfalls. Treating rockfalls as spatially and temporally independent is therefore problematic, as is experienced in other types of landform change. For example, Milan et al. (2007) found an increase in erosion and deposition volumes within a proglacial river channel when monitored using daily terrestrial laser scan (TLS) surveys as opposed to surveys separated by 8 days. This was attributed to the temporal length scales of discharge and sediment supply, with return periods of less than 8 days. The influence of monitoring or sampling interval on measured process rates is more widely considered in the Sadler effect, in which sediment accumu-

lation rates observed in stratigraphic sections exhibit a negative power-law dependence on measurement interval (Sadler, 1981; Wilkinson, 2015). Importantly, therefore, in settings that change little but often, the ability to capture true magnitude and frequency without higher-frequency monitoring is subject to an unknown degree of event superimposition and coalescence, as well as temporal coincidence. Our first aim, therefore, is to capture the influence of near-continuous monitoring on the magnitude–frequency distribution of rockfalls from an actively failing rock face, across which spatially contiguous rockfalls have previously been observed (Rosser et al., 2005). This draws upon a database of $\sim 10^3$ individual 3-D scans collected at ~ 1 h intervals, with each comprising $> 10^6$ points.

1.3 Scanning from a fixed position

The improvement in temporal resolution gained by monitoring from (semi-)permanent installations is weighed against a series of compromises in the quality of data generated. These ultimately arise from scanning a complex surface from only a single position, which has not previously required consideration due to the scarcity of near-continuous lidar monitoring in the geosciences. Scanning from a single position has a direct impact on the similarity of point distributions between successive surveys, which is critical in determining the accuracy of subsequent change detections. Scanning from a single position results in occlusion, leaving "holes" in the final point cloud in areas invisible to the scanner. At topographic edges, range measurements may be averaged from multiple returns recorded from separate surfaces, which are intersected within a single footprint. Since laser scanners never measure exactly the same point twice (Hodge et al., 2009); the perimeter of these holes and the position of topographic edges will move between successive point clouds, even despite no movement of the instrument. The likelihood of generating similar point distributions between surveys is also influenced by surface reflectance characteristics, such as moisture and colour, and by surface relief (Clark and Robson, 2004; Bae et al., 2005; Lichti et al., 2007; Kaasalainen et al., 2008, 2010; Pesci et al., 2008, 2011; Soudarissanane et al., 2011). Scan lines in most laser scanners result in non-uniformly distributed data, with heterogeneity often on a scale and orientation comparable to surface structure, leading to aliasing that is also inconsistent (Lichti and Jamtsho, 2006). The influence of these combined effects is exaggerated if the scanner view is oblique to the surface, which may be common due to logistical constraints when siting a semi-permanent instrument. Our second aim, therefore, is to minimise the errors that arise from near-continuous monitoring in order to reduce the minimum detectable movement. Kromer et al. (2015b) presented a 4-D smoothing technique to reduce the offset between successive point clouds, such as those from near-continuous monitoring. Similarly, the method presented here is optimised for handling large (10^2–10^4) numbers of high-resolution 3-D scans, critically without user intervention, but can also be applied to point clouds from non-continuous monitoring.

1.4 Uncertainty in near-continuous monitoring

The minimum detectable movement, or level of detection (LoD), is a fundamental parameter in the delineation and calculation of erosion volumes, here rockfalls. This involves masking regions of change that exceed a hard threshold at the LoD, which is estimated either locally (e.g. Wheaton et al., 2010; Lague et al., 2013) or across the entire point cloud (e.g. Rosser et al., 2005; Abellán et al., 2009). Methods that estimate spatially variable LoDs have enhanced the ability to identify volumetric loss as compared to the application of a single LoD, with the latter set to exceed a significant portion of the modelled uncertainty across the area of interest (AOI). The spatially variable uncertainties described in Sect. 1.3 raise the potential for real change to be masked when using a single LoD but, equally, the benefits of using a single LoD are primarily in the consistency in measurement across the AOI. For example, if the purpose of monitoring is to generate a rockfall inventory in which the relative magnitude of events is important, a single LoD ensures consistency in the minimum detectable rockfall across the AOI and minimises the potential for recording erroneous events. The application of a single LoD also becomes increasingly computationally efficient when dealing with a large number of surveys. While the LoD remains constant, however, the accuracy of volume estimates is also contingent upon the ability to accurately identify depth change within boundary cells that occupy the rockfall perimeter. Previous approaches have ignored cells with a depth change below the instrument precision, thereby assuming that erosion events with an aerial extent below the cell size cannot be detected (e.g. Dussauge et al., 2003; Rosser et al., 2005; Abellán et al., 2006). However, these often fail to model volumetric errors that arise from extrapolating measured depth changes across each cell at the rockfall perimeter. In accounting for this, the volumetric uncertainty increases as a proportion of estimated volume for smaller events and specific geometries. Our third and final aim, therefore, is to describe the impact that a changing magnitude–frequency distribution has on the overall uncertainty of eroded volume estimates through time.

2 Study site and data collection

Data are presented from a monitored coastal cliff in North Yorkshire, UK. Located at East Cliff, Whitby, the rock cliff is near vertical, reaching ~ 60 m in height, and is actively eroding. The erosion of this coast has previously been monitored and averages ~ 0.1 m a^{-1} (Rosser et al., 2005, 2007; Miller, 2007). Typical rockfalls include small-scale joint defined wedges and larger-scale failures released via rock bridge breakage, which can be inferred from the exposed fresh fracture surfaces visible after failure. Rockfalls have been mea-

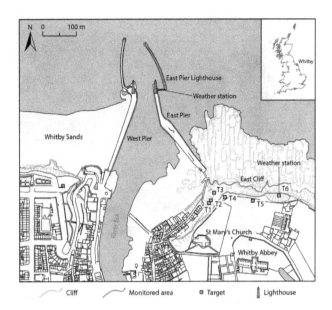

Figure 1. Map of Whitby with the area scanned delineated with red lines. A Riegl VZ-1000 scanner is installed within East Pier lighthouse. The targets installed for the SiteMonitor4D range correction factor estimation are illustrated (T1–T6) in addition to the weather stations. Whitby Abbey lies 180 m from the cliff top. Map produced using shape files from Ordnance Survey

sured up to 2.5×10^3 m^3, but the volume loss is dominated by smaller-scale rockfalls with median volumes approaching 1.0×10^{-3} m^3 (Rosser et al., 2013). The slope monitoring system presented surveys the cliff using a remotely controlled Riegl VZ-1000 laser scanner, housed inside the former lantern room at the top of East Pier lighthouse (Fig. 1), ~ 350 m seaward of the cliff face (Fig. 2a).

The viewpoint of the scanner results in some loss of spatial continuity in surface measurement due to occlusion, as a result of surface relief and the high incidence angle of parts of the rock face relative to the scanner (Fig. 2b). The closest point on the cliff is 342 m from the scanner with an incidence angle onto the strike of the cliff of $\sim 25°$, while the furthest monitored point is 533 m at an angle of $\sim 42°$. Range correction for atmospheric effects and precise point cloud alignment is automatically conducted every 3 h using very-high-resolution scanning (5.0×10^{-4} m point spacing) of six fixed 0.25 m^2 control targets, the precise relative positions ($\pm 5.0 \times 10^{-3}$ m) of which are known from a total station survey. Atmospheric range correction, derived from comparing scanned to surveyed target ranges, typically scales range measurements by $\sim 1.0 \times 10^{-5}$ with minimal diurnal fluctuation and is therefore largely inconsequential at this site, but would be significant for locations with greater ranges or areas subject to more extreme atmospheric conditions. The TLS survey was managed using SiteMonitor4D (3D Laser Mapping Ltd), which schedules scans, manages the atmospheric

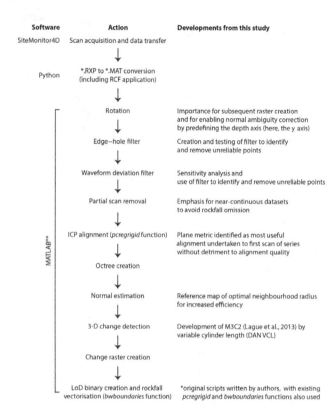

Figure 2. (a) Image of the cliff taken 1 h before high tide on 25 November 2015. Horizontally bedded strata are evident, with upper beds stained orange from downslope wash from glacial till of variable depth. The lower buttress comprises shales and some sandstone, while the near-vertical upper portion of the cliff comprises outcropping sandstone and sandstone interbedded with carbonaceous muds. **(b)** Slope model of the cliff showing the area covered by the TLS (light grey) draped over a 3-D model of the cliff, surveyed from multiple positions along the foreshore (dark grey). The total area measured is 8561 m^2, or 89 % of the cliff face area (9592 m^2). The cliff is $\sim$ 210 m across and $\sim$ 60 m high.

correction, and applies an affine rigid-body rotation matrix to compensate for tilt and yaw in the scanner position based upon the scanned control target positions in real time. The reported dataset has been collected using this setup between 5 March and 31 December 2015, totalling $\sim$ 9000 scans with $\sim$ 1 h intervals between surveys. Gaps in the dataset arise from system outages and are excluded from the analysis.

3   Method: optimising event extraction from near-continuously collected point clouds

Repeatability in change detection is dependent on point clouds that consistently describe the monitored surface. While the specification of successive scans is identical here, and despite no positional change in either the instrument or the surface, individual point clouds differ due to the inherent uncertainties described in Sect. 1.3. Minimising the impact of these differences on the resulting rockfall inventories is the primary objective of the preprocessing phases of our method, involving point cloud filtering and alignment, as well as the change detection phase. Once 3-D change has been calculated, the point clouds of change are interpolated and classified into 2.5-D datasets from which 3-D change geometry can be analysed (Fig. 3). While this paper does not seek to create a real-time system, the approach described is optimised for computational efficiency to allow data to be processed at a rate that is at least as quick as collection.

Figure 3. Flow diagram representing the stages of the optimised near-continuous monitoring change detection method. All stages following ASCII to MAT conversion were written in MATLAB®, with ICP alignment and rockfall vectorisation using the built-in functions *pcregrigid* and *bwboundaries*. Point clouds are initially rotated to become approximately planar across the $x - z$ plane, enabling the removal of points outside a tight bounding cuboid and rasterising of the point clouds of change. This rotation also enables an efficient solution to subsequent normal direction ambiguity, where the y component of each normal vector should always point out of the surface.

3.1   Point cloud filtering

Optimising a point cloud for change detection involves filtering points on features that cannot be consistently measured, such as edges and vegetation. For large numbers of scans obtained from near-continuous monitoring, an automated approach is required that finds and removes such points consistently for each cloud. Below we describe the application of filtering prior to change detection. Given that the change detection process draws on a neighbourhood of points to describe the surface morphology, filtering points in advance removes the need for change detection of erroneous measurements and ensures that such measurements do not impact the change detection of surrounding neighbours.

3.1.1 AOI extraction

Scan data collected outside of the AOI can provide useful information for scan-to-scan registration, particularly if they cover a wider geographical area, have a distribution that is less planar than the AOI itself, or if a large portion of the AOI is undergoing deformation. Here, given that the control target network provides range estimate control and the scanner remains unmoved, the spatial extent of the AOI is clipped at the beginning of the workflow in order to increase the speed of subsequent processing steps. For a small number of scans (< 20), AOI extraction can be undertaken manually. However, for the ~ 9000 scans used, points that lie outside a pre-defined cuboidal bounding box are automatically removed, which typically reduces the raw dataset from $\sim 1.9 \times 10^6$ to 1.1×10^6 points.

3.1.2 Edge and hole filter

Once the AOI has been extracted, morphological (this section) and radiometric (Sect. 3.1.3) filters are applied to remove points with high positional uncertainties. These specific filters have not previously been used to our knowledge; however, the reasoning behind their implementation is well documented as described in Sect. 1.3. We refer to edges as morphological breaks in gradient on the slope surface and holes as features that surround regions of occlusion in the point cloud, which frequently occur at edges due to high surface inclination. To detect edges, neighbouring points within a fixed radius of each (query) point, q, are identified and the central position of the neighbourhood points, CoG, calculated. The 3-D Euclidean distance, ED, between q and the CoG is then calculated. For a point at an edge, CoG tends away from the query point. The ED is therefore larger for query points that lie closer to an edge. Applying a threshold to ED needs to account for the varying point density across the cloud as in regions of low point density ED will always be larger. The edge–hole value EH assigned to each point is therefore reported as a ratio of the ED to the number of points in a spherical domain centred on each point:

$$\text{EH} = \frac{\text{ED}}{k}, \tag{1}$$

where k is the number of neighbouring points.

3.1.3 Waveform deviation filter

A limitation of many laser scanners is the inability to quantify the accuracy of each range measurement. With the common absence of more accurate data, assessing reliability is therefore challenging. In most TLS systems used for rock slope monitoring, range is estimated using the time of flight of a laser, where time is stamped based upon a characteristic of the measured reflection (e.g. intensity gate, maximum intensity amplitude) that varies between scanners. Some systems have the ability to capture the full energy–time distribution of the reflection. The energy of the received laser pulse structure depends on the spatial and temporal energy distribution of the emitted pulse, which are modified by the geometric and reflectance properties of the target surface (Stilla and Jutzi, 2008; Soudarissanane et al., 2011; Hartzell et al., 2015; Telling et al., 2017). This provides a means of estimating the relative quality of recorded measurements as a function of the number of separate reflections from a single pulse, the incidence angle of the laser beam with respect to the surface (the elongation through time of the reflected pulse relative to the emitted pulse), or the reflectance intensity (the integral of the reflection energy–time distribution). Here, the characteristics of the returned signal are used to remove vegetation (multiple returns per pulse) and edges (elongated reflections) in order to increase the consistency between successive point clouds. While the sensitivity of the waveform to target geometry has previously been highlighted (Williams et al., 2013), it has not previously been documented as a method to filter points acquired from terrestrial lidar.

The Riegl VZ-1000 TLS, with "full waveform" capture, records the intensity of each returned signal at 2.01×10^{-9} s intervals, providing 15–70 amplitude measurements per pulse. The "deviation", δ, of the waveform describes the change in shape of the received waveform relative to a modelled (emitted) Gaussian energy–time distribution according to

$$\delta = \sum_{i=1}^{N} |s_i - p_i|, \tag{2}$$

where N is the observations in pulse s_i, compared to the reference values, p_i. $\delta = 0$ represents identical emitted and received waveforms, as would be expected from a nadir-oriented planar specular surface. δ is less sensitive to target range than incidence angle.

3.1.4 Filtering of partially obscured point clouds

In addition to the filtering of individual points considered unreliable, entire surveys required removal from the overall inventory due to inclement weather conditions. In conventional monitoring with no scan schedule automation, scans that are partially or fully obscured due to rain or fog are manually removed and/or repeated. Due to the frequency and duration of near-continuous scanning, some scans will almost certainly be partially or fully obscured. Scans that are entirely obscured can be removed automatically with relative ease. Unobscured areas in partial scans still allow some accurate change detection and rockfall identification, and so may be valuable to retain. Given that change is detected between scan pairs, however, it is critical that these partial scans are removed prior to change detection and remain unused. Figure 4 describes a scenario in which a rockfall occurs between

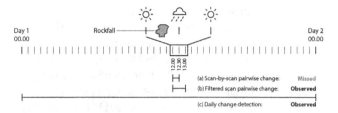

Figure 4. Conceptual illustration of the significance of removing partial scans. While parts of these scans provide accurate estimates of surface change, if a rockfall occurs in an area of no data, the failure will be missed using pairwise change. These scans should therefore be removed prior to change detection of the scan database.

12:00 and 12:30 (all times are GMT) during adverse weather, which partially obscures the impending scan at 12:30. While some areas of this scan allow accurate change detection of the surface, if the rockfall occurs in an obscured area, it can be omitted from the inventory entirely. However, if surfaces are compared between 12:00 and the following scan at 13:00, with both captured during fair conditions, the rockfall will be observed and included in the inventory.

Given the variability in the persistence of poor weather during a scan, a threshold based on the number of points can be difficult to define. At present, no automated method for detecting partial scans has been developed, though the removal of any scan that coincides with measured rainfall may represent a first step. Here, given that the same scan line patterns are applied during each survey, partially obscured point clouds were identified as those > 1 MB below the average file size. While the maximum possible number of change detections was 8986, this was reduced to 8596 because of complete obscuration and finally to 8270 because of partial scan removal. The point distribution of the remaining scans was manually examined by creating a video of every point cloud prior to reanalysis of the dataset. The reduction in the number of scans has a direct impact on the time interval between scans and hence deformation analysis prior to failures that occur during bad weather, which may result in some of the most active periods of rockfalls.

3.2   Precise alignment

While range correction factors automatically scale range estimates in response to atmospheric variation, point clouds collected from a single fixed position still require alignment due to small shifts in scanner inclination that have the potential to propagate to several centimetres over distances of several hundred metres. It is therefore assumed that successive point clouds are approximately but not perfectly aligned, and hence require adjustment. In this section we describe the protocol for scan alignment applied to the near-continuous dataset. As discussed by Abellán et al. (2014), aligning point clouds can be undertaken using common surveyed and modelled targets combined with measured global

coordinates (Teza et al., 2007; Olsen et al., 2009), feature-based registration based on the planarity and curvature of surfaces (e.g. Besl and Jain, 1988; Belton and Lichti, 2006; Rabbani et al., 2006), and point-to-point and point-to-surface methods, which use iterative closest point (ICP) alignment to progressively reduce the distance between two clouds (Besl and McKay, 1992; Chen and Medioni, 1992; Zhang, 1994). The accuracy of alignment is one of the key sources of error when detecting change between two point clouds (Teza et al., 2007).

Here, once filtered, point clouds are automatically registered to a reference point cloud to improve alignment. Registration of datasets into a global system has a significant impact on data file sizes due to multi-digit coordinates, which becomes problematic with large numbers of large point clouds from near-continuous monitoring. By retaining a local coordinate system in this study, resulting file sizes were halved. ICP registration was applied using MATLAB®'s *pcregrigid* function (Besl and McKay, 1992; Chen and Medioni, 1992), which searches for the closest point in the reference scan for each point in the moving scan and estimates the combination of rigid rotation and translation that best aligns them (Mitra et al., 2004). Pottmann and Hofer (2003) showed that for point clouds that are approximately aligned, as here, minimising the point-to-plane distance provided the best estimate of convergence. Given that the raw point cloud data commonly have systematic structure (vertical or horizontal scan lines), and where the AOI is approximately planar (a ∼2-D rock face rather than a fully ∼3-D scene), the success of the ICP can be dominated by aligning structure in the data, rather than the macro-scale cliff geometry. Point-to-point minimisation in ICP was therefore found to be less effective than point-to-plane alignment in this instance, using point clouds down-sampled to a fixed 0.25 m point spacing.

There is no established protocol for choosing the ideal reference scan for alignment. This reference scan may be the first available scan, a later single scan, or an average of a subset of previous scans. Schürch et al. (2011) aligned a series of three scans to the previous scan in a sequence, rather than to the first of the monitoring campaign, in order to gain more precise change estimates between successive surveys, at the cost of the overall positional accuracy. This procedure is advantageous as it ensures that the shape of a rapidly deforming or changing surface can be matched to the previous survey, rather than one captured considerably earlier. For the near-continuous monitoring dataset, however, the series of scans is aligned to the first survey. This is undertaken because ICP alignment minimises the point-to-plane distance of down-sampled point clouds, such that individual (small relative to the AOI) rockfall events do not impact upon the overall success of the alignment. Second, even with low alignment errors between scan pairs, the potential for the point clouds to drift over time increases with the number of scans sequentially aligned. While Schürch et al. (2011) assessed pairwise change between scans, high-frequency data allow for change detection to be conducted over multiple intervals that the time series enables, so aligning all scans with respect to each other is important. Third, given that all scans were aligned to the initial reference scan, these could be assigned to a single hierarchical structure created only for the reference scan. The partitioning of 3-D data, here into an octree structure, enables faster searching of point clouds with each point assigned a $3 \times n$ bit code, where n is the maximum octree level (Frisken and Perry, 2002; Girardeau-Montaut et al., 2005; Jaboyedoff et al., 2007, 2009; Elsberg et al., 2011, 2013; Hornung et al., 2013). Partitioning point clouds into a single predefined structure reduces computation time and ensures consistency in subsequent operations. For neighbourhood searches, such as during normal vector estimation, points from both the individual octree cube and the surrounding 26 cubes are used. As highlighted by Girardeau-Montaut et al. (2005), the subdivision level at which normal estimation and change detection is performed therefore influences only the computation time.

3.3 Normal vector estimation

The distance between successive clouds is measured along the normal vector of each point in the cloud. Accurate estimation of each normal vector is therefore critical in determining the magnitude and direction of change and should be derived from an appropriately sized neighbourhood of points that adds topological context (Riquelme et al., 2014). In this section, we describe the widely adopted method for normal estimation and gains in efficiency that have been made in this study, including the definition of a single reference map of optimal neighbourhood radii.

In order to calculate the normal direction of each neighbourhood, a tangent plane must be fitted to every point and its neighbours, with each being considered as a potential plane subset. Using eigenvectors calculated from principal component analysis, the eigenvector v_3 with the smallest associated eigenvalue is orthogonal to the plane, and therefore defines the normal (Hoppe et al., 1992). It so follows that the plane minimises the sum of squared distances to the neighbours of query point, p,

$$(p_i - \overline{p}) \cdot v_3 = 0, \tag{3}$$

and passes through the centroid, $\overline{p}$,

$$\overline{p} = \frac{\sum_{i=1}^{r} p_i}{r}, \tag{4}$$

where r is the number of neighbours in the neighbourhood, and p_i represents the Cartesian coordinates of each point within the neighbourhood (Pauly et al., 2002). The identification of a local surface normal using the third eigenvector is the equivalent of forming a total least-squares fitting plane. However, in a total least-squares fitting the entities in the covariance matrix are not divided by k, and the smallest eigenvalue is equal to the sum of the residuals squared (Pauly et al., 2002; Belton and Lichti, 2006).

The sign ambiguity of each vector is also corrected (Mitra and Nguyen, 2003; Ioannou et al., 2012). This can typically be resolved using the position of the query point, q, relative to the sensor position, s, by

$$\hat{s} = [X_s, Y_s, Z_s] - [X_q Y_q Z_q] \tag{5}$$

$$\text{In } \mathbb{R}^3 : \alpha = \arctan \left(\left\| \hat{s} \times \hat{n} \right\|_2 \right), \tag{6}$$

where $\times$ denotes the vector cross product and $\|$ denotes the Euclidean norm of the cross product. α denotes the angle between the unit-normal vector $\hat{n}$ at q and the vector between q and s, $\hat{s}$. If $\alpha > \frac{\pi}{2}$ or $\alpha < -\frac{\pi}{2}$, i.e. if the angle between the direction of the normal vector and the vector between the surface and the sensor is not within $\pm 90°$, the normal direction $\hat{n}$ is reversed:

$$\hat{n}_{\text{rev}} \langle u, v, w \rangle = \hat{n} \langle -u, -v, -w \rangle. \tag{7}$$

In order to minimise the computation time required to apply Eqs. (5) and (6), the axis orthogonal to the surface is introduced as a string of either "X" or "Y", similar to other point cloud processing packages such as MATLAB® and CloudCompare. With this information, the relevant component of the unit vector is used to determine whether the vector should be reversed or not. For example, if the approximate range in a near-nadir point cloud is measured along the y axis and the vector $\hat{n} = \langle -u, -v, -w \rangle$ then $\hat{n}$ is directed into the surface and should be reversed using Eq. (7). For each normal, this can provide a ~ 40–50% reduction in the time taken for sign correction relative to Eqs. (5) and (6).

The neighbourhood size strongly determines the direction of surface normals (Mitra and Nguyen, 2003; Lalonde et al., 2005; Bae et al., 2009; Lague et al., 2013; Riquelme et al., 2014). If the size of the neighbourhood is below the scale of surface roughness, the resulting normals will fluctuate in direction and are less likely to be consistent between successive point clouds. Here, by varying the size of the neighbourhood for each point between 0.1 and 2.5 m, the radius that produced the most planar surface is identified, which is ideally suited to normal estimation (Lague et al., 2013). An example of this is shown in Fig. 5a, with Fig. 5b illustrating surface planarity across East Cliff. This shows a clear similarity to the distribution of point density, such that the search radius is increased in regions of low point density. Importantly, identifying the optimum neighbourhood radius for 10^3–10^4 point clouds adds considerable computational cost in processing. As a compromise, the neighbourhood radius of each point in

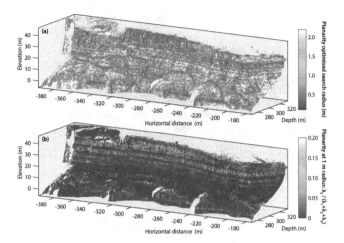

Figure 5. Search radii used for normal estimation across the rock face. **(a)** The radius for each point on the cliff at which the point clouds are most planar, with a mean value of 1.1 m, used to estimate the normal vector prior to change detection. This point cloud was used as a reference model, such that the normal radius of points in subsequent scans was assigned based on the radius of closest point in this scan. **(b)** Surface planarity at a radius of 1 m, where higher values indicate a more 3-D neighbourhood. These occur at inflections in slope profile and in areas of high local relief, such as the sandstone beds near the cliff top. Gaps in point cloud are zones of occlusion illustrated in Fig. 2b.

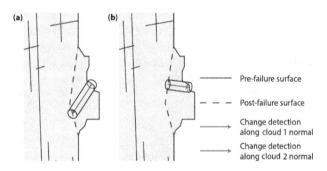

Figure 6. Conceptual variation in the distance along the normal for a rockfall. **(a)** Change detection along a surface-normal direction estimated from the pre-failure surface (Cloud 1). **(b)** The normal direction estimated using a planar, post-failure surface more accurately represents the direction of change than the post-failure surface vector due to the complexity of the pre-failure surface. With both change detections originating in the same approximate positions on the cliff face, the difference in vector lengths illustrates the sensitivity of the 3-D change measurements to the normal estimation.

this study is made equal to the distance to the closest point in the reference cloud in Fig. 5a. Notably, the normal for each point estimated uses the second cloud, such that change is accurately measured along the normal of a planar, post-failure surface, rather than the yet-to-fail surface (Fig. 6).

3.4 Change detection

The distance calculation used is based upon the structure of the M3C2 algorithm, developed by Lague et al. (2013). The algorithm is described first followed by a modification, which has been incorporated to improve the overall accuracy of change detection and to streamline the workflow when applied to large time series scan datasets.

Once the normal vector is estimated, a bounding cylinder with a user-defined radius is created along the normal running through the query point. In order to enforce the boundaries of this cylinder, the orthogonal distance between every point within the current and neighbouring 26 octree cubes and the normal vector was estimated:

$$\hat{d} = [X_n, Y_n, Z_n] - [X_p Y_p Z_p], \quad (8)$$

where $\hat{d}$ is a vector that connects each neighbour point p to a point on the normal vector $\hat{n}$, such as the query point, q. The projection of each point onto the normal P is therefore

$$P = q \times \hat{d} \quad (9)$$

$$, or \, P = q + \left(\frac{\hat{d} \cdot \hat{n}}{\hat{n} \cdot \hat{n}}\right) \times \hat{n} \quad (10)$$

and the orthogonal distance is

$$d_{orth} = \sqrt{(X_n - X_P)^2 + (Y_n - Y_P)^2 + (Z_n - Z_P)^2}. \quad (11)$$

Given that the position of each neighbouring point and its orthogonal distance to the normal vector are known, the cylinder boundaries can be enforced using the user-defined cylinder radius, r, retaining only points at which $d_{orth} \leq r$ (Fig. 7). Once the points, c, in the cylinder are isolated for both point clouds, the mean point CoG is estimated by

$$CoG = \left(\frac{\sum_{i=1}^{k_c} x}{k_c}, \frac{\sum_{i=1}^{k_c} y}{k_c}, \frac{\sum_{i=1}^{k_c} z}{k_c}, \right), \quad (12)$$

where k_c is the number of points in c. Both mean points are then projected onto the normal vector using Eqs. (9) and (10). The mean projected points of each sub-cloud, CP, are subtracted to give a distance vector, $\hat{v}$:

$$\hat{v} = CP_2 - CP_1. \quad (13)$$

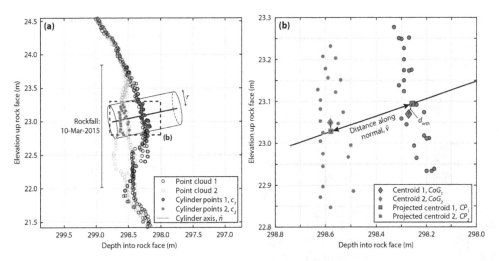

Figure 7. Approach to change detection used in this study based on empirical rockfall data. **(a)** A 2 m wide transect taken midway up East Cliff. The black points are taken from Cloud 1 and the grey from Cloud 2, with a 1.75 m high rockfall shown. Points within the cylinder radius, which intersects the two clouds, are shown as red and blue. The cylinder axis, which travels through the query point, is also shown. **(b)** Area of interest selected from **(a)** the centroids of each point cloud are determined and their orthogonal projection onto the normal vector (cylinder axis) is estimated (dashed lines). The distance measured in this study is between these projections, along the normal.

If the vector of change is along the direction of the normal vector (forward movement), the dot product of both vectors is > 0. If the vector of change is counter to the normal direction (backward movement), the dot product is < 0 and the vector is inverted.

M3C2 imposes a user-defined maximum cylinder length to decrease processing times. Cylinder length is critically important for determining the accuracy of change estimation, particularly at topographic edges within the point cloud. As described, edges are likely to be more prevalent in point clouds collected from single or off-nadir viewpoints. A method to reduce the effect of edge change uncertainty in change detection is therefore required. In Fig. 8, the influence of the choice of cylinder length is illustrated with respect to a jointed rock mass surface. The plots illustrate variation in measured change for a single point. When the cylinder extends 0.25 m in both directions, only points from this surface are included in the cylinder; as such, the centroid positions of each point cloud are both fitted onto that surface. The measured distance for this point, the distance between the two centroids, is $+0.0011$ m. With a cylinder extending ± 0.50 m, points that lie between surfaces are included in the change detection. Given that the distribution of points is rarely consistent between point clouds, the position of the centroid of each neighbourhood differs considerably from the centroids estimated using a shorter cylinder and the resulting change estimate is -0.1460 m. At a length of ± 10 m, the cylinder intersects multiple surfaces and the centroid positions are averaged between these surfaces. The inclusion of a greater number of points over a wider area increases the similarity of the mean position in both point clouds, but the resulting vector of change is $+0.0938$ m, a difference of 0.24 m from the 0.50 m

cylinder length and significantly higher than the true change estimate. To address this, a distance along the normal with variable cylinder length (DAN VCL) for each point is used. The approach begins with a cylinder that extends ± 0.10 m. If fewer than four points are found, the minimum number to estimate a centroid, the cylinder extends. This process is recursive and accepts a user-defined range of cylinder lengths.

3.5 Extracting discrete changes and quantifying volumetric error

The use of an increased number of scans over a consistent timescale increases the potential for error propagation and accumulation within the dataset. Here we aim to model the uncertainty of eroded volume estimates by focussing on the conversion of rockfall depth and area into volume for cells at the perimeter of a rockfall scar. Given that the proportion of rockfall cells at the edge of a scar decreases for a larger footprint, the uncertainty in rockfall volume estimates has the potential to vary with the temporal resolution of monitoring.

An LoD was identified between scan pairs in which no rockfalls occurred as 2 standard deviations of the 3-D change, after Abellán et al. (2009). This was of comparable magnitude to the LoD recorded for every scan pair in the dataset; hence, the maximum recorded LoD was applied to all scan pairs in the dataset. Similar to Kromer et al. (2017), these change estimates are assumed to include the registration error, which is reduced here through range correction using finely scanned targets and through ICP. For sites whose geometry creates a highly variable point spacing within a single survey, a spatially variable LoD would be more appropriate even for the purpose of compiling an inventory of geomorphic events, so long as a record of the LoD across the surface is kept. Open-pit highwalls, for example, typically comprise a series of benches to minimise the travel distance of rockfalls downslope. This design generates considerable variation in instrument–object distances across the slope and a spatially variable LoD. More broadly, spatially variable LoDs can be considered better suited to measuring total erosion budgets across a single surface than the relative contribution of individual events of varying sizes.

The resulting LoD was used to threshold 2.5-D rasters of the 3-D change data, created by linear interpolation of change values across the $x - z$ plane. The images produced included consistently located holes (no data) due to occlusion, which were identified and masked for each raster. Pixels that consistently exceeded the LoD within the first 100 point clouds, including in a non-systematic manner (e.g. both forward and backward movement), were also masked from all rasters. This prevented several (predominantly single) pixels of noise from being identified as detachments.

Once a change image is thresholded according to the LoD, the volume of each erosion event, V_E, can be calculated as

$$V_E = \sum_{i=1}^{N} d_i \times A_C, \tag{14}$$

where N is the number of cells that are classified as volume lost, d_i is the depth of change in cell i, and A_C is the cell area. Previous approaches have ignored cells with a depth change below the instrument precision and assumed that erosion events with an aerial extent $< A_C$ cannot be detected but often fail to quantify uncertainty in volume estimates derived using Eq. (14) for rockfalls with areas greater than A_C (e.g. Dussauge et al., 2003; Rosser et al., 2005; Abellán et al., 2006). Basic assumptions about how uncertainty in aerial extent propagates into volumetric uncertainty are needed, in particular for failures of varying geometry. This is of critical importance considering the relatively low spatial resolution of raster cells (here 0.15 m) relative to the accuracy of the change in depth within pixels recorded by TLS (here 1 in 10 000 to 1 in 100 000). Assuming any cell that lies on the boundary of an area of change can contain any fraction (0–1) of true change, the maximum area of change $A_{E_{\max}}$ is

$$A_{E_{\max}} = A_C \times N. \tag{15}$$

In reality, Eq. (15) represents the largest possible area because the likelihood that border cells are entirely covered by the true change is small. Conversely, the theoretical minimum area $A_{E_{\min}}$ approaches

$$A_{E_{\min}} = A_C(N - N_b), \tag{16}$$

where N_b is the number of boundary cells. The maximum range in uncertainty associated with the area of change is then

$$A_{\text{maxerror}} = A_{E_{\max}} - A_{E_{\min}}. \tag{17}$$

This value can be applied as a threshold to the rockfall inventory, such that failure areas below A_{maxerror} are removed. This threshold, however, represents the maximum possible error associated with the rockfall area. Jahne (2000) defined the variance σ_x^2 of the position of a single point in an image (cell), introduced by the cell size dx, as

$$\sigma_x^2 = \frac{1}{\Delta x} \int_{x_n - \Delta x/2}^{x_n + \Delta x/2} (x - x_n)^2 dx = \frac{(\Delta x)^2}{12}, \tag{18}$$

assuming a constant probability density function within the cell area, i.e. all positions are equally probable. The standard deviation σ_x is approximately $\frac{1}{\sqrt{12}} \approx 0.3$ times the cell size. Therefore, to accommodate for uncertainty in the position of the area of change within each boundary cell as a function of cell size, 2σ can be used as a threshold as follows:

$$A_{E_{\max}} = A_C \left(N + \frac{1}{\sqrt{12}} N_b \right) \tag{19}$$

$$A_{E_{\min}} = A_C \left(N - \frac{1}{\sqrt{12}} N_b \right) \tag{20}$$

$$A_{\text{error}} = A_{E_{\max}} - A_{E_{\min}}. \tag{21}$$

The volumetric error is hence

$$V_{\text{error}} = \sum_{i=1}^{N_b} d_i \times \frac{2}{\sqrt{12}} A_C. \tag{22}$$

Equation (22) shows that the number of border cells relative to the total number of cells within the area of change is critical in determining the net volumetric error. A higher ratio of border cells to the total number of cells results in a greater proportional area (and hence volume) error. While this volumetric error assessment is applied to rasters of 3-D-derived change, its use also extends to extraction of discrete events from DEMs of difference (DoDs).

4   Results

4.1   Processing of near-continuous monitoring data

To test the influence of the filtering phases on the subsequent change detection, comparisons were undertaken between two

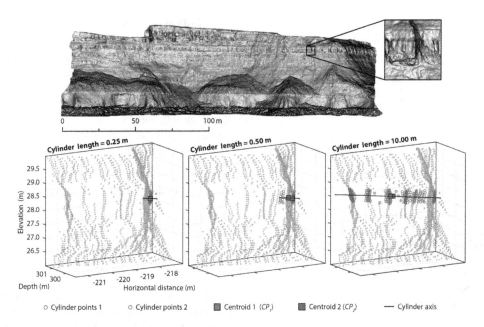

Figure 8. Inputs used for distance estimation with varying cylinder lengths. No appreciable change occurred between these two scans. As the cylinder length increases (from 0.25 to 0.50 to 10 m), the number of surfaces that the cylinder intersects increases (direction equal to the normal vector). All points within a 0.25 m radius would be included as cylinder points (circles) and the distance between their mean positions (squares) calculated. From top to bottom, this distance is 0.0011 to −0.1460 to 0.0938 m. Longer cylinders intersect multiple surfaces and therefore measure the distance between projected centroids that do not accurately represent the surface to which the query point belongs.

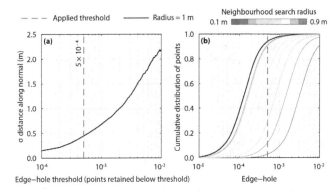

Figure 9. Sensitivity analysis for the edge–hole filter. **(a)** A change detection is undertaken between two point clouds where no observable movement occurred. The standard deviation for a single point cloud therefore indicates the level of noise between the two. When high EH values are retained, the standard deviation of change increases. **(b)** Cumulative proportion of EH values within an entire point cloud. Each line represents a different neighbourhood radius search, with a 1 m radius selected to ensure a minimum of four points, the minimum needed to estimate the CoG, would be found. An inflection in the number of points retained is used to define the threshold at 95 %, ensuring that artefacts such as holes are not introduced into the point cloud by over-removal of points. While the EH values change with the neighbourhood search radius, their distribution across the point cloud remains the same due to the normalisation by point density.

aligned point clouds following point removal based on various thresholds. The theoretical distance between the two point clouds is assumed to be zero given that no rockfalls were observed between their collection. Any offset therefore represents uncertainty, which is quantified for this purpose as the standard deviation of a 3-D change detection. This uncertainty is plotted against the applied edge and hole threshold in Fig. 9a. As the threshold is lowered, points with higher EH values are retained and the offset between the two point clouds (pre-alignment) increases. The distribution of EH values across the cloud is presented in Figs. 9b and S1 in the Supplement. Using a 1 m search radius, an inflection at the 95th percentile of points occurs at 5×10^{-4}. As a threshold, this value typically removes 5 % of points, which, as depicted by the dashed line in Fig. 9, generally account for uncertainties > 0.5 m. In addition to identifying edge features on the cliff face, this also helps to delineate areas of occlusion. The point density in Eq. (1), k, is used to filter spurious "floating points" in the dataset (for example, birds or dust). k values < 4, the minimum number to accurately define a centroid with an associated error, were removed.

Points removed using the waveform deviation filter (Fig. S2) occupied similar locations across the point cloud and, once removed, also resulted in decreased uncertainty in change estimates. In Fig. 10a, the mean absolute distance between points is used to the represent the positional uncertainty between point clouds. This uncertainty is ~ 0.02–0.03 m for points where $\delta \leq 25$. Conversely, points where $\delta > \sim 25$ exhibit more significant scatter, often approaching 2

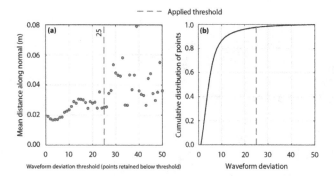

Figure 10. Sensitivity analysis for waveform deviation filter. **(a)** Mean absolute distance between two point clouds, attributed to points of each waveform deviation, from 1 to 50. Similar to Fig. 9, this indicates the comparison uncertainty between both scans. Error increases from ~ 0.03 to 0.06 m at values > 25. The variability in error also increases such that the selection of an appropriate threshold > 25 is not possible. **(b)** Cumulative distribution of waveform deviation values. A threshold of 25 removes 2 % of points.

to 3 times the level of uncertainty in the entire cloud. Removing points with $\delta > 25$ (Fig. 10b) retains 98 % of points, which accounted for a standard deviation of error between point clouds of 0.18 m prior to removal. The sequence with which the edge–hole and waveform deviation filters are applied appears to have little bearing on the outcome and the subsets of points removed by each have common members. When combined and applied to the dataset in this study, the filters reduced the standard deviation of change measurements between two stable point clouds from 0.078 to 0.055 m, thereby lowering the LoD that could be applied during rockfall or deformation identification by ~ 30 %. Although the mean offset between unregistered point clouds at these sites was 0.51 m, this was improved using the ICP alignment described in Sect. 3.2 (0.0053 m). Over the entire dataset, an average registration error of 0.005 m between point cloud pairs was obtained (n to $n + x$, where $1 > x > 8987$). Despite the alignment of all point clouds to the initial reference scan, no increase in error was observed through time. However, considerable alteration to the surface topography, via the occurrence of a single large event or the continued spalling of material over time, would require new reference scans to be assigned over shorter timescales.

Following filtering and alignment, the DAN VCL method significantly lowered the LoD (0.03 m) compared to that achieved using DoDs (1.04 m). These were created using the same pairs of point clouds rasterised at 0.25 m with each pixel containing > 1 point. In this instance, the LoD also shows a 5-fold improvement relative to the M3C2 algorithm applied using the same normal and fixed cylinder radius (LoD $= 0.165$ m). The LoD was derived for every sequential scan to ensure that no increase in registration or epistemic errors developed through the monitoring period. This value lay consistently between 0.01 and 0.03 m. The maxi-

mum LoD, 0.03 m, was therefore applied to each point cloud to prevent recording erroneous pixels in the resulting rockfall inventory. Combined with a cell size of 0.15 m, this provided a minimum detectable rockfall across the survey area of 6.75×10^{-4} m^3. More than 180 000 detachments were detected using the highest frequency of scans ($\sim$ hourly) over the 10-month monitoring period. The spatial and temporal distributions of rockfalls observed are shown in Fig. 11.

4.2 Understanding the influence of survey interval in magnitude–frequency relationships

In order to assess the influence of more frequent monitoring on the resultant volume frequency distribution, two inventories were compared. These were analysed over the same monitoring period, using scans separated by different intervals (T_{int}) of 1 h (hours) and 30 days. An increase in the number of small rockfalls and the proportional contribution of small events to the overall rockfall volume distribution is evident at $T_{int} < 1$ h (Fig. 12a), with the volume contribution from small (< 0.1 m^3) rockfalls increasing from 67 to 98 % of the total. The power-law scaling exponent, β, increases from 1.78 (30 days) to 2.27 (< 1 h). Notably, while a rollover occurs at $T_{int} = 30$ days, this is not apparent at $T_{int} < 1$ h. Given that both sets of scans were processed using the same LoD for change detection, the comparison at this site demonstrates that the observed rollover occurs due to superimposition and coalescence of events when longer survey intervals are used.

For all rockfalls observed at $T_{int} < 1$ h, volume error was modelled according to Eq. (22) (Fig. 12b). Larger rockfall volumes exhibit a smaller error in proportion to their volume. Importantly, however, as the vast majority of rockfall volumes are between 0.001 m^3 (a minimum of 2 pixels) and 0.01 m^3 (a minimum of 14 pixels), the uncertainty in volume ranges between 80 and 160 % of the estimate. A consequence is that the total volumetric uncertainty over 10 months of the $T_{int} < 1$ h rockfall inventory is greater than that collected at $T_{int} = 30$ days. High-frequency monitoring where change is dominated by a high frequency of low-magnitude events is not well suited to accurate measurement of total change through time. The error estimates demonstrate that the uncertainty in volume is greatest for the datasets in which T_{int} is low. For the highest-frequency dataset, the total estimated volume is 110.87 ± 52.44 m^3 (± 47 %), while the total estimated volume for the 30-day dataset is 72.37 ± 27.51 m^3 (± 38 %; Fig. 13).

5 Discussion

5.1 Processing techniques for near-continuous surface monitoring

Improvements in near-continuous point cloud acquisition currently outstrip the development of techniques for data treatment and analysis (Eitel et al., 2016; Kromer et al.,

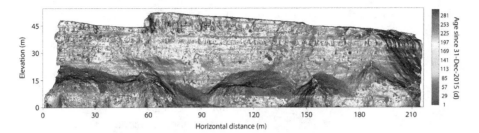

Figure 11. Distribution of rockfalls across East Cliff monitored at sub-hourly intervals between 5 March and 30 December 2015. Rockfalls are distributed across the entire cliff face, in particular in areas of exposed bedrock. Although the high water mark is below the portion of cliff shown in this figure, the largest and most frequent rockfalls occur at the base of the cliff. Accumulation and loss of material in the areas of unexposed bedrock on the cliff buttresses, which runs across the cliff face at ~ 17 m elevation, were removed. Colours represent the age since 31 December, where red represents the oldest rockfall.

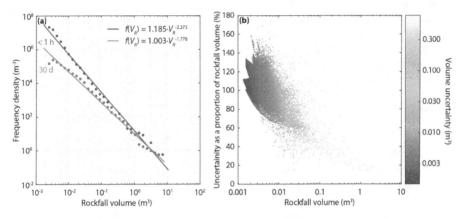

Figure 12. **(a)** Magnitude–frequency distribution for rockfall inventories acquired at varying T_{int} showing that a higher proportion of small events is established by monitoring at high frequencies. Rockfalls used range from 5 March to 30 November 2015 to enable direct comparison between $T_{int} < 1$ h and nine $T_{int} = 30$ days change detections. **(b)** Rockfall volumes from < 1 h rockfall inventory. Percentage volume error is estimated using the LoD, number of internal pixels, and number of edge pixels. Frequency densities (kernel density estimates) are appended to each axis, showing that rockfall volumes < 0.01 m^3 account for the greatest proportion of measured rockfalls (modal volume $= 0.0081$ m^3). As a result, errors range from ~ 60 to ~ 140 % for most rockfalls (modal error percentage $= 109$ %). Cumulative volume estimations using rockfalls of this size may vary by at least the actual volume.

2017). Near-continuous TLS has the potential to generate a considerable number of point clouds ($> 10^3$–10^4), representing a 2 to 3 order of magnitude increase in data volume over previous terrestrial lidar monitoring campaigns with lower temporal resolution (e.g. Teza et al., 2007; Abellán et al., 2010; Rosser et al., 2013; Royán et al., 2015). Key attributes of the techniques developed to process such datasets therefore relate to computational efficiency, the ability to automate processing, and minimising the accumulation of error between each survey pair. These have necessitated tailored approaches to lidar processing, such as those described here, which often differ from previous applications.

The relative gains of each processing step applied are evident in a gradual improvement to the applied LoD. A 30 % improvement with the application of radiometric and morphological filters occurs due to the removal of points considered to be both less accurate and less repeatable. This approach to the removal of unreliable points can also be used in alternative processing techniques, for example, as members of the fuzzy inference approach developed by Wheaton et al. (2010) to quantify spatially variable DoD uncertainty. The approach to change detection adopted in this study also yielded an improvement to the LoD, highlighting the importance of precise calculation of difference, here as a function of the cylinder length, for scenarios in which multiple surfaces may be intersected by the same normal vector in the resulting point cloud. Such surfaces can be characterised as increasingly three-dimensional on the scale of the applied cylinder length (e.g. Brodu and Lague, 2012) or as rough surfaces that are surveyed at oblique viewing angles (Hodge et al., 2009). This problem is exacerbated for near-continuous monitoring, given that scanning from a fixed position increases the width of occluded zones on surfaces inclined away from the scanner, yielding higher offsets between measured surfaces inclined towards the scanner. Critically, adaptive change detection techniques are necessary to account for the variability in point cloud quality across surfaces surveyed from a fixed position, where the installation location may

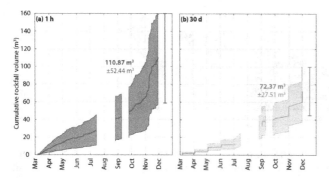

Figure 13. Cumulative rockfall volumes measured though the monitoring period, using data from all 11 monitoring intervals. The results show that far higher volumes of material, up to twice those recorded by 30 days of monitoring, are measured at sub-daily intervals. The times of pairwise change detections are recorded as the date of the first scan, rather than the second. As a result, although all scan intervals record a significantly increased rate of rockfall activity during November, this appears earlier on the plot for longer scan intervals. The total estimated volumes are not included for comparison as change detections cannot be recorded up to the final day of monitoring for longer time intervals (30 December).

yield unfavourable target geometries. These may take several forms, including spatially variable LoDs, varying cylinder lengths, or varying cylinder widths.

The cylinder radius determines the degree of spatial averaging over change measurements and, as such, should be informed by the style and scale of movements under investigation. In theory, the smaller the radius, the finer the spatial detail that can be established. However, this comes with a compromise in that the increase in accuracy by accounting for neighbouring points is reduced, the likelihood of intersecting points in the second point cloud is reduced, and the statistical significance of calculations is reduced by only drawing on a small number of points. Future development of a variable cylinder radius may draw upon the same point density estimated during normal estimation, varying in size until ~ 20 points, the minimum suggested by Lague et al. (2013), are found. Here, cylinder radii between 0.15 and 2.00 m were applied to scan pairs in which rockfalls had, and had not, occurred from East Cliff. For scans in which no change occurred, the standard deviation of differences across the point cloud was greatest for the 0.15 m radius (0.018 m), approaching the point spacing, but decreased and stabilised at a radius of 0.25 m (0.013 m). For scan pairs in which rockfalls had occurred, the size and shape of rockfalls were contrasted with the results of varying cylinder radii and a Hausdorff distance estimation. The Hausdorff distance measure itself is influenced considerably by the scan line spacing and the local point density but, for this purpose, it provides an indication of rockfall geometry with the smallest degree of spatial averaging. As the cylinder radius increases, the difference in shape relative to the Hausdorff distance also increases. While a radius of 0.15 m best approximates the size and shape of

the rockfall, this value is too close to the scan line spacing at all but the most proximal regions of the cliff face. A search radius of 0.25 m was therefore selected, providing similar rockfall geometries to the 0.15 m. This emphasises the potential to apply a variable cylinder radius across the point cloud based upon local point density, which could be helpful in future research.

The applied method removes scans undertaken during poor weather conditions (e.g. rain or fog) in order to preserve the accuracy of the resulting rockfall inventory. Given the potential for rockfalls to occur during poor weather conditions, this constitutes an important drawback of near-continuous TLS monitoring for rockfall inventory compilation. Techniques that can operate during inclement weather conditions, such as ground-based interferometric synthetic-aperture radar (InSAR), are therefore better suited to maintaining temporal consistency in rockfall datasets. However, while the precision and measurement frequency of InSAR monitoring surpasses that of lidar, highly precise change measurements are also spatially averaged across large pixel sizes, resulting in a minimum detectable rockfall volume several orders of magnitude higher ($\sim 1\,\text{m}^3$) compared to terrestrial lidar datasets. While small magnitude changes that occur across large areas can be accurately characterised at high frequency, detecting the frequency density of small erosion events is therefore compromised. Point cloud generation for slope monitoring has been supplemented in recent years by the development of new photogrammetric techniques, in particular structure from motion (SfM; Niethammer et al., 2011; Westoby et al., 2012; Lucieer et al., 2013; Turner et al., 2015; Carrivick et al., 2016). When imagery is acquired from unmanned aerial vehicles, SfM has the advantage of far lower operational costs than TLS, minimising areas of occlusion that occur from ground-level monitoring and providing highly dense point clouds due to the potentially small distances between the unmanned aerial vehicle and the slope. At present, however, the technique requires further development before it can be deployed for near-continuous monitoring.

5.2 Future developments in processing near-continuous TLS collection

Given that small events present the highest aerial, and hence volumetric, uncertainty, the recent development of 3-D volume estimation from point clouds (Carrea et al., 2012; Benjamin et al., 2016) may play an important role in near-continuous scanning. The uncertainty of these techniques is determined by the precision of the point cloud, thereby eliminating uncertainty in object aerial extents due to linear interpolation into a fixed grid. However, these techniques also contain uncertainties that arise in part from the meshing approach adopted (Soudarissanane et al., 2011; Hartzell et al., 2015; Telling et al., 2017). Due to the dependence of these techniques on a minimum of four points to create a closed hull, fully 3-D techniques are also limited in their ability to resolve small, single point detachments. The development of scanners with increasingly small angular step widths and increased rates of point acquisition, however, will decrease the minimum resolvable detachment. At present, the 3-D clustering required to isolate points belonging to geomorphic change, combined with subsequent meshing of these points, comes at a considerable computational cost. These techniques therefore remain to be applied for > 10 scans (e.g. Carrea et al., 2012; Benjamin et al., 2016; van Veen et al., 2017).

This study demonstrates the need to adjust the frequency of data collection and processing in accordance with the study objectives. Here, monitoring has been undertaken to detect near instantaneous discrete changes to the slope (rockfalls) where both the spatial and temporal resolution of monitoring are important. Longer-term total change is more prone to error when change accrues from many small events, and big changes can occur in both the short and long term. There is a lack of research into this trade-off in spatial and temporal resolution but approaches that allow this to occur would be helpful in the future. The collection of a high-frequency time series of scan data presents the opportunity to reduce uncertainty by averaging point positions through both time and space, as points are independent in neither space nor in time. This averaging can take the form of averaging the 3-D position of each point, as utilised here and in M3C2 (Lague et al., 2013), or the averaging of differences between points (Abellán et al., 2009; Kromer et al., 2015b).

5.3 Implications for rockfall magnitude–frequency

While the precise magnitude–frequency exponent reported is specific to East Cliff, the scale-invariant behaviour of rockfalls is similar to that observed in other rockfall inventories, albeit across a narrower range of magnitudes. Along the same stretch of coastline, previous monthly monitoring has yielded exponents of $\beta = 1.43$–1.91 (Rosser et al., 2007; Barlow et al., 2012), similar to that identified in this study using the $T_{int} = 30$ days inventory ($\beta = 1.78$). Both the exponent and presence of a rollover show a dependence upon monitoring interval on temporal scales considerably lower than the intervals used by Barlow et al. (2012). For rockfall distributions created from TLS surveys, Young et al. (2011) noted that the ability to resolve small-scale changes should not introduce a rollover because the smallest reported rockfall is larger than the minimum detectable event identifiable in change mapping. From a statistical perspective, this statement holds true as long as the frequency density is not estimated using a moving kernel, which enforces an extrapolation of density that extends one kernel half width beyond the range of the observations both below the minimum and above the maximum, introducing inflections in the frequency density at the tails (Lim et al., 2010). Here, a rollover in the magnitude–frequency distribution is identified for the $T_{int} = 30$ days inventory. However, this rollover

was not present in the $T_{int} < 1$ h dataset. Only rockfalls larger than 0.03 m (LoD) $\times$ 0.15 m $\times$ 0.15 m (the area of each cell, which exceeds the minimum point spacing) were analysed, which equates to a volume of 6.75×10^{-4} m^3. Critically, this implies that where events coincide, or coalesce, or where the mechanisms driving change are not spatially independent, event frequency is partially determined by survey interval. Given the development of spatially contiguous rockfall scars that has been observed in this setting (Fig. 11) and in other studies (Rosser et al., 2007, 2013; Stock et al., 2012; Kromer et al., 2015a; Rohmer and Dewez, 2015; Royán et al., 2015), the creation of magnitude–frequency distributions from near-continuous monitoring has the potential to generate improved understanding of the underlying mechanisms of rockfall failure.

Monitoring at lower frequencies may provide more accurate estimates of rates of total change over longer periods. This is related to both the longer and hence time-averaged conditions captured but also to the fact that the same level of change measured infrequently has less volumetric error than when measured frequently, particularly when change is accrued by many small, discrete events. A decrease in T_{int} that approaches near-continuous monitoring, 1 h, results in a shift in the exponent of the inverse power law of rockfall volumes from to 1.78 (30 days) to 2.27 (1 h). With a maximum plausible volume error, uncertainty in total rockfall volume ranged from 20 to 160 % of the measured volume. Although critical to measure the full rockfall volume distribution, high-frequency monitoring in this setting is not suitable for measuring net volume loss as a result of large numbers of small events. In summary, magnitude–frequency analysis of rockfall volumes (Fig. 12a) indicates that more frequent scanning detects a greater proportion of smaller rockfall events. Consequently, more frequent scanning also presents increased uncertainty in cumulative volume. Cumulatively, this error can be significant relative to the total flux over the monitoring period given that the size distribution of rockfall volumes adheres to a power law.

6 Conclusions

The magnitude–frequency distribution of geomorphic change is an important descriptor of the relative efficacy of event sizes and the nature of the hazard that they pose. Improvements in the ability to resolve the magnitude of events have surpassed the ability to constrain event frequency over short time intervals (< days). However, increasing the temporal resolution of monitoring of a changing surface increases the cumulative error over the same monitoring periods, particularly where change is dominated by numerous small events. In this study, we have discussed the practicalities and techniques to reduce this error for near-continuously acquired 3-D monitoring data, using one of the highest-temporal-resolution 3-D datasets collected to date.

The findings of our workflow are distilled here. Both morphological and radiometric filters can be effective in removing unreliable points, such as edges, surfaces of high incidence angle, and vegetation, the effects of which become increasingly prominent when scanning from a fixed position. Applying these filters lowered the standard deviation of change detected between two stable point clouds from 0.078 to 0.055 m. Scans with any degree of occlusion, arising from atmospheric conditions (e.g. rain) should be entirely removed to ensure that no rockfall, which are more probable during these conditions, are likely to be missed in pairwise change detection. The alignment of large numbers of scans (10^2–10^4) to the first scan of the dataset prevents drift without detriment to the overall alignment accuracy, though this may not apply in settings where the overall slope morphology changes through the monitoring period. Three-dimensional change detection improves the LoD in relation to 2.5-D change detection techniques (DoDs). However, in its application to an actively failing rock slope, the DAN VCL approach described here yielded a 5-fold decrease in uncertainty relative to M3C2, the effectiveness of which increases with the frequency of holes and edges in the dataset.

By comparing rockfall inventories collected at $T_{int} < 1$ h and $T_{int} = 30$ days, it is apparent that more frequent monitoring captures a higher proportion of small rockfalls, represented by a higher magnitude–frequency scaling coefficient. Importantly, both the size and shape of rasterised events determines the ability to accurately quantify their volume, with smaller events and events with a higher proportion of boundary cells producing a higher degree of volume uncertainty. The method proposed to quantify this uncertainty represents an important consideration for volume estimation during future near-continuous monitoring campaigns. Critically, a higher proportion of small rockfalls in an inventory increases volumetric uncertainty, which accumulates because of the increased frequency of these events. Net long-term eroded volume is therefore most accurately quantified between the first and last surveys of a monitoring period.

Competing interests. The authors declare that they have no conflict of interest.

Special issue statement. This article is part of the special issue "4-D reconstruction of earth surface processes: multi-temporal and multi-spatial high resolution topography". It is not associated with a conference.

Acknowledgements. This research formed part of a PhD studentship provided by the Department of Geography, Durham University, and ran alongside the Knowledge Transfer Partnership (KTP8878) awarded to Nick J. Rosser, Richard J. Hardy, Ashraf A. Afana, and 3-D Laser Mapping Ltd. We thank ICL Fertilizers (UK) Ltd for ongoing support of this project. Assistance in the development of this system was provided by Navstar Geomatics, and the maintenance and running of the system was supported by Samantha Waugh and Dave Hodgson. We thank Marc-Henri Derron, Álvaro Gómez-Gutiérrez, and Carlos Castillo for their constructive comments, which helped to improve the paper.

References

Abellán, A., Vilaplana, J. M., and Martínez, J.: Application of a long-range Terrestrial Laser Scanner to a detailed rockfall study at Vall de Núria (Eastern Pyrenees, Spain), Eng. Geol., 88, 136–148, 2006.

Abellán, A., Jaboyedoff, M., Oppikofer, T., and Vilaplana, J. M.: Detection of millimetric deformation using a terrestrial laser scanner: experiment and application to a rockfall event, Nat. Hazards Earth Syst. Sci., 9, 365–372, https://doi.org/10.5194/nhess-9-365-2009, 2009.

Abellán, A., Calvet, J., Vilaplana, J. M., and Blanchard, J.: Detection and spatial prediction of rockfalls by means of terrestrial laser scanner monitoring, Geomorphology, 119, 162–171, 2010.

Abellán, A., Oppikofer, T., Jaboyedoff, M., Rosser, N. J., Lim, M., and Lato, M. J.: Terrestrial laser scanning of rock slope instabilities, Earth Surf. Proc. Land., 39, 80–97, 2014.

Bae, K. H., Belton, D., and Lichti, D.D.: A framework for position uncertainty of unorganised three-dimensional point clouds from near-monostatic laser scanners using covariance analysis, Int. Arch. Photogramm., 36, 7–12, 2005.

Bae, K. H., Belton, D., and Lichti, D. D.: A closed-form expression of the positional uncertainty for 3D point clouds, IEEE T. Pattern. Anal., 31, 577–590, 2009.

Barlow, J., Lim, M., Rosser, N., Petley, D., Brain, M., Norman, E., and Geer, M.: Modeling cliff erosion using negative power law scaling of rockfalls, Geomorphology, 139–140, 416–424, 2012.

Belton, D. and Lichti, D. D.: Classification and segmentation of terrestrial laser scanner point clouds using local variance information, Int. Arch. Photogramm., 36, 44–49, 2006.

Benjamin, J., Rosser, N. J., and Brain, M. J.: Rockfall detection and volumetric characterisation using LiDAR, in: Landslides and Engineered Slopes, Experience, Theory and Practice: Proceedings of the 12th International Symposium on Landslides, edited by: Aversa, S., Cascini, L., Picarelli, L., and Scavia, C., Napoli, Italy, 12–19 June 2016, CRC Press, The Netherlands, 389–395, 2016.

Besl, P. J. and Jain, R. C.: Segmentation through variable-order surface fitting, IEEE T. Pattern. Anal., 10, 167–192, 1988.

Besl, P. J. and McKay, N. D.: Method for registration of 3-D shapes, IEEE T. Pattern Anal., 14, 239–256, 1992.

Brodu, N. and Lague, D.: 3D terrestrial lidar data classification of complex natural scenes using a multi-scale dimensionality criterion: Applications in geomorphology, ISPRS J. Photogramm., 68, 121–134, 2012.

Carrea, D., Abellán, A., Derron, M. H., Gauvin, N., and Jaboyedoff, M.: Using 3D surface datasets to understand landslide evolution: from analogue models to real case study, in: Landslides and Engineered Slopes: Protecting Society through Improved Understanding, edited by: Eberhardt, E., Froese, C., and Turner, K., London: CRC Press, 575–579, 2012.

Carrivick, J. L., Smith, M. W., and Quincey, D. J. (Eds.): Structure from Motion in the Geoscience, John Wiley & Sons, New York, USA, 2016.

Chen, Y. and Medioni, G.: Object modelling by registration of multiple range images, Image Vision Comput., 10, 145–155, 1992.

Clark, J. and Robson, S.: Accuracy of measurements made with a Cyrax 2500 laser scanner against surfaces of known colour, Surv. Rev., 37, 626–638, 2004.

Dussauge, C., Grasso, J. R., and Helmstetter, A.: Statistical analysis of rockfall volume distributions: Implications for rockfall dynamics, J. Geophys. Res.-Solid, 108, 2286–2296, https://doi.org/10.1029/2001JB000650, 2003.

Dussauge-Peisser, C., Helmstetter, A., Grasso, J.-R., Hantz, D., Desvarreux, P., Jeannin, M., and Giraud, A.: Probabilistic approach to rock fall hazard assessment: potential of historical data analysis, Nat. Hazards Earth Syst. Sci., 2, 15–26, https://doi.org/10.5194/nhess-2-15-2002, 2002.

Eitel, J. U., Höfle, B., Vierling, L. A., Abellán, A., Asner, G. P., Deems, J. S., Glennie, C. L., Joerg, P. C., LeWinter, A. L., Magney, T. S., and Mandlburger, G.: Beyond 3-D: The new spectrum of lidar applications for earth and ecological sciences, Remote Sens. Environ., 186, 372–392, 2016.

Elseberg, J., Borrmann, D., and Nüchter, A.: Efficient processing of large 3D point clouds, in: Proceedings of the 23rd International Symposium on Information, Communication and Automation Technologies (ICAT), Sarajevo, Bosnia and Herzegovina, 27–29 October 2011, 132–138, 2011.

Elseberg, J., Borrmann, D., and Nüchter, A.: One billion points in the cloud–an octree for efficient processing of 3D laser scans, ISPRS J. Photogramm., 76, 76–88, 2013.

Frisken, S. F. and Perry, R. N.: Simple and efficient traversal methods for quadtrees and octrees, J. Graphic. Tools, 7, 1–11, 2002.

Gintz, D., Hassan, M. A., and Schmidt, K. H.: Frequency and magnitude of bedload transport in a mountain river, Earth Surf. Proc. Land., 21, 433–445, 1996.

Girardeau-Montaut, D., Roux, M., Marc, R., and Thibault, G.: Change detection on points cloud data acquired with a ground laser scanner, Int. Arch. Photogramm., 36, 30–35, 2005.

Guthrie, R. H. and Evans, S. G.: Magnitude and frequency of landslides triggered by a storm event, Loughborough Inlet, British Columbia, Nat. Hazards Earth Syst. Sci., 4, 475–483, https://doi.org/10.5194/nhess-4-475-2004, 2004.

Guthrie, R. H. and Evans, S. G.: Work, persistence, and formative events: the geomorphic impact of landslides, Geomorphology, 88, 266–275, 2007.

Guzzetti, F., Malamud, B. D., Turcotte, D. L., and Reichenbach, P.: Power-law correlations of landslide areas in central Italy, Earth Planet. Sc. Lett., 195, 169–183, 2002.

Guzzetti, F., Reichenbach, P., Cardinali, M., Galli, M., and Ardizzone, F.: Probabilistic landslide hazard assessment at the basin scale, Geomorphology, 72, 272–299, 2005.

Hartzell, P. J., Glennie, C. L., and Finnegan, D. C.: Empirical waveform decomposition and radiometric calibration of a terrestrial full-waveform laser scanner, IEEE T. Geosci. Remote, 53, 162–172, 2015.

Hodge, R., Brasington, J., and Richards, K.: In situ characterization of grain-scale fluvial morphology using Terrestrial Laser Scanning, Earth Surf. Proc. Land., 34, 954–968, 2009.

Hooke, J. M.: Magnitude and distribution of rates of river bank erosion, Earth Surf. Proc. Land, 5, 143–157, 1980.

Hoppe, H., DeRose, T., Duchamp, T., McDonald, J., and Stuetzle, W.: Surface reconstruction from unorganized points, ACM SIG-GRAPH Computer Graphics, 26, 71–78, 1992.

Hornung, A., Wurm, K. M., Bennewitz, M., Stachniss, C., and Burgard, W.: OctoMap: An efficient probabilistic 3D mapping framework based on octrees, Auton. Robot., 34, 189–206, 2013.

Hovius, N., Stark, C. P., and Allen, P. A.: Sediment flux from a mountain belt derived by landslide mapping, Geology, 25, 231–234, 1997.

Hovius, N., Stark, C. P., Hao-Tsu, C., and Jiun-Chuan, L.: Supply and removal of sediment in a landslide-dominated mountain belt: Central Range, Taiwan, J. Geol., 108, 73–89, 2000.

Ioannou, Y., Taati, B., Harrap, R., and Greenspan, M.: Difference of normals as a multi-scale operator in unorganized point clouds, in: Proceedings of the 2nd International Conference on 3D Imaging, Modeling, Processing, Visualization and Transmission, Zurich, Switzerland, 13–15 October 2012, 501–508, 2012.

Jaboyedoff, M., Metzger, R., Oppikofer, T., Couture, R., Derron, M.H., Locat, J., and Turmel, D.: New insight techniques to analyze rock-slope relief using DEM and 3D-imaging cloud points: COLTOP-3D software, in: Rock Mechanics: Meeting Society's Challenges and Demands, edited by: Eberhardt, E., Stead, D., and Morrison, T., Proceedings of the 1st Canada-US Rock Mechanics Symposium, Vancouver, Canada, 27–31 May 2007, 61–68, 2007.

Jaboyedoff, M., Couture, R., and Locat, P.: Structural analysis of Turtle Mountain (Alberta) using digital elevation model: toward a progressive failure, Geomorphology, 103, 5–16, 2009.

Jahne, B. (Ed.): Computer vision and applications: a guide for students and practitioners, Orlando, FL: Academic Press, 679 pp., 2000.

Kaasalainen S., Kukko A., Lindroos T., Litkey P., Kaartinen H., Hyyppa J., and Ahokas E.: Brightness measurements and calibration with airborne and terrestrial laser scanners, IEEE T. Geosci. Remote, 46, 528–534, 2008.

Kaasalainen, S., Niittymaki, H., Krooks, A., Koch, K., Kaartinen, H., Vain, A., and Hyyppa, H.: Effect of target moisture on laser scanner intensity, IEEE T. Geosci. Remote, 48, 2128–2136, 2010.

Kromer, R. A., Hutchinson, D. J., Lato, M. J., Gauthier, D., and Edwards, T.: Identifying rock slope failure precursors using LiDAR for transportation corridor hazard management, Eng. Geol., 195, 93–103, 2015a.

Kromer, R. A., Abellán, A., Hutchinson, D. J., Lato, M., Edwards, T., and Jaboyedoff, M.: A 4D filtering and calibration technique for small-scale point cloud change detection with a terrestrial laser scanner, Remote Sens., 7, 13029–13052, 2015b.

Kromer, R. A., Abellán, A., Hutchinson, D. J., Lato, M., Chanut, M.-A., Dubios, L., and Jaboyedoff, M.: Automated Terrestrial Laser Scanning with Near Real-Time Change Detection – Monitoring of the Séchilienne Landslide, Earth Surf. Dynam., 5, 293–310, 2017.

Lague, D., Brodu, N., and Leroux, J.: Accurate 3D comparison of complex topography with terrestrial laser scanner: Application to the Rangitikei canyon (NZ), ISPRS J. Photogramm., 82, 10–26, 2013.

Lalonde, J. F., Unnikrishnan, R., Vandapel, N., and Hebert, M.: Scale selection for classification of point-sampled 3D surfaces, in: Proceedings of the 5th International Conference on 3-D Digital Imaging and Modeling, Ottawa, Canada, 13–16 June 2005, 285–292, 2005.

Li, G., West, A. J., Densmore, A. L., Hammond, D. E., Jin, Z., Zhang, F., Wang, J., and Hilton, R. G.: Connectivity of earthquake-triggered landslides with the fluvial network: Implications for landslide sediment transport after the 2008 Wenchuan earthquake, J. Geophys. Res.-Earth, 121, 703–724, 2016.

Lichti, D. D. and Jamtsho, S.: Angular resolution of terrestrial laser scanners, Photogramm. Rec., 21, 141–160, 2006.

Lim, M., Rosser, N. J., Allison, R. J., and Petley, D. N.: Erosional processes in the hard rock coastal cliffs at Staithes, North Yorkshire, Geomorphology, 114, 12–21, 2010.

Lucieer, A., de Jong, S., and Turner, D.: Mapping landslide displacements using Structure from Motion (SfM) and image correlation of multi-temporal UAV photography, Prog. Phys. Geog., 38, 97–116, 2014.

Malamud, B. D., Turcotte, D. L., Guzzetti, F., and Reichenbach, P.: Landslides, earthquakes, and erosion, Earth Planet. Sc. Lett., 229, 45–59, 2004.

Milan, D. J., Heritage, G. L., and Hetherington, D.: Application of a 3D laser scanner in the assessment of erosion and deposition volumes and channel change in a proglacial river, Earth Surf. Proc. Land., 32, 1657–1674, 2007.

Miller, P. E.: A robust surface matching technique for coastal geohazard monitoring, PhD thesis, School of Civil Engineering and Geosciences, Newcastle University, UK, 311 pp., 2007.

Mitra, N. J. and Nguyen, A.: Estimating surface normals in noisy point cloud data, in: Proceedings of the 19th Annual Symposium on Computational Geometry, San Diego, California, 8–10 June 2003, 322–328, 2003.

Mitra, N. J., Gelfand, N., Pottmann, H., and Guibas, L.: Registration of point cloud data from a geometric optimization perspective, in: Proceedings of the 2004 Eurographics/ACM SIGGRAPH Symposium on Geometry Processing, Nice, France, 8–10 July 2004, 22–31, 2004.

Nash, D. B.: Effective sediment-transporting discharge from magnitude-frequency analysis, J. Geol., 102, 79–95, 1994.

Niethammer, U., Rothmund, S., Schwaderer, U., Zeman, J., and Joswig, M.: Open source image-processing tools for low-cost UAV-based landslide investigations, Int. Arch. Photogramm., 38, 161–166, 2011.

Olsen, M. J., Kuester, F., Chang, B. J., and Hutchinson, T. C.: Terrestrial laser scanning-based structural damage assessment, J. Comput. Civil Eng., 24, 264–272, 2009.

Pauly, M., Gross, M., and Kobbelt, L. P.: Efficient simplification of point-sampled surfaces, in: Proceedings of the Conference on Visualization (VIS '02), Boston, MA, 27 October–1 November, 163–170, 2002.

Pelletier, J. D., Malamud, B. D., Blodgett, T., and Turcotte, D. L.: Scale-invariance of soil moisture variability and its implications for the frequency-size distribution of landslides, Eng. Geol., 48, 255–268, 1997.

Pesci, A., Teza, G., and Ventura, G.: Remote sensing of volcanic terrains by terrestrial laser scanner: preliminary reflectance and RGB implications for studying Vesuvius crater (Italy), Ann. Geophys.-Italy, 51, 633–653, 2008.

Pesci, A., Teza, G., and Bonali, E.: Terrestrial laser scanner resolution: numerical simulations and experiments on spatial sampling optimization, Remote Sens., 3, 167–184, 2011.

Pottmann, H. and Hofer, M.: Geometry of the Squared Distance Function to Curves and Surfaces, in: Visualization and Mathe-

matics III, Mathematics and Visualization, Springer, edited by: Hege, H. C. and Polthier, K., Berlin, 221–242, 2003.

Rabbani, T., Van Den Heuvel, F., and Vosselmann, G.: Segmentation of point clouds using smoothness constraint, Int. Arch. Photogramm., 36, 248–253, 2006.

Riquelme, A. J., Abellán, A., Tomás, R., and Jaboyedoff, M.: A new approach for semi-automatic rock mass joints recognition from 3D point clouds, Comput. Geosci., 68, 38–52, 2014.

Rohmer, J. and Dewez, T.: Analysing the spatial patterns of erosion scars using point process theory at the coastal chalk cliff of Mesnil-Val, Normandy, northern France, Nat. Hazards Earth Syst. Sci., 15, 349–362, https://doi.org/10.5194/nhess-15-349-2015, 2015.

Rosser, N. J., Petley, D. N., Lim, M., Dunning, S. A., and Allison, R. J.: Terrestrial laser scanning for monitoring the process of hard rock coastal cliff erosion, Q. J. Eng. Geol. Hydroge., 38, 363–375, 2005.

Rosser, N. J., Dunning, S. A., Lim, M., and Petley, D. N.: Terrestrial laser scanning for quantitative rockfall hazard assessment, in: Landslide Risk Management, A. T. Balkema, edited by: Hungr, O., Fell, R., Couture, R., and Eberhardt, E., Amsterdam, 2007.

Rosser, N. J., Brain, M. J., Petley, D. N., Lim, M., and Norman, E. C.: Coastline retreat via progressive failure of rocky coastal cliffs, Geology, 41, 939–942, 2013.

Royán, M. J., Abellán, A., and Vilaplana, J. M.: Progressive failure leading to the 3 December 2013 rockfall at Puigcercós scarp (Catalonia, Spain), Landslides, 12, 585–595, 2015.

Sadler, P. M.: Sediment accumulation rates and the completeness of stratigraphic sections, J. Geol., 89, 569–584, 1981.

Schürch, P., Densmore, A. L., Rosser, N. J., Lim, M., and McArdell, B. W.: Detection of surface change in complex topography using terrestrial laser scanning: application to the Illgraben debris-flow channel, Earth Surf. Proc. Land., 36, 1847–1859, 2011.

Soudarissanane, S., Lindenbergh, R., Menenti, M., and Teunissen, P. J. G.: Scanning geometry: Influencing factor on the quality of terrestrial laser scanning points, ISPRS J. Photogramm., 66, 389–399, 2011.

Stark, C. P. and Hovius, N.: The characterization of landslide size distributions, Geophys. Res. Lett., 28, 1091–1094, 2001.

Stilla, U. and Jutzi, B.: Waveform analysis for small-footprint pulsed laser systems, in: Topographic Laser Ranging and Scanning Principles and Processing, edited by: Shan, J. and Toth, C., CRC Press Taylor & Francis Group, Boca Raton, USA, 215–234, 2008.

Stock, G. M., Martel, S. J., Collins, B. D., and Harp, E. L.: Progressive failure of sheeted rock slopes: the 2009–2010 Rhombus Wall rock falls in Yosemite Valley, California, USA, Earth Surf. Proc. Land., 37, 546–561, 2012.

Telling, J., Lyda, A., Hartzell, P., and Glennie, C.: Review of Earth science research using terrestrial laser scanning, Earth Sci. Rev., 169, 35–68, 2017.

Teza, G., Galgaro, A., Zaltron, N., and Genevois, R.: Terrestrial laser scanner to detect landslide displacement fields: a new approach, Int. J. Remote Sens., 28, 3425–3446, 2007.

Turcotte, D. L., Malamud, B. D., Guzzetti, F., and Reichenbach, P.: Self-organization, the cascade model, and natural hazards, P. Natl. Acad. Sci. USA., 99, 2530–2537, 2002.

Turner, D., Lucieer, A., and de Jong, S. M.: Time series analysis of landslide dynamics using an unmanned aerial vehicle (UAV), Remote Sens., 7, 1736–1757, 2015.

van Veen, M., Hutchinson, D. J., Kromer, R., Lato, M., and Edwards, T.: Effects of sampling interval on the frequency-magnitude relationship of rockfalls detected from terrestrial laser scanning using semi-automated methods, Landslides, 14, 1579–1592, 2017.

Westoby, M. J., Brasington, J., Glasser, N. F., Hambrey, M. J., and Reynolds, J. M.: "Structure-fromMotion" photogrammetry: A low-cost, effective tool for geoscience applications, Geomorphology, 179, 300–314, 2012.

Wheaton, J. M., Brasington, J., Darby, S. E., and Sear, D.A.: Accounting for uncertainty in DEMs from repeat topographic surveys: improved sediment budgets, Earth Surf. Proc. Land., 35, 136–156, 2010.

Wilkinson, B. H.: Precipitation as meteoric sediment and scaling laws of bedrock incision: Assessing the Sadler effect, J. Geol., 123, 95–112, 2015.

Williams, J. G., Rosser, N. J., Afana, A., Hunter, G., and Hardy, R. J.: Can full-waveform technology enhance the use of terrestrial laser scanning to monitor rock slope deformation?, Proceedings of the 2013 Symposium on Slope Stability in Open Pit Mining and Civil Engineering, Brisbane, Australia, 25–27 September 2013, 763–774, 2013.

Wolman, M. G. and Gerson, R.: Relative scales of time and effectiveness of climate in watershed geomorphology, Earth Surf. Proc., 3, 189–208, 1978.

Wolman, M. G. and Miller, J. P.: Magnitude and frequency of forces in geomorphic processes, J. Geol., 68, 54–74, 1960.

Young, A. P., Guza, R. T., O'Reilly, W. C., Flick, R. E., and Gutierrez, R.: Short-term retreat statistics of a slowly eroding coastal cliff, Nat. Hazards Earth Syst. Sci., 11, 205–217, https://doi.org/10.5194/nhess-11-205-2011, 2011.

Zhang, Z.: Iterative point matching for registration of free-form curves and surfaces, Int. J. Comput. Vision, 13, 119–152, 1994.

PERMISSIONS

LIST OF CONTRIBUTORS

Sébastien Monnier
Instituto de Geografía, Pontificia Universidad Católica de Valparaíso, Valparaíso, Chile

Christophe Kinnard
Département des Sciences de l'Environnement, Université du Québec à Trois-Rivières, Trois-Rivières, Québec, Canada

Daniel N. Scott and Ellen E. Wohl
Department of Geosciences, Colorado State University, Fort Collins, CO 80521, USA

Gregory E. Tucker
Cooperative Institute for Research in Environmental Science (CIRES) and Department of Geological Sciences, University of Colorado, Boulder, CO 80305, USA

Scott W. McCoy
Department of Geological Sciences and Engineering, University of Nevada, Reno, NV 89557, USA

Daniel E. J. Hobley
School of Earth and Ocean Sciences, Cardiff University, Cardiff, CF10 3AT, Wales

Jasper R. F. W. Leuven, Sanja Selaković and Maarten G. Kleinhans
Faculty of Geosciences, Utrecht University, Princetonlaan 8A, 3584 CB, Utrecht, the Netherlands

Odin Marc, André Stumpf and Jean-Philippe Malet
École et Observatoire des Sciences de la Terre, Institut de Physique du Globe de Strasbourg, Centre National de la Recherche Scientifique UMR 7516, University of Strasbourg, 67084 Strasbourg CEDEX, France

Marielle Gosset
Géoscience Environnement Toulouse, Toulouse, France

Taro Uchida
National Institute for Land and Infrastructure Management, Research Center for Disaster Risk Management, Tsukuba, Japan

Shou-Hao Chiang
Center for Space and Remote Sensing Research, National Central University, Taoyuan City 32001, Taiwan

Sebastian G. Mutz, Todd A. Ehlers and Jingmin Li
Department of Geosciences, University Tübingen, 72074 Tübingen, Germany

Jingmin Li
Institute for Geography and Geology, University of Würzburg, Würzburg, 97074 Germany

Martin Werner, Gerrit Lohmann and Christian Stepanek
Department of Paleo climate Dynamics, Alfred Wegener Institute, Helmholtz Centre for Polar and Marine Research, 27570 Bremerhaven, Germany

Virginia Ruiz-Villanueva and Markus Stoffel
Institute for Environmental Sciences, University of Geneva, Boulevard Carl-Vogt 66, 1205 Geneva, Switzerland

Alexandre Badoux, Dieter Rickenmann, Martin Böckli, Nicolas Steeb and Christian Rickli
Swiss Federal Research Institute WSL, Zürcherstrasse 111, 8903 Birmensdorf, Switzerland

Salome Schläfli
Institute of Geological Sciences, University of Bern, Baltzerstrasse 1+3, 3012, Bern, Switzerland

Markus Stoffel
Department of Earth Sciences, University of Geneva, 13 rue des Maraîchers, 1205 Geneva, Switzerland
Department F.-A. Forel for Aquatic and Environmental Sciences, University of Geneva, 66 Boulevard Carl Vogt, 1205 Geneva, Switzerland

Ivar R. Lokhorst, Lisanne Braat, Jasper R. F. W. Leuven, Anne W. Baar, Sanja Selaković and Maarten G. Kleinhans
Faculty of Geosciences, Utrecht University, 3508 TC Utrecht, the Netherlands

Mijke van Oorschot
Department of Freshwater Ecology & Water Quality, Deltares, 2600 MH Delft, the Netherlands

Phillipe A. Wernette and Chris Houser
Department of Earth and Environmental Sciences, University of Windsor, Windsor, Ontario, N9B 3P4, Canada

Phillipe A. Wernette and Michael P. Bishop
Department of Geography, Texas A&M University, College Station, Texas, 77843, USA

Phillipe A. Wernette
anow at: Department of Earth and Environmental Sciences, University of Windsor, 401 Sunset Ave., Windsor, Ontario, N9B 3P4, Canada

Bradley A. Weymer
GEOMAR Helmholtz Centre for Ocean Research Kiel,
24148 Kiel, Germany

Mark E. Everett and Bobby Reece
Department of Geology and Geophysics, Texas A&M
University, College Station, Texas, 77843, USA

Liping Liao and Yunchuan Yang
College of Civil Engineering and Architecture, Guangxi
University, Nanning 530004, China
Key Laboratory of Disaster Prevention and Structural
Safety of Ministry of Education, Guangxi University,
Nanning 530004, China
Guangxi Key Laboratory of Disaster Prevention and
Engineering Safety, Guangxi University, Nanning
530004, China

Zhiquan Yang
Faculty of Land Resource Engineering, Kunming
University of Science and Technology, Kunming 650500,
China

Yingyan Zhu and Jin Hu
Institute of Mountain Hazards and Environment,
Chinese Academy of Sciences and Ministry of Water
Conservancy, Chengdu 610041, China

D. H. Steve Zou
Department of Civil and Resource Engineering,
Dalhousie University, Halifax, NS, B3H4K5, Canada

**Jack G. Williams, Nick J. Rosser, Richard J. Hardy
and Matthew J. Brain**
Department of Geography, Durham University, Lower
Mountjoy, South Road, Durham, UK

Ashraf A. Afana
National Trust, Kemble Drive, Swindon, UK

Index